N. Mohan Kumar

AN INTRODUCTION TO HARMONIC ANALYSIS

Yitzhak Katznelson
INSTITUTE OF MATHEMATICS
THE HEBREW UNIVERSITY OF JERUSALEM

Second Corrected Edition

DOVER PUBLICATIONS, INC.
NEW YORK

Published in Canada by General Publishing Company, Ltd., 30 Lesmill Road, Don Mills, Toronto, Ontario.
Published in the United Kingdom by Constable and Company, Ltd., 10 Orange Street, London WC 2.

This Dover edition, first published in 1976, is an unabridged and corrected republication of the work originally published by John Wiley & Sons, Inc., New York, in 1968.

International Standard Book Number: 0-486-63331-4
Library of Congress Catalog Card Number: 76-3075

Manufactured in the United States of America
Dover Publications, Inc.
180 Varick Street
New York, N.Y. 10014

ליונתן

Preface

Harmonic analysis is the study of objects (functions, measures, etc.), defined on topological groups. The group structure enters into the study by allowing the consideration of the translates of the object under study, that is, by placing the object in a translation-invariant space. The study consists of two steps. First: finding the "elementary components" of the object, that is, objects of the same or similar class, which exhibit the simplest behavior under translation and which "belong" to the object under study (harmonic or spectral *analysis*); and second: finding a way in which the object can be construed as a combination of its elementary components (harmonic or spectral *synthesis*).

The vagueness of this description is due not only to the limitation of the author but also to the vastness of its scope. In trying to make it clearer, one can proceed in various ways†; we have chosen here to sacrifice generality for the sake of concreteness. We start with the circle group **T** and deal with classical Fourier series in the first five chapters, turning then to the real line in Chapter VI and coming to locally compact abelian groups, only for a brief sketch, in Chapter VII. The philosophy behind the choice of this approach is that it makes it easier for students to grasp the main ideas and gives them a large class of concrete examples which are essential for the proper understanding of the theory in the general context of topological groups. The presentation of Fourier series and integrals differs from that in [1], [7], [8], and [28] in being, I believe, more explicitly aimed at the general (locally compact abelian) case.

The last chapter is an introduction to the theory of commutative

† Hence the indefinite article in the title of the book.

Banach algebras. It is biased, studying Banach algebras mainly as a tool in harmonic analysis.

This book is an expanded version of a set of lecture notes written for a course which I taught at Stanford University during the spring and summer quarters of 1965. The course was intended for graduate students who had already had two quarters of the basic "real-variable" course. The book is on the same level; the reader is assumed to be familiar with the basic notions and facts of Lebesgue integration, the most elementary facts concerning Borel measures, some basic facts about holomorphic functions of one complex variable, and some elements of functional analysis, namely: the notions of a Banach space, continuous linear functionals, and the three key theorems—"the closed graph," the Hahn-Banach, and the "uniform boundedness" theorems. All the prerequisites can be found in [23] and (except, for the complex variable) in [22]. Assuming these prerequisites, the book, or most of it, can be covered in a one-year course. A slower moving course or one shorter than a year may exclude some of the starred sections (or subsections). Aiming for a one-year course forced the omission not only of the more general setup (non-abelian groups are not even mentioned), but also of many concrete topics such as Fourier analysis on $\mathbf{R}^n$, $n > 1$, and finer problems of harmonic analysis in $\mathbf{T}$ or $\mathbf{R}$ (some of which can be found in [13]). Also, some important material was cut into exercises, and we urge the reader to do as many of them as he can.

The bibliography consists mainly of books, and it is through the bibliographies included in these books that the reader is to become familiar with the many research papers written on harmonic analysis. Only some, more recent, papers are included in our bibliography. In general we credit authors only seldom—most often for identification purposes. With the growing mobility of mathematicians, and the happy amount of oral communication, many results develop within the mathematical folklore and when they find their way into print it is not always easy to determine who deserves the credit. When I was writing Chapter III of this book, I was very pleased to produce the simple elegant proof of Theorem 1.6 there. I could swear I did it myself until I remembered two days later that six months earlier, "over a cup of coffee," Lennart Carleson indicated to me this same proof.

The book is divided into chapters, sections, and subsections. The chapter numbers are denoted by roman numerals and the sections and subsections, as well as the exercises, by arabic numerals. In cross references within the same chapter, the chapter number is omitted; thus Theorem III.1.6, which is the theorem in subsection 6 of Section 1 of Chapter III, is referred to as Theorem 1.6 within Chapter III, and Theorem III.1.6 elsewhere. The exercises are gathered at the end of the sections, and exercise V.1.1 is the first exercise at the end of Section 1, Chapter V. Again, the chapter number is omitted when an exercise is referred to within the same chapter. The ends of proofs are marked by a triangle (◀).

The book was written while I was visiting the University of Paris and Stanford University and it owes its existence to the moral and technical help I was so generously given in both places. During the writing I have benefitted from the advice and criticism of many friends; I would like to thank them all here. Particular thanks are due to L. Carleson, K. DeLeeuw, J.-P. Kahane, O. C. McGehee, and W. Rudin. I would also like to thank the publisher for the friendly cooperation in the production of this book.

YITZHAK KATZNELSON

Jerusalem
April 1968

Contents

Symbols

Page	*Symbol*
1	$\mathbf{R}, \mathbf{T}, \mathbf{Z}$
2	$L^1(\mathbf{T})$
3	$\hat{f}(n), S[f], \tilde{S}[f]$
5	$f * g$
12	$\mathbf{K}_n, \sigma_n(f), \sigma_n(f, t)$
13	$S_n(f), D_n$
14	$C(\mathbf{T}), C^n(\mathbf{T}), L^p(\mathbf{T})$
15	$\mathbf{V}_n(t), \mathbf{P}(r, t)$
16	$L^\infty(\mathbf{T}), \mathrm{Lip}_\alpha(\mathbf{T}), \mathrm{lip}_\alpha(\mathbf{T})$
25	$\omega(f, h), \Omega(f, h)$
31	$A(\mathbf{T})$
34	$S[\mu]$
36	$S_n(\mu), S_n(\mu, t), \sigma_n(\mu), \sigma_n(\mu, t)$
37	$M(\mathbf{T}), \delta, \delta_\tau$
47	$\boldsymbol{S}_n, \|\boldsymbol{S}_n\|^B, L_n$
48	$\tilde{f}, f^\flat$
50	$\Lambda_\varphi, \lambda_\varphi$
55	$S_n^*(f, t), S^*(f, t)$
65	$\mathfrak{m}_f(x)$
73	$M_f(t)$
82	$h_p(f, r)$
84	H^p, N
101	$\| \ \|_{\mathscr{F}\ell^p}$
104	$\mathrm{B}_\Lambda, (L^1(\mathbf{T}))_\Lambda$
112	$C^*\{M_n\}, C^\#\{M_n\}$
116	$\mathscr{M}(r)$
120	$L^1(\mathbf{R})$
121	$\hat{\mathbf{R}}$
123	$A(\hat{\mathbf{R}}), C_0(\hat{\mathbf{R}})$
124	$\mathbf{K}_\lambda, \mathbf{K}(x)$
126	$\mathbf{V}_\lambda, \mathbf{P}_\lambda$
127	$\mathscr{L}, \mathscr{L}_c$
131	$M(\mathbf{R})$
146	$\mathscr{F}L^p$
147	$S(\mathbf{R}), S^*(\mathbf{R})$
149	$\Sigma(\nu)$
155	$\Lambda(\varepsilon, f), AP(\mathbf{R})$
156	$W_0(f)$
157	$W(f)$
159	$\sigma(h)$
160	$M(f)$
162	$\langle f, g \rangle_M$
164	$f \underset{M}{*} g$
187	$\mathbf{D}$
194	$\mathbf{C}$
195	$C(X), C_0(X)$
205	$\mathfrak{M}$
207	$\hat{B}$
222	$\boldsymbol{h}(I)$
223	$\boldsymbol{k}(E)$
249	$A(E)$

AN INTRODUCTION TO HARMONIC ANALYSIS

Chapter I

Fourier Series on T

We denote by **R** the additive group of real numbers and by **Z** the subgroup consisting of the integers. The group **T** is defined as the quotient $\mathbf{R}/2\pi\mathbf{Z}$ where, as indicated by the notation, $2\pi\mathbf{Z}$ is the group of the integral multiples of 2π. There is an obvious identification between functions on **T** and 2π-periodic functions on **R**, which allows an implicit introduction of notions such as continuity, differentiability, etc. for functions on **T**. The *Lebesgue measure* on **T** also can be defined by means of the preceding identification: a function f is integrable on **T** if the corresponding 2π-periodic function, which we denote again by f, is integrable on $[0, 2\pi)$ and we set

$$\int_{\mathbf{T}} f(t)\,dt = \int_0^{2\pi} f(x)\,dx\,.$$

In other words, we consider the interval $[0, 2\pi)$ as a model for **T** and the Lebesgue measure dt on **T** is the restriction of the Lebesgue measure of **R** to $[0, 2\pi)$. The total mass of dt on **T** is equal to 2π and many of our formulas would be simpler if we normalized dt to have total mass 1, that is, if we replace it by $dx/2\pi$. Taking intervals on **R** as "models" for **T** is very convenient, however, and we choose to put $dt = dx$ in order to avoid confusion. We "pay" by having to write the factor $1/2\pi$ in front of every integral.

An all-important property of dt on **T** is its *translation invariance*, that is, for all $t_0 \in \mathbf{T}$ and f defined on **T**,

$$\int f(t - t_0)\,dt = \int f(t)\,dt\dagger$$

† Throughout this chapter, integrals with unspecified limits of integration are taken over **T**.

1. FOURIER COEFFICIENTS

1.1 We denote by $L^1(\mathbf{T})$ the space of all complex-valued, Lebesgue integrable functions on $\mathbf{T}$. For $f \in L^1(\mathbf{T})$ we put

$$\|f\|_{L^1} = \frac{1}{2\pi} \int_{\mathbf{T}} |f(t)|\, dt,$$

It is well known that $L^1(\mathbf{T})$, with the norm so defined, is a Banach space.

DEFINITION: *A trigonometric polynomial on* $\mathbf{T}$ is an expression of the form

$$P \sim \sum_{n=-N}^{N} a_n e^{int}. \tag{1.1}$$

The numbers n appearing in (1.1) are called *the frequencies of* P; the largest integer n such that $|a_n| + |a_{-n}| \neq 0$ is called *the degree of* P. The values assumed by the index n are integers so that each of the summands in (1.1) is a function on $\mathbf{T}$. Since (1.1) is a finite sum, it represents a function, which we denote again by P, defined for each $t \in \mathbf{T}$ by

$$P(t) = \sum_{n=-N}^{N} a_n e^{int}. \tag{1.2}$$

Let P be defined by (1.2). Knowing the function P we can compute the coefficients a_n by the formula

$$a_n = \frac{1}{2\pi} \int P(t) e^{-int}\, dt \tag{1.3}$$

which follows immediately from the fact that for integers j,

$$\frac{1}{2\pi} \int e^{ijt}\, dt = \begin{cases} 1 & \text{if } j = 0 \\ 0 & \text{if } j \neq 0. \end{cases}$$

Thus we see that the function P determines the expression (1.1) and there seems to be no point in keeping the distinction between the expression (1.1) and the function P; we shall consider trigonometric polynomials as both formal expressions and functions.

1.2 DEFINITION: *A trigonometric series on* **T** is an expression of the form

$$S \sim \sum_{n=-\infty}^{\infty} a_n e^{int}. \tag{1.4}$$

Again, n assumes integral values; however, the number of terms in (1.4) may be infinite and there is no assumption whatsoever about the size of the coefficients or about convergence. *The conjugate*† $\tilde{S}$ of the series (1.4) is, by definition, the series

$$\tilde{S} \sim \sum_{n=-\infty}^{\infty} -i\,\mathrm{sgn}\,(n) a_n e^{int}$$

where $\mathrm{sgn}\,(n) = 0$ if $n = 0$ and $\mathrm{sgn}\,(n) = n/|n|$ otherwise.

1.3 Let $f \in L^1(\mathbf{T})$. Motivated by (1.3) we define the nth *Fourier coefficient* of f by

$$\hat{f}(n) = \frac{1}{2\pi} \int f(t) e^{-int} dt. \tag{1.5}$$

DEFINITION: *The Fourier series* $S[f]$ of a function $f \in L^1(\mathbf{T})$ is the trigonometric series

$$S[f] \sim \sum_{n=-\infty}^{\infty} \hat{f}(n) e^{int}.$$

The series conjugate to $S[f]$ will be denoted by $\tilde{S}[f]$ and referred to as *the conjugate Fourier series* of f. We shall say that a trigonometric series is a *Fourier series* if it is the Fourier series of some $f \in L^1(\mathbf{T})$,

We turn to some elementary properties of Fourier coefficients.

1.4 ***Theorem***: *Let* $f, g \in L^1(\mathbf{T})$, *then*

(a) $\widehat{(f+g)}(n) = \hat{f}(n) + \hat{g}(n)$.

(b) *For any complex number* α,

$$\widehat{(\alpha f)}(n) = \alpha \hat{f}(n).$$

(c) *If* $\bar{f}$ *is the complex conjugate*‡ *of* f *then* $\hat{\bar{f}}(n) = \overline{\hat{f}(-n)}$.

(d) *Denote* $f_\tau(t) = f(t - \tau)$, $\tau \in \mathbf{T}$; *then*

$$\hat{f}_\tau(n) = \hat{f}(n) e^{-in\tau}.$$

† See chapter III for motivation of the terminology.

‡ Defined by: $\bar{f}(t) = \overline{f(t)}$ for all $t \in \mathbf{T}$.

(e) $|\hat{f}(n)| \leqq \frac{1}{2\pi}\int |f(t)|\,dt = \|f\|_{L^1}$.

The proofs of (a) through (e) follow immediately from (1.5) and the details are left to the reader.

1.5 *Corollary*: *Assume* $f_j \in L^1(\mathbf{T})$, $j = 0, 1, \ldots,$ *and* $\|f_j - f_0\|_{L^1} \to 0$. *Then* $\hat{f}_j(n) \to \hat{f}_0(n)$ *uniformly.*

1.6 *Theorem*: *Let* $f \in L^1(\mathbf{T})$, *assume* $\hat{f}(0) = 0$, *and define*

$$F(t) = \int_0^t f(\tau)\,d\tau .$$

Then F *is continuous,* 2π *periodic, and*

$$\hat{F}(n) = \frac{1}{in}\hat{f}(n), \qquad n \neq 0. \tag{1.6}$$

PROOF: The continuity (and, in fact, the absolute continuity) of F is evident. The periodicity follows from

$$F(t + 2\pi) - F(t) = \int_t^{t+2\pi} f(\tau)\,d\tau = 2\pi\hat{f}(0) = 0,$$

and (1.6) is obtained through integration by parts:

$$\hat{F}(n) = \frac{1}{2\pi}\int_0^{2\pi} F(t)e^{-int}dt = \frac{-1}{2\pi}\int_0^{2\pi} F'(t)\frac{1}{-in}e^{-int}dt = \frac{1}{in}\hat{f}(n). \blacktriangleleft$$

1.7 We now define the convolution operation in $L^1(\mathbf{T})$. The reader will notice the use of the group structure of $\mathbf{T}$ and of the invariance of dt in the subsequent proofs.

***Theorem*:** *Assume* $f, g \in L^1(\mathbf{T})$. *For almost all* t, *the function* $f(t-\tau)g(\tau)$ *is integrable (as a function of* τ *on* $\mathbf{T}$), *and, if we write*

$$h(t) = \frac{1}{2\pi}\int f(t-\tau)g(\tau)\,d\tau, \tag{1.7}$$

then $h \in L^1(\mathbf{T})$ *and*

$$\|h\|_{L^1} \leqq \|f\|_{L^1}\|g\|_{L^1}. \tag{1.8}$$

Moreover

$$\hat{h}(n) = \hat{f}(n)\hat{g}(n) \qquad \textit{for all } n. \tag{1.9}$$

PROOF: The functions $f(t-\tau)$ and $g(\tau)$, considered as functions of the two variables (t,τ), are clearly measurable, hence so is

$$F(t,\tau) = f(t-\tau)\,g(\tau)\,.$$

For almost all τ, $F(t,\tau)$ is just a constant multiple of f_τ, hence integrable, and

$$\frac{1}{2\pi}\int\left(\frac{1}{2\pi}\int |F(t,\tau)|\,dt\right)d\tau = \frac{1}{2\pi}\int |g(\tau)|\cdot\|f\|_{L^1}d\tau = \|f\|_{L^1}\|g\|_{L^1}.$$

Hence, by the theorem of Fubini, $f(t-\tau)g(\tau)$ is integrable (over $(0,2\pi)$) as a function of τ for almost all t, and

$$\frac{1}{2\pi}\int |h(t)|\,dt = \frac{1}{2\pi}\int\left|\frac{1}{2\pi}\int F(t,\tau)\,d\tau\right|dt \leqq \frac{1}{4\pi^2}\iint |F(t,\tau)|\,dt\,d\tau =$$

$$= \|f\|_{L^1}\|g\|_{L^1},$$

which establishes (1.8). In order to prove (1.9) we write

$$\hat{h}(n) = \frac{1}{2\pi}\int h(t)e^{-int}\,dt = \frac{1}{4\pi^2}\iint f(t-\tau)e^{-in(t-\tau)}g(\tau)e^{-in\tau}dt\,d\tau =$$

$$= \frac{1}{2\pi}\int f(t)e^{-int}dt\cdot\frac{1}{2\pi}\int g(\tau)e^{-in\tau}d\tau = \hat{f}(n)\hat{g}(n)\,.$$

As above the change in the order of integration is justified by Fubini's theorem. ◀

1.8 DEFINITION: *The convolution* $f * g$ of the ($L^1(\mathbf{T})$) functions f and g is the function h defined by (1.7).

Using the star notation for the convolution, we can write (1.9):

(1.10) $$\widehat{f*g}(n) = \hat{f}(n)\hat{g}(n)\,.$$

Theorem: *The convolution operation in* $L^1(\mathbf{T})$ *is commutative, associative, and distributive (with respect to the addition).*

PROOF: The change of variable $\vartheta = t-\tau$ gives

$$\frac{1}{2\pi}\int f(\vartheta)g(t-\vartheta)\,d\vartheta = \frac{1}{2\pi}\int g(t-\vartheta)f(\vartheta)\,d\vartheta\,,$$

that is,

$$f * g = g * f.$$

If $f_1, f_2, f_3 \in L^1(\mathbf{T})$, then

$$[(f_1 * f_2) * f_3](t) = \frac{1}{4\pi^2} \iint f_1(t-u-\tau) f_2(u) f_3(\tau) du\, d\tau =$$

$$= \frac{1}{4\pi^2} \iint f_1(t-\omega) f_2(\omega-\tau) f_3(\tau)\, d\omega\, d\tau = [f_1 * (f_2 * f_3)](t).$$

Finally, the distributive law

$$f_1 * (f_2 + f_3) = f_1 * f_2 + f_1 * f_3$$

is evident from (1.7). ◀

1.9 ***Lemma:*** *Assume* $f \in L^1(\mathbf{T})$ *and let* $\varphi(t) = e^{int}$ *for some integer* n. *Then*

$$(\varphi * f)(t) = \hat{f}(n) e^{int}.$$

PROOF:

$$(\varphi * f)(t) = \frac{1}{2\pi} \int e^{in(t-\tau)} f(\tau)\, d\tau = e^{int} \frac{1}{2\pi} \int f(\tau) e^{-in\tau}\, d\tau. \quad ◀$$

Corollary: *If* $f \in L^1(\mathbf{T})$ *and* $k(t) = \sum_{-N}^{N} a_n e^{int}$, *then*

(1.11) $$(k * f)(t) = \sum_{-N}^{N} a_n \hat{f}(n) e^{int}.$$

EXERCISES FOR SECTION 1

1. Compute the Fourier coefficients of the following functions (defined by their values on $[-\pi, \pi)$):

(a) $$f(t) = \begin{cases} \sqrt{2\pi} & |t| < \frac{1}{2} \\ 0 & \frac{1}{2} \leqq |t| \leqq \pi. \end{cases}$$

(b) $$\Delta(t) = \begin{cases} 1-|t| & |t| < 1 \\ 0 & 1 \leqq |t| < \pi. \end{cases}$$

What relation do you see between f and Δ?

(c) $$g(t) = \begin{cases} 1 & -1 < t \leqq 0 \\ -1 & 0 < t < 1 \\ 0 & 1 \leqq |t|. \end{cases}$$

What relation do you see between g and Δ?

(d) $h(t) = t \qquad -\pi \leqq t < \pi.$

2. Remembering Euler's formulas

$$\cos t = \tfrac{1}{2}(e^{it} + e^{-it}), \quad \sin t = \frac{1}{2i}(e^{it} - e^{-it}),$$

or

$$e^{it} = \cos t + i \sin t,$$

show that the Fourier series of a function $f \in L^1(\mathbf{T})$ is formally equal to

$$\frac{A_0}{2} + \sum_{n=1}^{\infty} (A_n \cos nt + B_n \sin nt)$$

where $A_n = \hat{f}(n) + \hat{f}(-n)$ and $B_n = i\{\hat{f}(n) - \hat{f}(-n)\}$. Equivalently:

$$A_n = \frac{1}{\pi}\int f(t) \cos nt\, dt$$

$$B_n = \frac{1}{\pi}\int f(t) \sin nt\, dt.$$

Show also that if f is real valued, then A_n and B_n are all real; if f is even, that is, if $f(t) = f(-t)$, then $B_n = 0$ for all n; and if f is odd, that is, if $f(t) = -f(-t)$, then $A_n = 0$ for all n.

3. Show that if $S \sim \sum a_j \cos jt$, then $\tilde{S} \sim \sum a_j \sin jt$.

4. Let $f \in L^1(\mathbf{T})$ and let $P(t) = \sum_{-N}^{N} a_n e^{int}$. Compute the Fourier coefficients of the function fP.

5. Let $f \in L^1(\mathbf{T})$, let m be a positive integer, and write

$$f_{(m)}(t) = f(mt).$$

Show

$$\hat{f}_{(m)}(n) = \begin{cases} \hat{f}\left(\dfrac{n}{m}\right) & \text{if } m \mid n. \\ 0 & \text{if } m \nmid n. \end{cases}$$

6. The trigonometric polynomial $\cos nt = \tfrac{1}{2}(e^{int} + e^{-int})$ is of degree n and has $2n$ zeros on **T**. Show that no trigonometric polynomial of degree $n > 0$ can have more than $2n$ zeros on **T**. *Hint*: Identify $\sum_{-n}^{n} a_j e^{ijt}$ on **T** with $z^{-n} \sum_{-n}^{n} a_j z^{n+j}$ on $|z| = 1$.

7. Denote by C^* the multiplicative group of complex numbers different from zero. Denote by T^* the subgroup of all $z \in C^*$ such that $|z| = 1$. Prove that if G is a subgroup of C^* which is compact (as a set of complex numbers), then $G \subseteq T^*$.

8. Let G be a compact proper subgroup of **T** Prove that G is finite and determine its structure. *Hint*: Show that G is discrete.

9. Let G be an infinite subgroup of **T**. Prove that G is dense in **T**. *Hint*: The closure of G in **T** is a compact subgroup.

10. Let α be an irrational multiple of 2π. Prove that $\{n\alpha(\text{mod}\,2\pi)\}_{n=0,\pm1,\pm2,\dots}$ is dense in **T**.

11. Prove that a continuous homomorphism of **T** into C^* is necessarily given by an exponential function. *Hint*: Use exercise 7 to show that the mapping is into T^*; determine the mapping on "small" rational multiples of 2π and use exercise 9.

12. If E is a subset of **T** and $\tau_0 \in \mathbf{T}$, we define $E+\tau_0=\{t+\tau_0;\ t \in E\}$; we say that E is *invariant under translation by* τ if $E=E+\tau$. Show that, given a set E, the set of $\tau \in \mathbf{T}$ such that E is invariant under translation by τ is a subgroup of **T**. Hence prove that if E is a measurable set on **T** and E is invariant under translation by infinitely many $\tau \in \mathbf{T}$, then either E or its complement has measure zero. *Hint*: A set of E of positive measure has points of density, that is, points τ such that $(2\varepsilon)^{-1}|E \cap (\tau-\varepsilon,\tau+\varepsilon)| \to 1$ as $\varepsilon \to 0$. ($|E_0|$ denotes the Lebesgue measure of E_0.)

13. If E and F are subsets of **T**, we write

$$E+F=\{t+\tau; t \in E,\ \tau \in F\}$$

and call $E+F$ the *algebraic sum* of E and F. Similarly we define the sum of any finite number of sets. A set E is called a *basis* for **T** if there exists an integer N such that $E+E+\cdots+E$ (N times) is **T**. Prove that every set E of positive measure on **T** is a basis. *Hint*: Prove that if E contains an interval it is a basis. Using points of density prove that if E has positive measure then $E+E$ contains intervals.

14. Show that measurable proper subgroups of **T** have measure zero.

15. Show that measurable homomorphisms of **T** into C^* map it into T^*.

16. Let f be a measurable homomorphism of **T** into T^*. Show that for all values of n, except possibly one value, $\hat{f}(n)=0$.

2. SUMMABILITY IN NORM AND HOMOGENEOUS BANACH SPACES ON T

2.1 We have defined the Fourier series of a function $f \in L^1(\mathbf{T})$ as a certain (formal) trigonometric series. The reader may wonder what is the point in the introduction of such formal series. After all, there is no more information in the (formal) expression $\sum_{n=-\infty}^{\infty} \hat{f}(n)\, e^{int}$ than there is in the simpler one $\{\hat{f}(n)\}_{n=-\infty}^{\infty}$ or the even simpler $\hat{f}$ with the understanding that the function $\hat{f}$ is defined on the integers. As we shall see, both expressions, $\sum \hat{f}(n) e^{int}$ and $\hat{f}$, have their advantages; the main advantages of the series notation being that it indicates the way in which f can be reconstructed from $\hat{f}$. Much of this chapter and all of chapter II will be devoted to clarifying the sense in which $\sum \hat{f}(n) e^{int}$ represents f. In this section we establish some of the main facts; we shall see that $\hat{f}$ determines f uniquely and we show how we can find f if we know $\hat{f}$.

Two very important properties of the Banach space $L^1(\mathbf{T})$ are the following:

(H-1′) *If* $f \in L^1(\mathbf{T})$ *and* $\tau \in \mathbf{T}$, *then*

$$f_\tau(t) = f(t-\tau) \in L^1(\mathbf{T}) \quad \textit{and} \quad \|f_\tau\|_{L^1} = \|f\|_{L^1}$$

(H-2′) *The* $L^1(\mathbf{T})$*-valued function* $\tau \to f_\tau$ *is continuous on* $\mathbf{T}$, *that is, for* $f \in L^1(\mathbf{T})$ *and* $\tau_0 \in \mathbf{T}$

$$\lim_{\tau \to \tau_0} \|f_\tau - f_{\tau_0}\|_{L^1} = 0. \tag{2.1}$$

We shall refer to (H-1′) as the translation invariance of $L^1(\mathbf{T})$; it is an immediate consequence of the translation invariance of the measure dt. In order to establish (H-2′) we notice first that (2.1) is clearly valid if f is a continuous function. Remembering that the continuous functions are dense in $L^1(\mathbf{T})$, we now consider an arbitrary $f \in L^1(\mathbf{T})$ and $\varepsilon > 0$. Let g be a continuous function on $\mathbf{T}$ such that $\|g - f\|_{L^1} < \varepsilon/2$; thus

$$\|f_\tau - f_{\tau_0}\|_{L^1} \leqq \|f_\tau - g_\tau\|_{L^1} + \|g_\tau - g_{\tau_0}\|_{L^1} + \|g_{\tau_0} - f_{\tau_0}\|_{L^1} =$$

$$= \|(f-g)_\tau\|_{L^1} + \|g_\tau - g_{\tau_0}\|_{L^1} + \|(g-f)_{\tau_0}\|_{L^1} < \varepsilon + \|g_\tau - g_{\tau_0}\|_{L^1}.$$

Hence $\overline{\lim}_{\tau \to \tau_0} \|f_\tau - f_{\tau_0}\|_{L^1} < \varepsilon$ and, ε being an arbitrary positive number, (H-2′) is established.

2.2 DEFINITION: A *summability kernel* is a sequence $\{k_n\}$ of continuous 2π-periodic functions satisfying:

$$\frac{1}{2\pi} \int k_n(t)\, dt = 1 \tag{S-1}$$

(S-2) $$\frac{1}{2\pi}\int |k_n(t)|\,dt \leqq \text{const}$$

(S-3) For all $0 < \delta < \pi$,

$$\lim_{n\to\infty}\int_\delta^{2\pi-\delta} |k_n(t)|\,dt = 0.$$

A *positive summability kernel* is one such that $k_n(t) \geqq 0$ for all t and n.

For positive kernels the assumption (S-2) is clearly redundant.

Sometimes we consider families k_r depending on a continuous parameter r instead of the discrete n. Thus the Poisson kernel $\mathbf{P}(r, t)$, which we shall define at the end of this section, is defined for $0 \leqq r < 1$ and we replace in (S-3), as well as in the applications, the limit "$\lim_{n\to\infty}$" by "$\lim_{r\to 1}$".

The following lemma is stated in terms of vector-valued integrals. We refer to the Appendix for the definition and relevant properties.

***Lemma*:** *Let B be a Banach space, φ a continuous B-valued function on $\mathbf{T}$, and $\{k_n\}$ a summability kernel. Then*:

$$\lim_{n\to\infty}\frac{1}{2\pi}\int k_n(\tau)\,\varphi(\tau)\,d\tau = \varphi(0).$$

PROOF: By (S-1) we have, for $0 < \delta < \pi$,

$$\begin{aligned}\frac{1}{2\pi}\int k_n(\tau)\varphi(\tau)\,d\tau - \varphi(0) &= \frac{1}{2\pi}\int k_n(\tau)(\varphi(\tau)-\varphi(0))\,d\tau = \\ &= \frac{1}{2\pi}\int_{-\delta}^{\delta} k_n(\tau)(\varphi(\tau)-\varphi(0))\,d\tau + \frac{1}{2\pi}\int_\delta^{2\pi-\delta} k_n(\tau)(\varphi(\tau)-\varphi(0))\,d\tau.\end{aligned} \tag{2.2}$$

Now

$$\left\| \frac{1}{2\pi}\int_{-\delta}^{\delta} k_n(\tau)(\varphi(\tau)-\varphi(0))\,d\tau\right\|_B \leqq \max_{|\tau|\leqq\delta} \|\varphi(\tau)-\varphi(0)\|_B \|k_n\|_{L^1} \tag{2.3}$$

and

$$\begin{aligned}&\left\|\frac{1}{2\pi}\int_\delta^{2\pi-\delta} k_n(\tau)(\varphi(\tau)-\varphi(0))d\tau\right\|_B \leqq \\ &\leqq \max\|\varphi(\tau)-\varphi(0)\|_B\,\frac{1}{2\pi}\int_\delta^{2\pi-\delta}|k_n(\tau)|d\tau.\end{aligned} \tag{2.4}$$

By (S-2) and the continuity of $\varphi(\tau)$ at $\tau = 0$, given $\varepsilon > 0$ we can find $\delta > 0$ so that (2.3) is bounded by ε, and keeping this δ, it results from (S-3) that (2.4) tends to zero as $n \to \infty$ so that (2.2) is bounded by 2ε for large n. ◀

2.3 For $f \in L^1(\mathbf{T})$ we put $\varphi(\tau) = f_\tau(t) = f(t - \tau)$, then by (H-1′) and (H-2′) φ is a continuous $L^1(\mathbf{T})$-valued function on **T** and $\varphi(0) = f$. Applying lemma 2.2 we obtain

Theorem: *Let $f \in L^1(\mathbf{T})$ and $\{k_n\}$ be a summability kernel; then*

$$f = \lim_{n\to\infty} \frac{1}{2\pi} \int k_n(\tau) f_\tau \, d\tau \tag{2.5}$$

in the $L^1(\mathbf{T})$ norm.

2.4 The integrals in (2.5) have the formal appearance of a convolution although the operation involved, that is, vector integration, is different from the convolution as defined in section 1. The ambiguity, however, is harmless.

Lemma: *Let k be a continuous function on* **T** *and $f \in L^1(\mathbf{T})$. Then*

$$\frac{1}{2\pi} \int k(\tau) f_\tau \, d\tau = k * f. \tag{2.6}$$

PROOF: Assume first that f is continuous on **T**. We have (see Appendix)

$$\frac{1}{2\pi} \int k(\tau) f_\tau \, d\tau = \frac{1}{2\pi} \lim \sum_j (\tau_{j+1} - \tau_j) k(\tau_j) f_{\tau_j},$$

the limit being taken in the $L^1(\mathbf{T})$ norm as the subdivision $\{\tau_j\}$ of $[0, 2\pi)$ becomes finer and finer. On the other hand,

$$\frac{1}{2\pi} \lim \sum (\tau_{j+1} - \tau_j) k(\tau_j) f(t - \tau_j) = (k * f)(t)$$

uniformly and the lemma is proved for continuous f. For arbitrary $f \in L^1(\mathbf{T})$, let $\varepsilon > 0$ be arbitrary and let g be a continuous function on **T** such that $\| f - g \|_{L^1} < \varepsilon$. Then, since (2.6) is valid for g,

$$\frac{1}{2\pi} \int k(\tau) f_\tau \, d\tau - k * f = \frac{1}{2\pi} \int k(\tau)(f - g)_\tau \, d\tau + k * (g - f)$$

and consequently

$$\left\| \frac{1}{2\pi} \int k(\tau) f_\tau \, d\tau - k * f \right\|_{L^1} \leqq 2 \| k \|_{L^1} \varepsilon . \qquad ◀$$

Using lemma 2.4 we can rewrite (2.5):

$$(2.5') \qquad f = \lim_{n \to \infty} k_n * f \qquad \text{in the } L^1(\mathbf{T}) \text{ norm.}$$

2.5 One of the most useful summability kernels and probably the best known is *Fejér's kernel* (which we denote by $\{\mathbf{K}_n\}$) defined by

$$(2.7) \qquad \mathbf{K}_n(t) = \sum_{j=-n}^{n} \left(1 - \frac{|j|}{n+1}\right) e^{ijt}.$$

The fact that $\mathbf{K}_n$ satisfies (S-1) is obvious from (2.7); that $\mathbf{K}_n(t) \geqq 0$ and that (S-3) is satisfied is clear from

Lemma:

$$\mathbf{K}_n(t) = \frac{1}{n+1} \left\{ \frac{\sin \frac{n+1}{2} t}{\sin \frac{1}{2} t} \right\}^2 .$$

PROOF: Recall that

$$(2.8) \qquad \sin^2 \frac{t}{2} = \frac{1}{2}(1 - \cos t) = -\frac{1}{4} e^{-it} + \frac{1}{2} - \frac{1}{4} e^{it}.$$

A direct computation of the coefficients in the product shows that

$$\left(-\frac{1}{4} e^{-it} + \frac{1}{2} - \frac{1}{4} e^{it}\right) \sum_{j=-n}^{n} \left(1 - \frac{|j|}{n+1}\right) e^{ijt} =$$

$$= \frac{1}{n+1} \left(-\frac{1}{4} e^{-i(n+1)t} + \frac{1}{2} - \frac{1}{4} e^{i(n+1)t}\right). \qquad ◀$$

We shall adhere to the generally used notation and write $\sigma_n(f) = \mathbf{K}_n * f$ and $\sigma_n(f, t) = (\mathbf{K}_n * f)(t)$. It follows from corollary 1.9 that

$$(2.9) \qquad \sigma_n(f, t) = \sum_{-n}^{n} \left(1 - \frac{|j|}{n+1}\right) \hat{f}(j) e^{ijt}.$$

2.6 The fact that $\sigma_n(f) \to f$ in the $L^1(\mathbf{T})$ norm for every $f \in L^1(\mathbf{T})$, which is a special case of (2.5′), and the fact that $\sigma_n(f)$ is a trigonometric polynomial imply that trigonometric polynomials are dense in $L^1(\mathbf{T})$.

Other immediate consequences are the following two important theorems.

2.7 ***Theorem***: (*The Uniqueness Theorem*): *Let* $f \in L^1(\mathbf{T})$ *and assume* $\hat{f}(n) = 0$ *for all* n. *Then* $f = 0$.

PROOF: By (2.9) $\sigma_n(f) = 0$ for all n. Since $\sigma_n(f) \to f$, it follows that $f = 0$. ◀

An equivalent form of the uniqueness theorem is: Let $f, g \in L^1(\mathbf{T})$ and assume $\hat{f}(n) = \hat{g}(n)$ for all n, then $f = g$.

2.8 ***Theorem***: (*The Riemann-Lebesgue Lemma*): *Let* $f \in L^1(\mathbf{T})$, *then* $\lim_{|n| \to \infty} \hat{f}(n) = 0$.

PROOF: Let $\varepsilon > 0$ and let P be a trigonometric polynomial on $\mathbf{T}$ such that $\| f - P \|_{L^1} < \varepsilon$. If $|n| >$ degree of P, then

$$|\hat{f}(n)| = |(\widehat{f - P})(n)| \leqq \| f - P \|_{L^1} < \varepsilon. \quad ◀$$

Remark: If K is a compact set in $L^1(\mathbf{T})$ and $\varepsilon > 0$, there exist a finite number of trigonometric polynomials $P_1, \dots, P_N$ such that for every $f \in K$ there exists a $j, 1 \leqq j \leqq N$, such that $\| f - P_j \|_{L^1} < \varepsilon$. If $|n|$ is greater than $\max_{1 \leqq j \leqq N}$ (degree of P_j) then $|\hat{f}(n)| < \varepsilon$ for all $f \in K$. Thus, the Riemann-Lebesgue lemma holds uniformly on compact subsets of $L^1(\mathbf{T})$.

2.9 For $f \in L^1(\mathbf{T})$ we denote by $S_n(f)$ the nth partial sum of $S[f]$, that is,

$$(2.10) \qquad (S_n(f))(t) = S_n(f, t) = \sum_{-n}^{n} \hat{f}(j) e^{ijt}.$$

If we compare (2.9) and (2.10) we see that

$$(2.11) \qquad \sigma_n(f) = \frac{1}{n+1} (S_0(f) + S_1(f) + \cdots + S_n(f)),$$

in other words, the $\sigma_n(f)$ are the arithmetic means of $S_n(f)$. It follows that if $S[f]$ converges† in $L^1(\mathbf{T})$, then the limit is necessarily f.

From corollary 1.9 it follows that $S_n(f) = D_n * f$ where D_n is the *Dirichlet kernel* defined by

$$(2.12) \qquad D_n(t) = \sum_{-n}^{n} e^{ijt} = \frac{\sin (n + \frac{1}{2}) t}{\sin \frac{1}{2} t}.$$

† That is, if $S_n(f)$ converge as $n \to \infty$.

It is important to notice that $\{D_n\}$ is not a summability kernel in our sense. It does satisfy condition (S-1); however, it does not satisfy either (S-2) or (S-3). This explains why the problem of convergence for Fourier series is so much harder than the problem of summability. We shall discuss convergence in chapter II.

2.10 DEFINITION: *A homogeneous Banach space on* $\mathbf{T}$ is a linear subspace B of $L^1(\mathbf{T})$ having a norm $\| \ \|_B \geqq \| \ \|_{L^1}$ under which it is a Banach space, and having the following properties:

(H-1) If $f \in B$ and $\tau \in \mathbf{T}$, then $f_\tau \in B$ and $\|f_\tau\|_B = \|f\|_B$ (where $f_\tau(t) = f(t-\tau)$).

(H-2) For all $f \in B$, $\tau, \tau_0 \in \mathbf{T}$, $\lim_{\tau \to \tau_0} \|f_\tau - f_{\tau_0}\|_B = 0$.

Remarks. Condition (H-1) is referred to as translation invariance and (H-2) as continuity of the translation. We could simplify (H-2) somewhat by requiring continuity at one specific $\tau_0 \in \mathbf{T}$, say $\tau_0 = 0$ rather than at every $\tau_0 \in \mathbf{T}$, since by (H-1)

$$\|f_\tau - f_{\tau_0}\|_B = \|f_{\tau - \tau_0} - f\|_B .$$

Also, the method of the proof of (H-2′) (see 2.1) shows that if we have a space B satisfying (H-1) and we want to show that it satisfies (H-2) as well, it is sufficient to check the continuity of the translation on a dense subset of B. An almost equivalent statement is

Lemma: *Let* $B \subset L^1(\mathbf{T})$ *be a Banach space satisfying* (H-1). *Denote by* B_c *the set of all* $f \in B$ *such that* $\tau \to f_\tau$ *is a continuous* B*-valued function. Then* B_c *is a closed subspace of* B.

Examples of homogeneous Banach spaces on $\mathbf{T}$.

(a) $C(\mathbf{T})$—the space of all continuous 2π-periodic functions with the norm

$$\|f\|_\infty = \max_t |f(t)| \tag{2.13}$$

(b) $C^n(\mathbf{T})$—the subspace of $C(\mathbf{T})$ of all n-times continuously differentiable functions (n being a rational integer) with the norm

$$\|f\|_{C^n} = \sum_{j=0}^{n} \frac{1}{j!} \max_t |f^{(j)}(t)| \tag{2.13′}$$

(c) $L^p(\mathbf{T})$, $1 \leqq p < \infty$—the subspace of $L^1(\mathbf{T})$ consisting of all the functions f for which $\int |f(t)|^p \, dt < \infty$ with the norm

(2.14)
$$\| f \|_{L^p} = \left(\frac{1}{2\pi} \int |f(t)|^p \, dt \right)^{1/p}.$$

The validity of (H-1) for all three examples is obvious. The validity of (H-2) for (a) and (b) is equivalent to the statement that continuous functions on **T** are uniformly continuous. The proof of (H-2) for (c) is identical to that of (H-2′) (see 2.1).

We now generalize theorem 2.3 to homogeneous Banach spaces on **T**.

2.11 ***Theorem:*** *Let B be a homogeneous Banach space on* **T**, *let* $f \in B$ *and let* $\{k_n\}$ *be a summability kernel. Then*

$$\| k_n * f - f \|_B \to 0 \qquad \text{as} \quad n \to \infty.$$

PROOF: Since $\| \ \|_B \geqq \| \ \|_{L^1}$, the B-valued integral $\frac{1}{2\pi} \int k_n(\tau) f_\tau \, d\tau$ is the same as the $L^1(\mathbf{T})$-valued integral which, by lemma 2.4, is equal to $k_n * f$. The theorem now follows from lemma 2.2. ◀

2.12 ***Theorem:*** *Let B be a homogeneous Banach space on* **T**. *Then the trigonometric polynomials in B are everywhere dense.*

PROOF: For every $f \in B$, $\sigma_n(f) \to f$. ◀

Corollary (*Weierstrass Approximation Theorem*): *Every continuous* 2π*-periodic function can be approximated uniformly by trigonometric polynomials.*

2.13 We finish this section by mentioning two important summability kernels.

(1) The de la Vallée Poussin kernel:

(2.15)
$$\mathbf{V}_n(t) = 2\mathbf{K}_{2n+1}(t) - \mathbf{K}_n(t)$$

(S-1), (S-2) and (S-3) are obvious from (2.15). $\mathbf{V}_n$ is a polynomial of degree $2n+1$ having the property that $\hat{\mathbf{V}}_n(j) = 1$ if $|j| \leqq n+1$; it is therefore very useful when we want to approximate a function f by polynomials having the same Fourier coefficients as f over prescribed intervals (namely $\mathbf{V}_n * f$).

(2) The Poisson kernel: for $0 \leqq r < 1$ put

(2.16)
$$\mathbf{P}(r,t) = 1 + 2 \sum_{j=1}^{\infty} r^j \cos jt = \frac{1 - r^2}{1 - 2r\cos t + r^2}.$$

It follows from corollary 1.9 and from the fact that the series in (2.16) converges uniformly, that

(2.17) $\mathbf{P}(r,t) * f = \sum \hat{f}(n) r^{|n|} e^{int}.$

Thus $\mathbf{P} * f$ is the Abel mean of $S[f]$ and theorem 2.11 (with Poisson's kernel) states that for $f \in B$, $S[f]$ is Abel summable to f in the B norm. Compared to the Fejér kernel, the Poisson kernel has the disadvantage of not being a polynomial; however, being essentially the real part of the Cauchy kernel (precisely: $\mathbf{P}(r,t) = \mathrm{Re}\left(\dfrac{1 + re^{it}}{1 - re^{it}}\right)$), the Poisson kernel links the theory of trigonometric series with the theory of analytic functions. We shall make much use of that in chapter III. Another important property of $\mathbf{P}(r,t)$ is that it is a decreasing function of t for $0 \leqq t \leqq \pi$.

EXERCISES FOR SECTION 2

1. Show that every measurable homomorphism of $\mathbf{T}$ into T^* has the form $t \to e^{int}$ where n is a rational integer. *Hint:* Use Exercise 1.16.

2. Show that in the following examples (H-1) is satisfied but (H-2) is not satisfied:

(a) $L^\infty(\mathbf{T})$—the space of essentially bounded functions in $L^1(\mathbf{T})$ with the norm

$$\|f\|_\infty = \operatorname*{ess\ sup}_t |f(t)|$$

(b) $\mathrm{Lip}_\alpha(\mathbf{T}), 0 < \alpha \leqq 1$—the subspace of $C(\mathbf{T})$ consisting of the functions f for which

$$\sup_{\substack{t \\ h \neq 0}} \frac{|f(t+h) - f(t)|}{|h|^\alpha} < \infty$$

with the norm

$$\|f\|_{Lip_\alpha} = \sup_t |f(t)| + \sup_{\substack{t \\ h \neq 0}} \frac{|f(t+h) - f(t)|}{|h|^\alpha}.$$

3. Show that for $B = L^\infty(\mathbf{T})$, B_c (see lemma 2.10) is $C(\mathbf{T})$.

4. Assume $0 < a < 1$; show that for $B = \mathrm{Lip}_\alpha(\mathbf{T})$

$$B_c = \mathrm{lip}_\alpha(\mathbf{T}) = \left\{ f; \lim_{h \to 0} \sup_t \frac{|f(t+h) - f(t)|}{|h|^\alpha} = 0 \right\}.$$

5. Show that for $B = \mathrm{Lip}_1(\mathbf{T})$, $B_c = C^1(\mathbf{T})$.

6. Let B be a Banach space on $\mathbf{T}$, satisfying (H-1). Prove that B_c is the closure of the set of trigonometric polynomials in B.

7. Use exercise 1.1 and the fact that step functions are dense in $L^1(\mathbf{T})$ to prove the Riemann-Lebesgue lemma.

8. (Fejér's lemma). If

$$f \in L^1(\mathbf{T}) \text{ and } g \in L^\infty(\mathbf{T}),$$

then

$$\lim_{n\to\infty} \frac{1}{2\pi} \int f(t)g(nt)\,dt = \hat{f}(0)\hat{g}(0).$$

Hint: Approximate f in the $L^1(\mathbf{T})$ norm by polynomials.

9. Show that for $f \in L^1(\mathbf{T})$ the norm of the operator $f: g \to f * g$ on $L^1(\mathbf{T})$ is $\|f\|_{L^1}$. *Hint:* $\|\mathbf{K}_n\|_{L^1} = 1$, $\|\mathbf{K}_n * f\|_{L^1} \to \|f\|_{L^1}$.

10. Defining the *support* of a function $f \in L^1(\mathbf{T})$ as the smallest closed set S such that $f(t) = 0$ almost everywhere in the complement of S, show that the support of $f * g$ for $f, g \in L^1(\mathbf{T})$ is included in the algebraic sum support (f) + support(g).

11. For $n = 1, 2, \ldots$, let k_n be a nonnegative, infinitely differentiable function on $\mathbf{T}$ having the properties (i) $\int k_n(t)\,dt = 1$, (ii) $k_n(t) = 0$ if $|t| > 1/n$. Show that $\{k_n\}$ is a summability kernel and deduce that if B is a homogeneous Banach space on $\mathbf{T}$ and $f \in B$, then f can be approximated in the B norm by infinitely differentiable functions with supports arbitrarily close to the support of f.

12. (Bernstein).† Let P be a trigonometric polynomial of degree n. Show that $\sup_t |P'(t)| \leq 2n \sup_t |P(t)|$. *Hint*: Show that $P' = -P * 2n\,\mathbf{K}_{n-1}(t) \sin nt$ and use the fact that $\|2n\,\mathbf{K}_{n-1}(t) \sin nt\|_{L^1(\mathbf{T})} < 2n$.

13. Let B be a homogeneous Banach space on $\mathbf{T}$. Show that if $g \in L^1(\mathbf{T})$ and $f \in B$ then $g * f \in B$, and

$$\|g * f\|_B \leq \|g\|_{L^1} \|f\|_B.$$

14. Let B be a homogeneous Banach space on $\mathbf{T}$. Let $H \subseteq B$ be a closed, translation-invariant subspace. Show that H is spanned by the exponentials it contains and deduce that a function $f \in B$ is in H if, and only if, for every $n \in \mathbf{Z}$ such that $\hat{f}(n) \neq 0$, there exists $g \in H$ such that $\hat{g}(n) \neq 0$.

3. POINTWISE CONVERGENCE OF $\sigma_n(f)$

We saw in section 2 that if $f \in L^1(\mathbf{T})$, then $\sigma_n(f)$ converges to f in the topology of any homogeneous Banach space that contains f. In particular, if $f \in C(\mathbf{T})$ then $\sigma_n(f)$ converges to f uniformly. However, if f is not continuous, we cannot usually deduce pointwise convergence of $\sigma_n(f)$ from its convergence in norm, nor can we relate the limit

† Bernstein's inequality is: $\sup |p'| \leq n \sup |p|$, and can be proved similarly.

of $\sigma_n(f, t_0)$, in case it exists, to $f(t_0)$, We therefore have to reexamine the integrals defining $\sigma_n(f)$ for pointwise convergence.

3.1 ***Theorem*** *(Fejér): Let* $f \in L^1(\mathbf{T})$.

(a) Assume that $\lim_{h\to 0}(f(t_0+h)+f(t_0-h))$ *exists (we allow the values* $-\infty$ *and* $+\infty$*); then*

$$\sigma_n(f, t_0) \to \tfrac{1}{2}\lim_{h\to 0}\,(f(t_0+h)+f(t_0-h)). \tag{3.1}$$

In particular, if t_0 *is a point of continuity of* f, *then* $\sigma_n(f, t_0) \to f(t_0)$.

(b) *If every point of a closed interval* I *is a point of continuity for* f, $\sigma_n(f, t)$ *converges to* $f(t)$ *uniformly on* I.

(c) *If for all* t, $m \leqq f(t)$, *then* $m \leqq \sigma_n(f, t)$; *if for all* $t, f(t) \leqq M$, *then* $\sigma_n(f, t) \leqq M$.

Remark: The proof will be based on the fact that $\{\mathbf{K}_n(t)\}$ is a positive summability kernel which has the following properties:

(3.2) For $0 < \vartheta < \pi$

$$\lim_{n\to\infty}\Big(\sup_{\vartheta < t < 2\pi - \vartheta} \mathbf{K}_n(t)\Big) = 0,$$

$$\mathbf{K}_n(t) = \mathbf{K}_n(-t). \tag{3.3}$$

The statement of the theorem remains valid if we replace $\sigma_n(f)$ by $k_n * f$, where $\{k_n\}$ is a positive summability kernel satisfying (3.2) and (3.3). For example: the Poisson kernel satisfies all the above requirements and the statement of the theorem remains valid if we replace $\sigma_n(f)$ by the Abel means of the Fourier series of f.

PROOF OF FEJÉR'S THEOREM: We assume for simplicity that $\check{f}(t_0) = \lim_{h\to 0} \dfrac{f(t_0+h)+f(t_0-h)}{2}$ is finite, the modifications needed for the cases $\check{f}(t_0) = +\infty$ or $\check{f}(t_0) = -\infty$ being obvious. Now

$$\sigma_n(f, t_0) - \check{f}(t_0) = \frac{1}{2\pi}\int_{\mathbf{T}} \mathbf{K}_n(\tau)(f(t_0-\tau) - \check{f}(t_0))\,d\tau =$$

$$= \frac{1}{2\pi}\left(\int_{-\vartheta}^{\vartheta} + \int_{\vartheta}^{2\pi-\vartheta}\right)\mathbf{K}_n(\tau)(f(t_0-\tau) - \check{f}(t_0))\,d\tau = \tag{3.4}$$

$$= \frac{1}{\pi}\left(\int_0^{\vartheta} + \int_{\vartheta}^{\pi}\right)\mathbf{K}_n(\tau)\left[\frac{f(t_0-\tau)+f(t_0+\tau)}{2} - \check{f}(t_0)\right]d\tau.$$

(Notice that the last equality in (3.4) depends on (3.3).)

Given $\varepsilon > 0$, we choose $\vartheta > 0$ so small that

$$|\tau| < \vartheta \Rightarrow \left| \frac{f(t_0 + \tau) + f(t_0 - \tau)}{2} - \check{f}(t_0) \right| < \varepsilon, \tag{3.5}$$

and then n_0 so large that $n > n_0$ implies

$$\sup_{\vartheta < \tau < 2\pi - \vartheta} \mathbf{K}_n(\tau) < \varepsilon. \tag{3.6}$$

From (3.4), (3.5), and (3.6) we obtain

$$|\sigma_n(f, t_0) - \check{f}(t_0)| < \varepsilon + \varepsilon \, \|f - \check{f}(t_0)\|_{L^1} \tag{3.7}$$

which proves part (a).

Part (b) follows from the uniform continuity of f on I; we can pick ϑ so that (3.5) is valid for all $t_0 \in I$, and n_0 depends only on ϑ (and ε).

Part (c) depends only on the fact that $\mathbf{K}_n(t) \geqq 0$; if $m \leqq f$ then

$$\sigma_n(f, t) - m = \frac{1}{2\pi} \int \mathbf{K}_n(\tau)(f(t - \tau) - m)\, d\tau \geqq 0,$$

the integrand being nonnegative. If $f \leqq M$ then

$$M - \sigma_n(f, t) = \frac{1}{2\pi} \int \mathbf{K}_n(\tau)(M - f(t - \tau))\, d\tau \geqq 0,$$

for the same reason. ◀

Corollary: *If t_0 is a point of continuity of f and if the Fourier series of f converges at t_0 then its sum is $f(t_0)$* (cf. 2.9).

3.2 Fejér's condition

$$\check{f}(t_0) = \lim_{h \to 0} \frac{f(t_0 + h) + f(t_0 - h)}{2}$$

implies that

$$\lim_{h \to 0} \frac{1}{h} \int_0^h \left| \frac{f(t_0 + \tau) + f(t_0 - \tau)}{2} - \check{f}(t_0) \right| d\tau = 0. \tag{3.8}$$

Requiring the existence of a number $\check{f}(t)$ such that (3.8) is valid is far less restrictive than Fejér's condition and more natural for summable functions. It does not change if we modify f on a set of measure zero and, although for some function f Fejér's condition may hold for no value t_0, (3.8) holds with $\check{f}(t_0) = f(t_0)$ for almost all t_0 (cf. [28], Vol. 1, p. 65).

Theorem *(Lebesgue): If (3.8) holds, then $\sigma_n(f, t_0) \to \check{f}(t_0)$. In particular $\sigma_n(f, t) \to f(t)$ almost everywhere.*

PROOF: As in the proof of Féjer's theorem,

$$(3.9)\quad \sigma_n(f, t_0) - \check{f}(t_0) = \frac{1}{\pi}\left(\int_0^{\vartheta} + \int_{\vartheta}^{\pi}\right) \mathbf{K}_n(\tau)\left[\frac{f(t_0 - \tau) + f(t_0 + \tau)}{2} - \check{f}(t_0)\right] d\tau.$$

Remembering that $\mathbf{K}_n(\tau) = \dfrac{1}{n+1}\left(\dfrac{\sin(n+1)\tau/2}{\sin \tau/2}\right)^2$ and that $\sin\dfrac{\tau}{2} > \dfrac{\tau}{\pi}$ for $0 < \tau < \pi$, we obtain

$$(3.10)\qquad \mathbf{K}_n(\tau) \leqq \min\left(n+1, \frac{\pi^2}{(n+1)\tau^2}\right)$$

In particular we see that the second integral in (3.9) tends to zero provided $(n+1)\vartheta^2$ tends to ∞. We pick $\vartheta = n^{-1/4}$ and turn to evaluate the first integral.

Denote

$$\Phi(h) = \int_0^h \left| \frac{f(t_0 + \tau) + f(t_0 - \tau)}{2} - \check{f}(t_0) \right| d\tau;$$

then

$$\left| \frac{1}{\pi} \int_0^{\vartheta} \mathbf{K}_n(\tau) \left[\frac{f(t_0 + \tau) + f(t_0 - \tau)}{2} - \check{f}(t_0)\right] d\tau \right| \leqq \frac{1}{\pi}\left|\int_0^{1/n}\right| + \frac{1}{\pi}\left|\int_{1/n}^{\vartheta}\right|$$

$$\leqq \frac{n+1}{\pi}\Phi\left(\frac{1}{n}\right) + \frac{\pi}{n+1}\int_{1/n}^{\vartheta} \left| \frac{f(t_0 + \tau) + f(t_0 - \tau)}{2} - \check{f}(t_0) \right| \frac{d\tau}{\tau^2}.$$

The term $\dfrac{n+1}{\pi}\Phi\left(\dfrac{1}{n}\right)$ tends to zero by (3.8). Integrating by parts we obtain

$$(3.11)\qquad \begin{aligned} &\frac{\pi}{n+1}\int_{1/n}^{\vartheta} \left| \frac{f(t_0 + \tau) + f(t_0 - \tau)}{2} - \check{f}(t_0) \right| \frac{d\tau}{\tau^2} = \\ &= \frac{\pi}{n+1}\left[\frac{\Phi(\tau)}{\tau^2}\right]_{1/n}^{\vartheta} + \frac{2\pi}{n+1}\int_{1/n}^{\vartheta} \frac{\Phi(\tau)}{\tau^3}\, d\tau. \end{aligned}$$

For $\varepsilon > 0$ and $n > n(\varepsilon)$ we have by (3.8)

$$\Phi(\tau) < \varepsilon\tau \qquad \text{in} \quad 0 < \tau < \vartheta = n^{-1/4}$$

hence (3.11) is bounded by

$$\frac{\pi\varepsilon n}{n+1} + \frac{2\pi\varepsilon}{n+1}\int_{1/n}^{\vartheta} \frac{d\tau}{\tau^2} < 3\pi\varepsilon. \qquad ◀$$

Corollary: *If the Fourier series of* $f \in L^1(\mathbf{T})$ *converges on a set* E *of positive measure, its sum coincides with* f *almost everywhere on* E. *In particular, if a Fourier series converges to zero almost everywhere, all its coefficients must vanish.*

Remark: This last result is not true for all trigonometric series. There are examples of trigonometric series converging to zero almost everywhere† without being identically zero.

3.3 The need to impose in theorem 3.2 the strict condition (3.8) rather than the weaker condition

$$(3.8') \qquad \Psi(h) = \int_0^h \left(\frac{f(t_0+\tau)+f(t_0-\tau)}{2} - \check{f}(t_0)\right)d\tau = o(h)$$

comes from the fact that in order to carry the integration by parts we have to replace $\mathbf{K}_n(t)$ by the monotonic majorant $\min\left(n+1, \dfrac{\pi^2}{(n+1)t^2}\right)$. If we want to prove the analogous result for $\mathbf{P}(r,t)$ rather that $\mathbf{K}_n(t)$, the condition (3.8′) is sufficient. Thus we obtain:

Theorem (*Fatou*): *If* (3.8′) *holds, then*

$$\lim_{r\to 1} \sum_{-\infty}^{\infty} \hat{f}(j) r^{|j|} e^{ijt_0} = \check{f}(t_0).$$

The condition (3.8′) with $\check{f}(t_0) = f(t_0)$ is satisfied at every point t_0 where f is the derivative of its integral (hence almost everywhere).

EXERCISES FOR SECTION 3

1. Let $0<\alpha\le 1$ and let $f\in L^1(\mathbf{T})$, Assume that at the point $t_0 \in \mathbf{T}$, f satisfies a Lipschitz condition of order α, that is, $|f(t_0+\tau)-f(t_0)| < K|\tau|^\alpha$ for $|\tau|\le\pi$. Prove that for $\alpha<1$

$$|\sigma_n(f,t_0)-f(t_0)| \le \frac{\pi+1}{1-\alpha} K n^{-\alpha}$$

while for $\alpha=1$

$$|\sigma_n(f,t_0)-f(t_0)| \le 2\pi K \frac{\log n}{n}.$$

Hint: Use (3.10) and (3.4) with $\vartheta = 1/n$.

2. If $f \in \mathrm{Lip}_\alpha(\mathbf{T})$, $0<\alpha\le 1$, then

† However, a trigonometric series converging to zero everywhere is identically zero (see [13], Chapter 5).

$$\|\sigma(f)-f\|_\infty \leqq \begin{cases} \text{const}\,\|f\|_{\text{Lip}_\alpha}\, n^{-\alpha}, & 0<\alpha<1 \\ \text{const}\,\|f\|_{\text{Lip}_1}\dfrac{\log n}{n}, & \alpha=1\,. \end{cases}$$

3. Let $f\in L^\infty(\mathbf{T})$ and assume $|\hat{f}(n)| \leqq K|n|^{-1}$. Prove that for all n and t, $|S_n(f,t)| \leqq \|f\|_\infty + 2K$. *Hint*:

$$S_n(f,t) = \sigma_n(f,t) + \sum_{-n}^{n} \frac{|j|}{n+1}\hat{f}(j)e^{ijt}.$$

4. Show that for all n and t, $\left|\sum_1^n j^{-1}\sin jt\right| \leqq \frac{1}{2}\pi + 1$. *Hint*: Consider $f(t)=t/2$ in $[0,2\pi)$.

4. THE ORDER OF MAGNITUDE OF FOURIER COEFFICIENTS

The only things we know so far about the size of Fourier coefficients $\{\hat{f}(n)\}$ of a function $f\in L^1(\mathbf{T})$ is that they are bounded by $\|f\|_{L^1}$ (1.4(e)) and that $\lim_{|n|\to\infty}\hat{f}(n) = 0$ (the Riemann-Lebesgue lemma). In this section we discuss the following three questions:

(a) Can the Riemann-Lebesgue lemma be improved to provide a certain rate of vanishing of $\hat{f}(n)$ as $|n|\to\infty$?

We show that the answer to (a) is negative; $\hat{f}(n)$ can go to zero arbitrarily slowly (see 4.1).

(b) In view of the negative answer to (a), is it true that any sequence $\{a_n\}$ which tends to zero as $|n|\to\infty$ is the sequence of Fourier coefficients of some $f\in L^1(\mathbf{T})$?

The answer to (b) is again negative (see 4.2).

(c) How are properties like boundedness, continuity, smoothness, etc. of a function f reflected by $\{\hat{f}(n)\}$?

Question (c) is one of the main problems of harmonic analysis. In the second half of this section we show how various smoothness conditions affect the size of the Fourier coefficients. The effect of square integrability will be discussed in the following section.

4.1. Theorem: *Let $\{a_n\}_{n=-\infty}^{\infty}$ be an even sequence of nonnegative numbers tending to zero at infinity. Assume that for $n>0$*

$$a_{n-1} + a_{n+1} - 2a_n \geqq 0. \tag{4.1}$$

Then there exists a nonnegative function $f\in L^1(\mathbf{T})$ such that $\hat{f}(n) = a_n$.

PROOF: We remark first that the convexity condition (4.1) implies that $(a_n - a_{n+1})$ is monotonically decreasing with n, hence

$$\lim_{n\to\infty} n(a_n - a_{n+1}) = 0,$$

and consequently

$$\sum_{n=1}^{N} n(a_{n-1} + a_{n+1} - 2a_n) = a_0 - a_N - N(a_N - a_{N+1})$$

converges to a_0 as $N \to \infty$. Put

$$(4.2) \qquad f(t) = \sum_{n=1}^{\infty} n(a_{n-1} + a_{n+1} - 2a_n)\, \mathbf{K}_{n-1}(t),$$

$\mathbf{K}_n$ denoting, as usual, the Fejér kernel of order n. Since $\| \mathbf{K}_n \|_{L^1} = 1$, the series (4.2) converges in $L^1(\mathbf{T})$ and, all its terms being nonnegative, its limit f is nonnegative. Now

$$\hat{f}(j) = \sum_{n=1}^{\infty} n(a_{n-1} + a_{n+1} - 2a_n)\hat{\mathbf{K}}_{n-1}(j) =$$
$$= \sum_{n=|j|+1}^{\infty} n(a_{n-1} + a_{n+1} - 2a_n)\left(1 - \frac{|j|}{n}\right) = a_{|j|},$$

and the proof is complete. ◀

4.2 Comparing theorem 4.1 to our next theorem shows the basic difference between sine-series $(a_{-n} = -a_n)$ and cosine-series $(a_{-n} = a_n)$.

Theorem: *Let $f \in L^1(\mathbf{T})$ and assume that $\hat{f}(|n|) = -\hat{f}(-|n|) \geqq 0$. Then*

$$\sum_{n\neq 0} \frac{1}{n}\hat{f}(n) < \infty.$$

PROOF: Without loss of generality we assume $\hat{f}(0) = 0$. Put $F(t) = \int_0^t f(\tau)\,d\tau$; then $F \in C(\mathbf{T})$ and, by theorem 1.6,

$$\hat{F}(n) = \frac{1}{in}\,\hat{f}(n), \qquad n \neq 0.$$

Since F is continuous, we can apply Fejér's theorem for $t_0 = 0$ and obtain

$$(4.3) \qquad \lim_{N\to\infty} 2\sum_{n=1}^{N}\left(1 - \frac{n}{N+1}\right)\frac{\hat{f}(n)}{n} = i(F(0) - \hat{F}(0))$$

and since $\dfrac{\hat{f}(n)}{n} \geqq 0$, the theorem follows. ◀

Corollary: *If* $a_n > 0$, $\sum a_n/n = \infty$, *then* $\sum_{n=1}^{\infty} a_n \sin nt$ *is not a Fourier series. Hence there exist trigonometric series with coefficients tending to zero which are not Fourier series.*

By theorem 4.1, the series $\sum_{n=2}^{\infty} \dfrac{\cos nt}{\log n} = \sum_{|n| \geq 2} \dfrac{e^{int}}{2\log|n|}$ is a Fourier series while, by theorem 4.2, its conjugate series $\sum_{n=2}^{\infty} \dfrac{\sin nt}{\log n}$ $= -i \sum_{|n| \geq 2} \dfrac{\operatorname{sgn}(n)}{2\log|n|} e^{int}$ is not.

4.3 We turn now to some simple results about the order of magnitude of Fourier coefficients of functions satisfying various smoothness conditions.

Theorem: *If* $f \in L^1(\mathbf{T})$ *is absolutely continuous, then* $\hat{f}(n) = o(1/n)$.

PROOF: By theorem 1.6 we have $\hat{f}(n) = (1/in)\widehat{f'}(n)$ and by the Riemann-Lebesgue lemma $\widehat{f'}(n) \to 0$. ◀

Remark: By repeated application of theorem 1.6 (i.e., by repeated integration by parts) we see that if f is k-times differentiable and $f^{(k)} \in L^1(\mathbf{T})$, then

$$\hat{f}(n) = o\left(\frac{1}{n^k}\right) \quad \text{as } |n| \to \infty. \tag{4.4}$$

4.4 We can obtain a somewhat more precise estimate than the asymptotic (4.4). All that we have to do is notice that if $0 \leq j \leq k$, then $\hat{f}(n) = (in)^{-j}\widehat{f^{(j)}}(n)$ and hence

$$|\hat{f}(n)| \leq |n|^{-j} \|f^{(j)}\|_{L^1}. \tag{4.5}$$

We thus obtain

Theorem: *If* f *is* k*-times differentiable and* $f^{(k)} \in L^1(\mathbf{T})$, *then*

$$|\hat{f}(n)| \leq \min_{0 \leq j \leq k} \frac{\|f^{(j)}\|_{L^1}}{|n|^j}.$$

If f *is infinitely differentiable, then*

$$|\hat{f}(n)| \leq \min_{0 \leq j} \frac{\|f^{(j)}\|_{L^1}}{|n|^j}.$$

4.5 ***Theorem***: *If* f *is of bounded variation on* $\mathbf{T}$, *then*

$$|\hat{f}(n)| \leqq \frac{\mathrm{Var}(f)}{2\pi|n|}.$$

PROOF: We integrate by parts using Stieltjes integrals

$$|\hat{f}(n)| = \left| \frac{1}{2\pi}\int e^{-int}f(t)\,dt \right| = \left| \frac{1}{2\pi in}\int e^{-int}\,df(t) \right| \leqq \frac{\mathrm{Var}(f)}{2\pi|n|}. \quad ◀$$

4.6 For $f \in C(\mathbf{T})$ we denote by $\omega(f,h)$ the *modulus of continuity* of f, that is,

$$\omega(f,h) = \max_{|y| \leq h, t} |f(t+y) - f(t)|.$$

For $f \in L^1(\mathbf{T})$ we denote by $\Omega(f,h)$ the *integral modulus of continuity* of f, that is,

$$\Omega(f,h) = \| f(t+h) - f(t) \|_{L^1}. \tag{4.6}$$

We clearly have $\Omega(f,h) \leqq \omega(f,h)$.

Theorem: *For* $n \neq 0$, $|\hat{f}(n)| \leqq \frac{1}{2}\Omega\left(f, \frac{\pi}{|n|}\right)$.

PROOF: $\hat{f}(n) = \frac{1}{2\pi}\int f(t)e^{-int}\,dt = \frac{-1}{2\pi}\int f(t)\,e^{-in(t+\pi/n)}\,dt$;

by a change of variable,

$$\hat{f}(n) = \frac{1}{4\pi}\int \left(f\left(t+\frac{\pi}{n}\right) - f(t)\right) e^{-int}\,dt,$$

hence

$$|\hat{f}(n)| < \frac{1}{2}\Omega\left(f, \frac{\pi}{|n|}\right). \quad ◀$$

Corollary: *If* $f \in \mathrm{Lip}_\alpha(\mathbf{T})$, *then* $\hat{f}(n) = O(n^{-\alpha})$.

4.7. ***Theorem***: *Let* $1 < p \leqq 2$ *and let* q *be the conjugate exponent* $\left(\text{i.e., } q = \frac{p}{p-1}\right)$. *If* $f \in L^p(\mathbf{T})$ *then* $\sum |\hat{f}(n)|^q < \infty$.

The case $p = 2$ will be proved in the following section. The case $1 < p < 2$ will be proved in chapter IV.

Remark: Theorem 4.7 cannot be extended to $p > 2$. Thus, if $f \in L^p(\mathbf{T})$ with $p > 2$, then $f \in L^2(\mathbf{T})$ and consequently $\sum |\hat{f}(n)|^2 < \infty$; this is all that we can assert. There exist continuous functions f such that $\sum |\hat{f}(n)|^{2-\varepsilon} = \infty$ for all $\varepsilon > 0$.

EXERCISES FOR SECTION 4

1. Given a sequence $\{\omega_n\}$ of positive numbers such that $\omega_n \to 0$ as $|n| \to \infty$, show that there exists a sequence $\{a_n\}$ satisfying the conditions of theorem 4.1 and

$$a_n > \omega_n \qquad \text{for all } n.$$

2. Show that if $\sum |\hat{f}(n)|\,|n|^l < \infty$, then f is l-times continuously differentiable. Hence, if $\hat{f}(n) = O(|n|^{-k})$ where $k > 2$, and if

$$l = \begin{cases} k-2 & k \text{ integer} \\ [k]-1 & \text{otherwise} \end{cases}$$

then f is l-times continuously differentiable.

Remarks: Properly speaking the elements of $L^1(\mathbf{T})$ are equivalence classes of functions any two of which differ only on a set of measure zero. Saying that a function $f \in L^1(\mathbf{T})$ is continuous or differentiable etc. is a convenient and innocuous abuse of language with obvious meaning.

Exercise 2 is all that we can state as a converse to theorem 4.4 if we look for continuous derivatives. It can be improved if we allow square summable derivatives (see exercise 5.5).

3. A function f is *analytic* on $\mathbf{T}$ if in a neighborhood of every $t_0 \in \mathbf{T}$, $f(t)$ can be represented by a power series (of the form $\sum_{n=0}^{\infty} a_n(t-t_0)^n$). Show that f is analytic if, and only if, f is infinitely differentiable on $\mathbf{T}$ and there exists a number R such that

$$\sup_t |f^{(n)}(t)| \leqq n!R^n, \qquad n > 0.$$

4. Show that f is analytic on $\mathbf{T}$ if, and only if, there exist constants $K > 0$ and $a > 0$ such that $|\hat{f}(j)| \leqq Ke^{-a|j|}$. Hence show that f is analytic on $\mathbf{T}$ if, and only if, $\sum \hat{f}(j)\, e^{ijz}$ converges for $|\mathrm{Im}(z)| < a$ for some $a > 0$.

5. Let f be analytic on $\mathbf{T}$ and let $g(e^{it}) = f(t)$. What is the relation between the Laurent expansion of g about 0 (which converges in an annulus containing the circle $|z| = 1$) and the Fourier series of f?

6. Let f be infinitely differentiable on $\mathbf{T}$ and assume that for $n \geqq 1$ $\sup_t |f^{(n)}(t)| < Kn^{\alpha n}$ with $\alpha > 0$, Show that

$$|\hat{f}(j)| \leqq K\exp\left(-\frac{\alpha}{e}|j|^{1/\alpha}\right).$$

7. Assume $|\hat{f}(j)| \leqq K\exp(-|j|^{1/\alpha})$. Show that f is infinitely differentiable and

$$|f^{(n)}(t)| < K_1\, e^{an} n^{\alpha n}$$

for some constants a and K_1. *Hint*: $|f^{(n)}(t)| \leqq 2K\sum |j|^n \exp(-|j|^{1/\alpha})$. Compare this last sum to $\int_0^\infty x^n \exp(-x^{1/\alpha})\,dx$ and change the variable of integration putting $y = x^{1/\alpha}$.

8. Prove that if $0 < \alpha \leqq 1$, then $f(t) = \sum_1^\infty \frac{\cos 3^n t}{3^{n\alpha}}$ belongs to $\text{Lip}_\alpha(\mathbf{T})$; hence corollary 4.6 cannot be improved.

9. Show that the series $\sum_{n=2}^\infty \frac{\sin nt}{\log n}$ converges for all $t \in \mathbf{T}$.

5. FOURIER SERIES OF SQUARE SUMMABLE FUNCTIONS

In some respects the greatest success in representing functions by means of their Fourier series happens for square summable functions. The reason is that $L^2(\mathbf{T})$ is a Hilbert space, its inner product being defined by

(5.1)
$$\langle f, g\rangle = \frac{1}{2\pi}\int f(t)\overline{g(t)}\,dt,$$

and in this Hilbert space the exponentials form a complete orthogonal system. We start this section with a brief review of the basic properties of orthonormal and complete systems in abstract Hilbert space and conclude with the corresponding statements about Fourier series in $L^2(\mathbf{T})$.

5.1 Let $\mathscr{H}$ be a complex Hilbert space. Let $f, g \in \mathscr{H}$. We say that f is *orthogonal* to g if $\langle f, g\rangle = 0$. This relation is clearly symmetric. If E is a subset of $\mathscr{H}$ we say that $f \in \mathscr{H}$ is *orthogonal to E* if f is orthogonal to every element of E. A set $E \subset \mathscr{H}$ is *orthogonal* if any two vectors in E are orthogonal to each other. A set $E \subset \mathscr{H}$ is an *orthonormal system* if it is orthogonal and the norm of each vector in E is one, that is, if, whenever $f, g \in E$, $\langle f, g\rangle = 0$ if $f \neq g$ and $\langle f, f\rangle = 1$.

Lemma: *Let $\{\varphi_n\}_{n=1}^N$ be a finite orthonormal system. Let $a_1, \ldots, a_N$ be complex numbers. Then*

$$\left\| \sum_1^N a_n\varphi_n \right\|^2 = \sum_{n=1}^N |a_n|^2.$$

PROOF:

$$\left\| \sum_1^N a_n\varphi_n \right\|^2 = \left\langle \sum_1^N a_n\varphi_n, \sum_1^N a_n\varphi_n \right\rangle = \sum_1^N a_n \left\langle \varphi_n, \sum_1^N a_m\varphi_m \right\rangle$$
$$= \sum a_n \bar{a}_n = \sum |a_n|^2 \quad ◀$$

Corollary: *Let $\{\varphi_n\}_{n=1}^{\infty}$ be an orthonormal system in $\mathscr{H}$ and let $\{a_n\}_{n=1}^{\infty}$ be a sequence of complex numbers such that $\sum |a_n|^2 < \infty$. Then $\sum_{n=1}^{\infty} a_n\varphi_n$ converges in $\mathscr{H}$.*

PROOF: Since $\mathscr{H}$ is complete, all that we have to show is that the partial sums $S_N = \sum_1^N a_n\varphi_n$ from a Cauchy sequence in $\mathscr{H}$. Now, for $N > M$,

$$\| S_N - S_M \|^2 = \Big\| \sum_{M+1}^{N} a_n\varphi_n \Big\|^2 = \sum_{M+1}^{N} |a_n|^2 \to 0 \quad \text{as } M \to \infty. \qquad \blacktriangleleft$$

5.2 ***Lemma:*** *Let $\mathscr{H}$ be a Hilbert space. Let $\{\varphi_n\}_{n=1}^{N}$ be a finite orthonormal system in $\mathscr{H}$. For $f \in \mathscr{H}$ write $a_n = \langle f, \varphi_n \rangle$. Then*

$$(5.2) \qquad 0 \leqq \Big\| f - \sum_1^N a_n\varphi_n \Big\|^2 = \| f \|^2 - \sum_1^N |a_n|^2.$$

PROOF:

$$\Big\| f - \sum_1^N a_n\varphi_n \Big\|^2 = \Big\langle f - \sum_1^N a_n\varphi_n, f - \sum_1^N a_n\varphi_n \Big\rangle =$$

$$= \| f \|^2 - \sum_1^N \bar{a}_n\langle f, \varphi_n\rangle - \sum_1^N a_n\langle \varphi_n, f\rangle + \sum_1^N |a_n|^2 = \| f \|^2 - \sum_1^N |a_n|^2. \qquad \blacktriangleleft$$

Corollary *(Bessel's inequality): Let $\mathscr{H}$ be a Hilbert space and $\{\varphi_\alpha\}$ an orthonormal system in $\mathscr{H}$. For $f \in \mathscr{H}$ write $a_\alpha = \langle f, \varphi_\alpha \rangle$. Then*

$$(5.3) \qquad \sum |a_\alpha|^2 \leqq \| f \|^2.$$

The family $\{\varphi_\alpha\}$ in the statement of Bessel's inequality need not be finite nor even countable. The inequality (5,3) is equivalent to saying that for every finite subset of $\{\varphi_\alpha\}$ we have (5.2). In particular $a_\alpha = 0$ except for countably many values of α and the series $\sum |a_\alpha|^2$ converges.

If $\mathscr{H} = L^2(\mathbf{T})$ all orthonormal systems in $\mathscr{H}$ are finite or countable (cf. exercise 2 at the end of this section) and we write them as sequences $\{\varphi_n\}$.

5.3 DEFINITION: A *complete orthonormal system* in $\mathscr{H}$ is an orthonormal system having the additional property that the only vector in $\mathscr{H}$ orthogonal to it is the zero vector.

Lemma: *Let $\{\varphi_n\}$ be an orthonormal system in $\mathscr{H}$. Then the following statements are equivalent*:

(a) *$\{\varphi_n\}$ is complete.*

(b) *For every $f \in \mathscr{H}$ we have*

$$\|f\|^2 = \sum |\langle f, \varphi_n \rangle|^2 . \tag{5.4}$$

(c) $$f = \sum \langle f, \varphi_n \rangle \varphi_n .$$

PROOF: That (b) and (c) are equivalent follows immediately from (5.2). If f is orthogonal to $\{\varphi_n\}$ and if (5.4) is valid, then $\|f\|^2 = 0$, hence $f = 0$. Thus (b) $\Rightarrow$ (a). We complete the proof by showing (a) $\Rightarrow$ (c). From Bessel's inequality and corollary 5.1 it follows that $\sum \langle f, \varphi_n \rangle \varphi_n$ converges in $\mathscr{H}$. If we denote $g = \sum \langle f, \varphi_n \rangle \varphi_n$ we see that $\langle g, \varphi_n \rangle = \langle f, \varphi_n \rangle$ or, equivalently, $g - f$ is orthogonal to $\{\varphi_n\}$. Thus if $\{\varphi_n\}$ is complete $f = g$. ◀

5.4 ***Lemma*** (*Parseval*): *Let $\{\varphi_n\}_{n=1}^{\infty}$ be a complete orthonormal system in $\mathscr{H}$. Let $f, g \in \mathscr{H}$. Then*

$$\langle f, g \rangle = \sum_{n=1}^{\infty} \langle f, \varphi_n \rangle \langle \varphi_n, g \rangle . \tag{5.5}$$

PROOF: If f is a finite linear combination of $\{\varphi_n\}$, (5.5) is obvious. In the general case

$$\langle f, g \rangle = \lim_{N \to \infty} \left\langle \sum_1^N \langle f, \varphi_n \rangle \varphi_n, g \right\rangle = \lim_{N \to \infty} \sum_1^N \langle f, \varphi_n \rangle \langle \varphi_n, g \rangle . \quad ◀$$

5.5 For $\mathscr{H} = L^2(\mathbf{T})$ the exponentials $\{e^{int}\}_{n=-\infty}^{\infty}$ form a complete orthonormal system. The orthonormality is evident:

$$\langle e^{int}, e^{imt} \rangle = \frac{1}{2\pi} \int e^{i(n-m)t}\, dt = \delta_{n,m} .$$

The completeness is somewhat less evident; it follows from theorem 2.7 since

$$\langle f, e^{int} \rangle = \frac{1}{2\pi} \int f(t) \overline{e^{int}}\, dt = \hat{f}(n) .$$

The general results about complete orthonormal systems in Hilbert space now yield

Theorem: (a) *Let $f \in L^2(\mathbf{T})$. Then*

$$\sum |\hat{f}(n)|^2 = \frac{1}{2\pi}\int |f(t)|^2\, dt$$

(b) $$f = \lim_{N\to\infty} \sum_{-N}^{N} \hat{f}(n)\, e^{int} \qquad \textit{in the } L^2(\mathbf{T}) \textit{ norm.}$$

(c) *Given any sequence* $\{a_n\}_{n=-\infty}^{\infty}$ *of complex numbers satisfying* $\sum |a_n|^2 < \infty$, *then there exists a unique* $f \in L^2(\mathbf{T})$ *such that* $a_n = \hat{f}(n)$.

(d) *Let* $f, g \in L^2(\mathbf{T})$. *Then*

$$\frac{1}{2\pi}\int f(t)\overline{g(t)}\, dt = \sum_{n=-\infty}^{\infty} \hat{f}(n)\overline{\hat{g}(n)}.$$

We denote by ℓ^2 the space of sequences $\{a_n\}_{-\infty}^{\infty}$ such that $\sum |a_n|^2 < \infty$. With pointwise addition and scalar multiplication and with the norm $(\sum |a_n|^2)^{1/2}$ or equivalently the inner product $\langle\{a_n\},\{b_n\}\rangle = \sum_{-\infty}^{\infty} a_n \bar{b}_n$, ℓ^2 is a Hilbert space. Theorem 5.5 is equivalent to the statement that the correspondence $f \leftrightarrow \{\hat{f}(n)\}$ is an isometry between $L^2(\mathbf{T})$ and ℓ^2.

EXERCISES FOR SECTION 5

1. Let $\{\varphi_n\}_{n=1}$ be an orthogonal system in a Hilbert space $\mathscr{H}$. Let $f \in \mathscr{H}$. Show that

$$\min_{a_1,\ldots,a_N} \left\| f - \sum_{1}^{N} a_j \varphi_j \right\|$$

is attained at the point $a_j = \langle f, \varphi_j\rangle$, $j = 1, \ldots, N$, and only there.

2. A Hilbert space $\mathscr{H}$ is *separable* if it contains a dense countable subset. Show that an orthonormal system in a separable Hilbert space is either finite or countable. *Hint*: The distance between two orthogonal vectors of norm 1 is $\sqrt{2}$.

3. Prove that an orthonormal system $\{\varphi_n\}$ in $\mathscr{H}$ is complete if, and only if, the set of finite linear combinations of $\{\varphi_n\}$ is dense in $\mathscr{H}$.

4. Let f be absolutely continuous on $\mathbf{T}$ and assume $f' \in L^2(\mathbf{T})$; prove that

$$\sum |\hat{f}(n)| \leqq \| f \|_{L^1} + \sqrt{2 \textstyle\sum_1^{\infty} n^{-2}}\; \|f'\|_{L^2}.$$

Hint: $|\hat{f}(0)| \leqq \|f\|_{L^1}$ and $\sum |n\hat{f}(n)|^2 = \|f'\|_{L^2}^2$; apply the Cauchy-Schwarz inequality to the last identity,

5. Assume $f \in L^1(\mathbf{T})$ and $\hat{f}(n) = O(|n|^{-k})$. Show that f is m-times differentiable with $f^{(m)} \in L^2(\mathbf{T})$ provided $k - m > \frac{1}{2}$.

6. ABSOLUTELY CONVERGENT FOURIER SERIES

We shall study absolutely convergent Fourier series in some detail later on: here we mention only some elementary facts.

6.1. We denote by $A(\mathbf{T})$ the space of (continuous) functions on **T** having an absolutely convergent Fourier series, that is, the functions f for which $\sum_{-\infty}^{\infty}|\hat{f}(n)| < \infty$. The mapping $f \to \{\hat{f}(n)\}_{n\in\mathbf{Z}}$ of $A(\mathbf{T})$ into ℓ^1 (the Banach space of absolutely convergent sequences) is clearly linear and one-to-one. If $\sum_{-\infty}^{\infty}|a_n| < \infty$ the series $\sum_{-\infty}^{\infty} a_n e^{int}$ converges uniformly on **T** and, denoting its sum by g, we have $a_n = \hat{g}(n)$. It follows that the mapping above is an isomorphism of $A(\mathbf{T})$ onto ℓ^1. We introduce a norm to $A(\mathbf{T})$ by

$$\|f\|_{A(\mathbf{T})} = \sum_{-\infty}^{\infty} |\hat{f}(n)|. \tag{6.1}$$

With this norm $A(\mathbf{T})$ is a Banach space isometric to ℓ^1; we now claim it is an algebra.

Lemma: *Assume that* $f, g \in A(\mathbf{T})$. *Then* $fg \in A(\mathbf{T})$ *and*

$$\|fg\|_{A(\mathbf{T})} \leqq \|f\|_{A(\mathbf{T})}\|g\|_{A(\mathbf{T})}.$$

PROOF: We have $f(t) = \sum \hat{f}(n)e^{int}$, $g(t) = \sum \hat{g}(n)e^{int}$ and since both series converge absolutely:

$$f(t)g(t) = \sum_k \sum_m \hat{f}(k)\hat{g}(m)e^{i(k+m)t}$$

Collecting the terms for which $k + m = n$ we obtain

$$f(t)g(t) = \sum_n \sum_k \hat{f}(k)\hat{g}(n-k)e^{int}$$

so that $\widehat{fg}(n) = \sum_k \hat{f}(k)\hat{g}(n-k)$; hence $\sum|\widehat{fg}(n)| \leqq \sum_n\sum_k|\hat{f}(k)||\hat{g}(n-k)| = \sum_k|\hat{f}(k)|\sum_n|\hat{g}(n)|$. ◀

6.2 Not every continuous function on **T** has an absolutely convergent Fourier series, and those that have cannot† be characterized by smoothness conditions (see exercise 5 of this section). Some smoothness conditions are sufficient, however, to imply the absolute convergence of the Fourier series.

† See, however, exercise 7.8.

Theorem: *Let f be absolutely continuous on* $\mathbf{T}$ *and assume* $f' \in L^2(\mathbf{T})$. *Then* $f \in A(\mathbf{T})$ *and*

$$\|f\|_{A(\mathbf{T})} \leqq \|f\|_{L^1} + \left(2\sum_1^{\infty} n^{-2}\right)^{1/2} \|f'\|_{L^2}. \tag{6.2}$$

PROOF: This is exercise 4 of the previous section and the hint given there is essentially the whole proof.

***6.3** We refer to exercise 2.2 for the definitions of $\mathrm{Lip}_\alpha(\mathbf{T})$ and of its norm.

Theorem (*Bernstein*): *Assume* $f \in \mathrm{Lip}_\alpha(\mathbf{T})$ *for some* $\alpha > \frac{1}{2}$. *Then* $f \in A(\mathbf{T})$ *and*

$$\|f\|_{A(\mathbf{T})} \leqq c_\alpha \|f\|_{\mathrm{Lip}_\alpha} \tag{6.3}$$

where the constant c_α *depends only on* α.

PROOF:

$$f(t-h) - f(t) \sim \sum (e^{-inh} - 1)\hat{f}(n)\, e^{int}.$$

If we take $h = 2\pi/(3.2^m)$ and $2^m \leqq n \leqq 2^{m+1}$, we have $|e^{-inh} - 1| \geqq \sqrt{3}$ and consequently

$$\sum_{2^m \leqq |n| < 2^{m+1}} |\hat{f}(n)|^2 \leqq \sum_n |e^{inh} - 1|^2 |\hat{f}(n)|^2 = \|f_h - f\|_{L^2}^2 \leqq \tag{6.4}$$

$$\leqq \|f_h - f\|_\infty^2 \leqq \left(\frac{2\pi}{3 \cdot 2^m}\right)^{2\alpha} \|f\|_{\mathrm{Lip}_\alpha}^2.$$

Noticing that the sum on the left of (6.4) consists of at most 2^{m+1} terms, we obtain by the Cauchy-Schwarz inequality

$$\sum_{2^m \leqq |n| < 2^{m+1}} |\hat{f}(n)| \leqq 2^{(m+1)/2} \left(\frac{2\pi}{3 \cdot 2^m}\right)^{\alpha} \|f\|_{\mathrm{Lip}_\alpha}. \tag{6.5_m}$$

Since $\alpha > \frac{1}{2}$, we can sum the inequalities $(6.5)_m$ for $m = 0, 1, \ldots$, and remembering that $|\hat{f}(0)| \leqq \|f\|_{\mathrm{Lip}_\alpha}$ we obtain (6.3). ◀

Bernstein's theorem is sharp; there exist functions in $\mathrm{Lip}_{1/2}(\mathbf{T})$ the Fourier series of which does not converge absolutely. A classical example is the Hardy-Littlewood series $\sum_{n=1}^{\infty} \frac{e^{in \log n}}{n} e^{int}$ (see [28], Vol. 1, p. 197). Another example is given in exercise 6.6.

***6.4** The Lipschitz condition in 6.3 can be relaxed if we assume that f is of bounded variation.

Theorem (*Zygmund*): *Let f be of bounded variation on* **T** *and assume $f \in \mathrm{Lip}_\alpha(\mathbf{T})$ for some $\alpha > 0$. Then $f \in A(\mathbf{T})$.*

We refer to [28], Vol. 1, p. 241, for the proof.

***6.5** *Remark*: There is a change of scene in this section compared with the rest of the chapter. We no longer talk about functions summable on **T** and their Fourier series—we discuss functions summable on **Z** (i.e., absolutely convergent sequences) and their "Fourier transforms" which happen to be continuous functions on **T**. Lemma 6.1, for instance, is completely analogous to theorem 1.7 with the roles of **T** and **Z** reversed.

EXERCISES FOR SECTION 6

1. For $n = 1, 2, \ldots$, let $f_n \in A(\mathbf{T})$ and $\|f_n\|_{A(\mathbf{T})} \leqq 1$. Assume that f_n converge to f uniformly on **T**. Show that $f \in A(\mathbf{T})$ and $\|f\|_{A(\mathbf{T})} \leqq 1$.

2. Show that the conditions in exercise 1 are not sufficient to imply $\lim_{n\to\infty} \|f - f_n\|_{A(\mathbf{T})} = 0$; however, if we add the assumption that $\|f\|_{A(\mathbf{T})} = \lim \|f_n\|_{A(\mathbf{T})}$, then $\|f - f_n\|_{A(\mathbf{T})} \to 0$.

3. For $0 < a \leqq \pi$ define

$$\Delta_a(t) = \begin{cases} 1 - a^{-1}|t| & \text{for } |t| \leqq a \\ 0 & \text{for } a \leqq |t| \leqq \pi \end{cases}$$

Show that $\Delta_a \in A(\mathbf{T})$ and $\|\Delta_a\|_{A(\mathbf{T})} = 1$. *Hint*: $\hat{\Delta}_a(n) \geqq 0$ for all n.

4. Let $f \in C(\mathbf{T})$ be even on $(-\pi, \pi)$, decreasing on $[0, \pi]$ and convex there (i.e., $f(t+2h) + f(t) \geqq 2f(t+h)$ for $0 \leqq t < t + 2h \leqq \pi$). Show that $f \in A(\mathbf{T})$ and, if $f \geq 0$, $\|f\|_{A(\mathbf{T})} = f(0)$. *Hint*: f can be approximated uniformly by positive combinations of Δ_a. Compare with theorem 4.1.

5. Let φ be a "modulus of continuity," that is, an increasing concave function on $[0,1]$ with $\varphi(0) = 0$, Show that if the sequence of integers $\{\lambda_n\}$ increases fast enough and if $f(t) = \sum \frac{1}{n^2} e^{i\lambda_n t}$ then $\omega(f,h) \neq O(\varphi(h))$ as $h \to 0$. $\omega(f,h)$ is the modulus of continuity of f (defined in 4.6).

6. (Rudin, Shapiro.) We define the trigonometric polynomials P_m and Q_m inductively as follows: $P_0 = Q_0 = 1$ and

$$P_{m+1}(t) = P_m(t) + e^{i2^m t} Q_m(t)$$
$$Q_{m+1}(t) = P_m(t) - e^{i2^m t} Q_m(t).$$

(a) Show that

$$|P_{m+1}(t)|^2 + |Q_{m+1}(t)|^2 = 2(|P_m(t)|^2 + |Q_m(t)|^2)$$

hence

$$|P_m(t)|^2 + |Q_m(t)|^2 = 2^{m+1}$$

and

$$\|P_m\|_{C(\mathbf{T})} \leqq 2^{(m+1)/2}.$$

(b) For $|n| < 2^m$, $\hat{P}_{m+1}(n) = \hat{P}_m(n)$, hence there exists a sequence $\{\varepsilon_n\}_{n=0}^{\infty}$ such that ε_n is either 1 or -1 and such that $P_m(t) = \sum_0^{2m-1} \varepsilon_n e^{int}$.
(c) Write $f_m = P_m - P_{m-1} = e^{i2^{m-1}t} Q_{m-1}$ and $f = \sum_1^{\infty} 2^{-m} f_m$. Show that $f \in \mathrm{Lip}_{1/2}(\mathbf{T})$ and $f \notin A(\mathbf{T})$. *Hint*: For $2^{-k} < h \leqq 2^{1-k}$ write

$$f(t+h) - f(t) = \left(\sum_1^k + \sum_{k+1}^{\infty}\right) 2^{-m}(f_m(t+h) - f_m(t)).$$

By part (a) the sum $\sum_{k+1}^{\infty}$ is bounded by $2\sum_{k+1}^{\infty} 2^{-m} 2^{m/2} < 5h^{1/2}$. Using exercise 2.12, part (a), and the fact that f_m is a trigonometric polynomial of degree $2^m - 1$, one obtains a similar estimate for $\sum_1^k$.

7. Let $f, g \in L^2(\mathbf{T})$. Show that $f * g \in A(\mathbf{T})$.

7. FOURIER COEFFICIENT OF LINEAR FUNCTIONALS

We consider a homogeneous Banach space B on $\mathbf{T}$ and assume, for simplicity, that $e^{int} \in B$ for all n. As usual, we denote by B^* the dual space of B.

7.1 The Fourier coefficients of a functional $\mu \in B^*$ are, by definition:

$$\hat{\mu}(n) = \overline{\langle e^{int}, \mu\rangle}, \qquad n \in \mathbf{Z}; \tag{7.1}$$

and we call the trigonometric series

$$S[\mu] \sim \sum_{-\infty}^{\infty} \hat{\mu}(n) e^{int},$$

the Fourier series† of μ. Clearly

$$|\hat{\mu}(n)| \leqq \|\mu\|_{B^*} \|e^{int}\|_B.$$

The notation (7.1) is consistent with our definition of Fourier coefficients in case that μ is identified naturally with a summable function. For instance, if $B = L^p(\mathbf{T})$, $1 < p < \infty$, B^* is canonically identified with $L^q(\mathbf{T})$ where $q = p/(p-1)$. To the function $g \in L^q(\mathbf{T})$ corresponds the linear functional

† We keep, however, the convention of 1.6 that *a Fourier series*, without complements, is a Fourier series of a summable function.

$$f \to \langle f, g \rangle = \frac{1}{2\pi} \int f(t)\overline{g(t)}\,dt, \qquad f \in L^p(\mathbf{T})$$

and

$$\overline{\langle e^{int}, g \rangle} = \frac{1}{2\pi} \int \overline{e^{\,int}\,\overline{g(t)}\,dt} = \frac{1}{2\pi} \int e^{-int} g(t)\,dt$$

thus $\hat{g}(n)$ as defined in (7.1) for the functional g coincides with the nth Fourier coefficient of the function g.

Theorem *(Parseval's formula): Let $f \in B$, $\mu \in B^*$; then*

(7.2) $$\langle f, \mu \rangle = \lim_{N \to \infty} \sum_{-N}^{N} \left(1 - \frac{|n|}{N+1}\right) \hat{f}(n)\overline{\hat{\mu}(n)}.$$

PROOF:

(a) For polynomials $P(t) = \sum_{-N}^{N} \hat{P}(n)\,e^{int}$ we clearly have $\langle P, \mu \rangle = \sum_{-N}^{N} \hat{P}(n)\overline{\hat{\mu}(n)}$.

(b) Since, by theorem 2.11, $f = \lim_{N \to \infty} \sigma_N(f)$ in the B norm, it follows from (a) and the continuity of μ that

$$\langle f, \mu \rangle = \lim \langle \sigma_N(f), \mu \rangle = \lim_{N \to \infty} \sum_{-N}^{N} \left(1 - \frac{|n|}{N+1}\right) \hat{f}(n)\overline{\hat{\mu}(n)}. \quad ◀$$

Remark: The fact that the limit in (7.2) exists is an implicit part of the theorem. It is equivalent to the C-1 summability of the series $\sum \hat{f}(n)\overline{\hat{\mu}(n)}$. If this last series converges then clearly

(7.3) $$\langle f, \mu \rangle = \sum_{-\infty}^{\infty} \hat{f}(n)\overline{\hat{\mu}(n)}.$$

We shall sometimes refer to (7.3) as Parseval's formula, keeping in mind that if the series on the right does not converge then (7.3) is simply an abbreviation for (7.2).

Corollary *(Uniqueness theorem): If $\hat{\mu}(n) = 0$ for all n, then $\mu = 0$.*

7.2 We shall often write $\mu \sim \sum \hat{\mu}(n)\,e^{int}$; if the series converges in some sense (which should be clear from the context), we may write $\mu = \sum \hat{\mu}(n)\,e^{int}$. This is an abuse of language which, if used with caution, presents no risk of misunderstanding and obviates tedious repetitions.

In accordance with our abuse of language we define, for $\mu \in B^*$, the elements $S_n(\mu)$ of B^* by

$$(7.4)\qquad \begin{aligned} S_n(\mu) &= \sum_{-n}^{n} \hat{\mu}(j)e^{ijt} \\ \sigma_n(\mu) &= \sum_{-n}^{n} \left(1 - \frac{|j|}{n+1}\right)\hat{\mu}(j)e^{ijt}. \end{aligned}$$

We shall also write

$$(7.5)\qquad \begin{aligned} S_n(\mu, t) &= \sum_{-n}^{n} \hat{\mu}(j)e^{ijt} \\ \sigma_n(\mu, t) &= \sum_{-n}^{n} \left(1 - \frac{|j|}{n+1}\right)\hat{\mu}(j)e^{ijt}. \end{aligned}$$

The correspondence between the functionals (7.4) and the functions (7.5) is clearly

$$\langle f, S_n(\mu)\rangle = \frac{1}{2\pi}\int f(t)\overline{S_n(\mu, t)}\, dt = \sum_{-n}^{n} \hat{f}(j)\,\overline{\hat{\mu}(j)}$$

for all $f \in B$; similarly for $\sigma_n(\mu)$.

The mapping $\mathbf{S}_n: f \to S_n(f)$ on B is clearly a bounded linear operator, and so is $\mathbf{S}_n: \mu \to S_n(\mu)$ on B^*. It follows from Parseval's formula that $\mathbf{S}_n$ on B^* is the adjoint of $\mathbf{S}_n$ on B and consequently has the same norm. Similarly, $\boldsymbol{\sigma}_n: \mu \to \sigma_n(\mu)$ on B^* is the adjoint of $\boldsymbol{\sigma}_n: f \to \sigma_n(f)$ on B and consequently† $\|\boldsymbol{\sigma}_n\|^{B^*} = 1$.

We remark that by Parseval's formula, for every $\mu \in B^*, \sigma_n(\mu)$ converges weak-star to μ.

7.3 Parseval's formula enables us to characterize sequences of Fourier coefficients of linear functionals.

Theorem: *Let B be a homogeneous Banach space on* $\mathbf{T}$. *Assume that $e^{int} \in B$ for all n. Let $\{a_n\}_{n=-\infty}^{\infty}$ be a sequence of complex numbers. Then the following two conditions are equivalent:*

(a) *There exists $\mu \in B^*, \|\mu\| \leq C$, such that $\hat{\mu}(n) = a_n$ for all n.*

(b) *For all trigonometric polynomials P*

$$\left|\sum \hat{P}(n)\overline{a_n}\right| \leq C\|P\|_B.$$

† $\|\boldsymbol{\sigma}_n\|^{B^*}$ denotes the norm of $\boldsymbol{\sigma}_n$ as operator on B^*.

PROOF: The implication (a) $\Rightarrow$ (b) follows immediately from Parseval's formula. If we assume (b) then

$$P \rightarrow \sum \hat{P}(n)\overline{a_n} \tag{7.6}$$

is a linear functional on the space of all trigonometric polynomials, bounded in the B norm, and therefore (theorem 2.12) admits a unique extension μ of norm $\leqq C$ to B. Since μ extends (7.6) we have

$$\hat{\mu}(n) = \overline{\langle e^{int}, \mu \rangle} = a_n . \quad ◀$$

***Corollary**: A trigonometric series $S \sim \sum a_n e^{int}$ is the Fourier series of some $\mu \in B^*$, $\| \mu \| \leqq C$, if, and only if, $\| \sigma_N(S) \|_{B^*} \leqq C$ for all N.*

Here $\sigma_N(S)$ denotes the element in B^* the Fourier series of which is $\sum_{-N}^{N}(1 - |j|/(N+1))\, a_j e^{ijt}$.

PROOF: The necessity follows from 7.2; the sufficiency from the foregoing theorem and the observation that for trigonometric polynomials P

$$\sum \hat{P}(n)\overline{a_n} = \lim_{N \to \infty} \langle P, \sigma_N(S) \rangle . \quad ◀$$

7.4 In the case $B = C(\mathbf{T})$ the dual space B^* is identified with the space $M(\mathbf{T})$ of all (Borel) measures on $\mathbf{T}$ (we set $\langle f, \mu \rangle = \int f \overline{d\mu}$) We shall refer to Fourier coefficients of measures as *Fourier-Stieltjes coefficients* and to Fourier series of measures as *Fourier-Stieltjes series.* The mapping $f \rightarrow (1/2\pi) f(t)\, dt$ is an isometric embedding of $L^1(\mathbf{T})$ in $M(\mathbf{T})$. The Fourier coefficients of $(1/2\pi) f(t)\, dt$ are precisely $\hat{f}(n)$, hence *a Fourier series is a Fourier-Stieltjes series.*

An example of a measure that is not obtained as $(1/2\pi) f(t)\, dt$ is the so-called Dirac measure; it is the measure δ of mass one concentrated at $t = 0$. δ can also be defined by $\langle f, \delta \rangle = f(0)$ for all $f \in C(\mathbf{T})$. We denote by δ_τ, $\tau \in \mathbf{T}$, the unit mass concentrated at τ. Thus $\delta = \delta_0$ and $\langle f, \delta_\tau \rangle = f(\tau)$ for all $\tau \in \mathbf{T}$. From (7.1) it follows that $\hat{\delta}_\tau(n) = e^{-in\tau}$ and in particular $\hat{\delta}(n) = 1$. This shows that Fourier-Stieltjes coefficients need not tend to zero at infinity (note, however, that by 7.1 $|\hat{\mu}(n)| \leqq \| \mu \|_{M(\mathbf{T})}$).

7.5 We recall that a measure μ is *positive* if $\mu(E) \geqq 0$ for every measurable set E, or equivalently, if $\int f\, d\mu \geqq 0$ whenever $f \in C(\mathbf{T})$ is nonnegative. If μ is absolutely continuous, that is, if $\mu = (1/2\pi)\, g(t)\, dt$ with $g \in L^1(\mathbf{T})$, then μ is positive if and only if $g(t) \geqq 0$ almost everywhere.

Lemma: *A necessary and sufficient condition for a series* $S \sim \sum a_n e^{int}$ *to be the Fourier-Stieltjes series of a positive measure is that, for all* n, $\sigma_n(S,t) = \sum_{-n}^{n}(1 - |j|/(n+1))a_j e^{ijt} \geqq 0$ *on* $\mathbf{T}$.

PROOF: If $S = S[\mu]$ for a positive $\mu \in M(\mathbf{T})$ and if $f \in C(\mathbf{T})$ is nonnegative, we have

$$\frac{1}{2\pi}\int f(t)\overline{\sigma_n(S,t)}\,dt = \sum_{-n}^{n}\left(1 - \frac{|j|}{n+1}\right)\hat{f}(j)\overline{\hat{\mu}(j)} = \int \sigma_n(f)\,\overline{d\mu} \geqq 0$$

since $\mu \geqq 0$ and, by 3 1, $\sigma_n(f,t) \geqq 0$. Since this is true for arbitrary nonnegative f, $\sigma_n(S,t) \geqq 0$ on $\mathbf{T}$.

Assuming $\sigma_n(S,t) \geqq 0$ we obtain

$$\| \sigma_n(S) \|_{M(\mathbf{T})} = \frac{1}{2\pi}\int \sigma_n(S,t)\,dt = a_0$$

and, by corollary 7.3, $S = S[\mu]$ for some $\mu \in M(\mathbf{T})$. For arbitrary nonnegative $f \in C(\mathbf{T})$, $\int f\,d\mu = \lim_{n\to\infty}(1/2\pi)\int f(t)\sigma_n(S,t)\,dt \geqq 0$ and it follows that μ is a positive measure. ◀

Remark: The condition "$\sigma_n(S,t) \geqq 0$ for all n" can clearly be replaced by "$\sigma_n(S,t) \geqq 0$ for infinitely many n's."

7.6 We are now able to characterize Fourier-Stieltjes coefficients of positive measures as *positive definite sequences.*

DEFINITION: A numerical sequence $\{a_n\}_{n=-\infty}^{\infty}$ is *positive definite* if for any sequence $\{z_n\}$ having only a finite number of terms different from zero we have

$$\sum_{n,m} a_{n-m} z_n \overline{z_m} \geqq 0.$$

Theorem: (*Herglotz*): *A numerical sequence* $\{a_n\}_{n=-\infty}^{\infty}$ *is positive definite if, and only if, there exists a positive measure* μ *such that* $a_n = \hat{\mu}(n)$ *for all* n.

PROOF: Assume $a_n = \hat{\mu}(n)$ with positive μ. Then

$$\sum_{n,m} a_{n-m} z_n \overline{z_m} = \sum_{n,m}\int e^{-int}e^{imt} z_n \overline{z_m}\,d\mu = \int \left|\sum_n z_n e^{-int}\right|^2 d\mu \geqq 0.$$

If, on the other hand, we assume that $\{a_n\}$ is positive definite, we write $S \sim \sum a_n e^{int}$ and, for arbitrary N and $t \in \mathbf{T}$ we choose

$$z_n = \begin{cases} e^{int} & |n| \leqq N \\ 0 & |n| > N \end{cases}$$

We have $\sum_{n,m} a_{n-m} z_n \overline{z_m} = \sum_j C_{j,N} a_j e^{ijt}$ where $C_{j,N}$ is the number of ways to write j in the form $n - m$ where $|n| \leqq N$ and $|m| \leqq N$, that is, $C_{j,N} = \max(0, 2N+1-|j|)$. It follows that

$$\sigma_{2N}(S,t) = \frac{1}{2N+1} \sum_j C_{j,N} a_j e^{ijt} \geqq 0$$

and the theorem follows from 7.5. ◀

7.7 An important property of Fourier-Stieltjes coefficients is that of being "universal multipliers." More precisely:

Theorem: *Let B be a homogeneous Banach space on* **T** *and let $\mu \in M(\mathbf{T})$. There exists a unique linear operator $\boldsymbol{\mu}$ on B having the following properties*:

(i) $\|\boldsymbol{\mu}\| \leqq \|\mu\|_{M(\mathbf{T})}$

(ii) $\widehat{\boldsymbol{\mu} f}(n) = \hat{\mu}(n)\hat{f}(n)$ for all $f \in B$.

PROOF: If an operator $\boldsymbol{\mu}$ satisfies (ii), then for $f = \sum_{-N}^{N} \hat{f}(n) e^{int}$ we have $\boldsymbol{\mu} f = \sum_{-N}^{N} \hat{\mu}(n)\hat{f}(n) e^{int}$, that is, $\boldsymbol{\mu}$ is completely determined on the polynomials in B. If $\boldsymbol{\mu}$ is bounded it is completely determined. In order to show the existence of $\boldsymbol{\mu}$ it is sufficient to show that if we define

$$\boldsymbol{\mu} f = \sum \hat{\mu}(n)\hat{f}(n) e^{int} \tag{7.7}$$

for all polynomials f then

$$\|\boldsymbol{\mu} f\|_B \leqq \|\mu\|_{M(\mathbf{T})} \|f\|_B, \tag{7.8}$$

since $\boldsymbol{\mu}$ would then have a unique extension of norm $\leqq \|\mu\|_{M(\mathbf{T})}$ to all of B. If $\mu = (1/2\pi)\, g(t)\, dt$ with $g \in C(\mathbf{T})$, it is clear that $\boldsymbol{\mu} f$, as defined in (7.7), is simply $g * f$ which we can write as a B-valued integral (see 2.4)

$$\boldsymbol{\mu} f = g * f = \frac{1}{2\pi} \int g(\tau) f_\tau \, d\tau \tag{7.9}$$

and deduce the estimate

$$\|\boldsymbol{\mu} f\|_B \leqq \|f\|_B \frac{1}{2\pi} \int |g(\tau)|\, d\tau = \|\mu\|_{M(\mathbf{T})} \|f\|_B .$$

For arbitrary $\mu \in M(\mathbf{T})$, $\sigma_n(\mu)$ has the form $(1/2\pi)\, g_n(t)\, dt$, where $g_n(t) = \sum_{-n}^{n}(1 - |j|/(n+1))\, \hat{\mu}(j)\, e^{ijt}$ and

$$\frac{1}{2\pi}\int |g_n(t)|\, dt = \|\sigma_n(\mu)\|_{M(\mathbf{T})} \leqq \|\mu\|_{M(\mathbf{T})}.$$

By our previous remark $\|g_n * f\|_B \leqq \|\mu\|_{M(\mathbf{T})}\|f\|_B$, and since $\boldsymbol{\mu} f = \lim_{n\to\infty} g_n * f$ (remember that f is a trigonometric polynomial) we obtain (7.8). ◀

Corollary: *Let $f \in B$ and $\mu \in M(\mathbf{T})$. Then $\{\hat{\mu}(n)\hat{f}(n)\}$ is the sequence of Fourier coefficients of some function in B.*

In view of (7.9) we shall write $\mu * f$ instead of $\boldsymbol{\mu} f$, and refer to it as the *convolution* of μ and f. With this notation, our earlier condition (ii) becomes a (formal) extension of (1.10).

7.8 For $\mu \in M(\mathbf{T})$ we define $\mu^{\#} \in M(\mathbf{T})$ by

(7.10) $$\mu^{\#}(E) = \overline{\mu(-E)}$$

for every Borel set E (recall that $-E = \{t; -t \in E\}$), or equivalently, by

(7.11) $$\int f(t)\, d\mu^{\#} = \int f(-t)\, \overline{d\mu}$$

for all $f \in C(\mathbf{T})$. It follows from (7.11) that

(7.12) $$\widehat{\mu^{\#}}(n) = \overline{\hat{\mu}(n)}.$$

7.9 By Parseval's formula, the adjoint of $\boldsymbol{\mu}$ is the operator which assigns to a $\nu \in B^*$ the element of B^* whose Fourier series is $\sum \overline{\hat{\mu}(n)}\hat{\nu}(n)\, e^{int} = \sum \widehat{\mu^{\#}}(n)\hat{\nu}(n)\, e^{int}$. We extend the notation of 7.7, write this element of B^* as $\mu^{\#} * \nu$, and refer to it as the convolution of $\mu^{\#}$ and ν. We summarize:

Theorem: *Let B be a homogeneous Banach space on $\mathbf{T}$ and B^* its dual. Let $\mu \in M(\mathbf{T})$, $\nu \in B^*$; then $\sum \hat{\mu}(n)\hat{\nu}(n)\, e^{int}$ is the Fourier series of an element $\mu * \nu \in B^*$. Moreover $\|\mu * \nu\|_{B^*} \leqq \|\mu\|_{M(\mathbf{T})}\|\nu\|_{B^*}$.*

The norm estimate follows from (7.8), the fact that the norm of the adjoint is the same as the norm of the operator, and $\|\mu^{\#}\|_{M(\mathbf{T})} = \|\mu\|_{M(\mathbf{T})}$.

It follows, in particular, that if $\mu, \nu \in M(\mathbf{T})$, $\sum \hat{\mu}(n)\hat{\nu}(n)\, e^{int}$ is the Fourier-Stieltjes series of the measure $\mu * \nu$.

7.10 We have introduced the convolution of $\mu, \nu \in M(\mathbf{T})$ by its Fourier-Stieltjes series. It can, of course, be done directly. With μ and ν given, and for $f \in C(\mathbf{T})$, the double integral

$$I(f) = \iint f(t+\tau)\, d\mu(t)\, d\nu(\tau)$$

is well defined, is clearly linear in f, and satisfies

$$|I(f)| \leq \|f\|_\infty \|\mu\|_{M(\mathbf{T})} \|\nu\|_{M(\mathbf{T})}.$$

By the Riesz representation theorem, which identifies $M(\mathbf{T})$ as the dual of $C(\mathbf{T})$, there exists a measure $\lambda \in M(\mathbf{T})$ such that $I(f) = \int f(t)\, d\lambda$. Taking $f(t) = e^{-int}$ we obtain $\hat{\lambda}(n) = \hat{\mu}(n)\hat{\nu}(n)$, that is, $\lambda = \mu * \nu$. In other words

$$\int f\, d(\mu * \nu) = \iint f(t+\tau)\, d\mu(t)\, d\nu(\tau). \tag{7.13}$$

Taking a sequence of functions f which converge to the characteristic function of an arbitrary closed† set E, we see that (7.13) is equivalent to (denoting $E - \tau = \{t; t+\tau \in E\}$)

$$(\mu * \nu)(E) = \int \mu(E-\tau)\, d\nu(\tau). \tag{7.14}$$

7.11 We recall that a measure $\mu \in M(\mathbf{T})$ is *discrete* if $\mu = \sum a_j \delta_{\tau_j}$ where a_j are complex numbers; we then have $\|\mu\|_{M(\mathbf{T})} = \sum |a_j|$. A measure μ is *continuous* if $\mu(\{t\}) = 0$ for every $t \in \mathbf{T}$ ($\{t\}$ is the set whose only member is the point t). Equivalently μ is continuous if $\lim_{\eta \to 0} \int_{t-\eta}^{t+\eta} |d\mu| = 0$ for all $t \in \mathbf{T}$. Every $\mu \in M(\mathbf{T})$ can be uniquely decomposed to a sum $\mu = \mu_c + \mu_d$ where μ_c is continuous and μ_d is discrete.

It is clear from (7.14) that if μ is a continuous measure and $\nu \in M(\mathbf{T})$ then $\mu * \nu$ is continuous. Also, it is clear that $\delta_\tau * \delta_{\tau'} = \delta_{\tau+\tau'}$, and consequently if $\mu = \sum a_j \delta_{\tau_j}$ and $\nu = \sum b_k \delta_{\tau'_k}$ then $\mu * \nu = \sum_{j,k} a_j b_k \delta_{\tau_j + \tau'_k}$. If $\mu = \mu_c + \mu_d$ is the decomposition of μ into continuous and discrete parts, then $\mu_c^{\#}$ is the continuous part of $\mu^{\#}$ and $\mu_d^{\#}$ is its discrete part. Thus

$$\mu * \mu^{\#} = (\mu_c * \mu_c^{\#} + \mu_c * \mu_d^{\#} + \mu_d * \mu_c^{\#}) + \mu_d * \mu_d^{\#},$$

the sum of the first three terms being continuous and the last term

† Hence, by regularity, (7.14) holds for every Borel set E.

being discrete. If $\mu_d = \sum a_j \delta_{\tau_j}$ then $\mu_d^{\#} = \sum \overline{a_j} \delta_{-\tau_j}$, and consequently the mass at $\tau = 0$ of the measure $\mu * \mu^{\#}$ is $\sum |a_j|^2$. We have proved:

Lemma: *Let $\mu \in M(\mathbf{T})$. Then $\sum_\tau |\mu(\{\tau\})|^2 = (\mu * \mu^{\#})(\{0\})$. In particular, a necessary and sufficient condition for the continuity of μ is that $(\mu * \mu^{\#})(\{0\}) = 0$.*

The discrete part of a measure μ can be "recovered" from its Fourier-Stieltjes series.

Theorem: *Let $\mu \in M(\mathbf{T})$, $\tau \in \mathbf{T}$, Then*

$$\mu(\{\tau\}) = \lim_{N \to \infty} \frac{1}{2N+1} \sum_{-N}^{N} \hat{\mu}(n) e^{in\tau}.$$

PROOF: The functions $\varphi_N(t) = \dfrac{1}{2N+1} D_N(t-\tau) = \dfrac{1}{2N+1} \sum_{-N}^{N} e^{-in\tau} e^{int}$ are bounded by 1 and tend to zero uniformly outside any neighborhood of $t = \tau$. Remembering that

$$\lim_{\vartheta \to 0} \int_{\tau-\vartheta}^{\tau+\vartheta} |d(\mu - \mu(\{\tau\})\delta_\tau)| = 0$$

we obtain

$$\lim_{N \to \infty} \langle \varphi_N, \mu - \mu(\{\tau\})\delta_\tau \rangle = 0. \tag{7.15}$$

Now

$$\overline{\langle \varphi_N, \mu - \mu(\{\tau\})\delta_\tau \rangle} = \frac{1}{2N+1} \sum_{-N}^{N} \hat{\mu}(n) e^{in\tau} - \mu(\{\tau\})$$

and the theorem follows from (7.15). ◀

Corollary (*Wiener*): *Let $\mu \in M(\mathbf{T})$. Then*

$$\sum_\tau |\mu(\{\tau\})|^2 = \lim_{N \to \infty} \frac{1}{2N+1} \sum_{-N}^{N} |\hat{\mu}(n)|^2.$$

In particular, a necessary and sufficient condition for the continuity of μ is

$$\lim_{N \to \infty} \frac{1}{2N+1} \sum_{-N}^{N} |\hat{\mu}(n)|^2 = 0.$$

EXERCISES FOR SECTION 7

1. Let B be homogeneous on $\mathbf{T}$ and B^* its dual. Show that $S \sim \sum a_n e^{int}$ is the Fourier series of some $\mu \in B^*$ if and only if $\|\sigma_N(S)\|_{B^*}$ is bounded as $N \to \infty$.

2. Denote $\mathbf{K}_{n,\tau}(t) = \mathbf{K}_n(t-\tau)$. Show that for every $\mu \in M(\mathbf{T})$

$$\sigma_n(\mu,\tau) = \overline{\langle \mathbf{K}_{n,\tau}, \mu \rangle}.$$

Deduce that $\sigma_n(\mu,\tau) \geqq 0$ if μ is positive.

3. Show that a trigonometric series $\sum a_n e^{int}$ such that $\sum_{-N}^{N} a_n e^{int} \geqq 0$ for all N and $t \in \mathbf{T}$ is a Fourier-Stieltjes series of a positive measure.

4. We shall prove later that if $f \in L^p(\mathbf{T})$ with $1 < p \leqq 2$, then

$$\left(\sum |\hat{f}(n)|^q\right)^{1/q} \leqq \|f\|_{L^p(\mathbf{T})} \qquad \left(q = \frac{p}{p-1}\right).$$

Assuming this, show that if $\{a_n\}$ is a numerical sequence satisfying $\sum |a_n|^p < \infty$, there exists a function $g \in L^q(\mathbf{T})$ such that $\hat{g}(n) = a_n$ and $\|g\|_{L^q(\mathbf{T})} \leqq (\sum |a_n|^p)^{1/p}$.

5. The elements of the dual space of $C^m(\mathbf{T})$ are called *distributions of order m* on $\mathbf{T}$. We denote by $\mathscr{D}^m(\mathbf{T}) = (C^m(\mathbf{T}))^*$ the space of distributions of order m on $\mathbf{T}$. Since $C^{m+1}(\mathbf{T}) \subset C^m(\mathbf{T})$ we have $\mathscr{D}^m(\mathbf{T}) \subset \mathscr{D}^{m+1}(\mathbf{T})$; we write $\mathscr{D}(\mathbf{T}) = \bigcup_m \mathscr{D}^m(\mathbf{T})$, $C^\infty(\mathbf{T}) = \bigcap_m C^m(\mathbf{T})$.

(a) Prove that if $\mu \in \mathscr{D}^m$ then

$$|\hat{\mu}(n)| \leqq \text{const}\,|n|^m \qquad n \neq 0.$$

(b) Given a numerical sequence $\{a_n\}$ satisfying $a_n = O(|n|^m)$, there exists a distribution $\mu \in \mathscr{D}^{m+1}$ such that $a_n = \hat{\mu}(n)$ for all n. (*Hint*: If $f \in C^{m+1}(\mathbf{T})$ then $\sum |n^m \hat{f}(n)| < \infty$.)

Thus a trigonometric series $\sum a_n e^{int}$ is the Fourier series of a distribution on $\mathbf{T}$ if and only if, for some m, $a_n = O(|n|^m)$, $n \neq 0$.

Let $\mu \in \mathscr{D}$ and let O be an open subset of $\mathbf{T}$. We say that μ *vanishes on O* if $\langle \varphi\ \mu \rangle = 0$ for all $\varphi \in C^\infty(\mathbf{T})$ such that the support of φ (i.e., the closure of the set $\{t; \varphi(t) \neq 0\}$) is contained in O.

(c) Prove that if μ vanishes on the open sets O_1 and O_2, then it vanishes on $O_1 \cup O_2$. (*Hint*: Show that if the support of $\varphi \in C^\infty(\mathbf{T})$ is contained in $O_1 \cup O_2$ then there exist $\varphi_1, \varphi_2 \in C^\infty(\mathbf{T})$, with supports contained in O_1, O_2 respectively, such that $\varphi = \varphi_1 + \varphi_2$.

(d) Extend the result of (c) to any finite union of open sets; hence, using the compactness of the support of the test functions φ, show that if μ vanishes in the open sets O_α, α running over some index set I, then μ vanishes on $\bigcup_{\alpha \in I} O_\alpha$.

Thus the union of all the open subsets of $\mathbf{T}$ on which μ vanishes is again such a set. This is clearly the largest open set on which μ vanishes.

Definition: The *support of* μ is the complement in $\mathbf{T}$ of the largest open set $O \subset \mathbf{T}$ on which μ vanishes.

(e) Show that if $\mu \in \mathscr{D}^m$ and if $f \in C^m(\mathbf{T})$ vanishes on a neighborhood of the support of μ, then $\langle f, \mu \rangle = 0$. The same conclusion holds if for some homogeneous Banach space B, the distribution μ belongs to B^* and $f \in B$ (see exercise 2.11).

(f) We define the derivative μ' of a distribution $\mu \in \mathscr{D}^m$ by

$$\langle f, \mu' \rangle = -\langle f', \mu \rangle \qquad \text{for } f \in C^{m+1}(\mathbf{T}).$$

Show that $\mu' \in \mathscr{D}^{m+1}$ and $\widehat{\mu'}(n) = in\hat{\mu}(n)$.

(g) Show that support$(\mu') \subseteq$ support(μ).

(h) Show that the mapping $\mu \to \mu'$ maps $\mathscr{D}^m$ onto the subspace $\mathscr{D}^{m+1}$ consisting of all $\mu \in \mathscr{D}^{m+1}$ satisfying $\hat{\mu}(0) = 0$.

Hence, every $\mu \in \mathscr{D}^m$ can be written in the form $\hat{\mu}(0)dt + \mu_1$ where μ_1 is the mth derivative of a measure.

(i) A distribution μ on $\mathbf{T}$ is *positive* if $\langle f, \mu \rangle \geqq 0$ for every nonnegative $f \in C^\infty(\mathbf{T})$. Show that a positive distribution is a measure.

6. The dual space of $A(\mathbf{T})$ is commonly denoted by $PM(\mathbf{T})$ and its elements referred to as pseudo-measures. Show that with the natural identifications $M(\mathbf{T}) \subset PM(\mathbf{T}) \subset \mathscr{D}^1(\mathbf{T})$, $PM(\mathbf{T})$ consisting precisely of those $\mu \in \mathscr{D}^1(\mathbf{T})$ for which $\{\hat{\mu}(n)\}$ is bounded. Moreover, the correspondence $\mu \leftrightarrow \{\hat{\mu}(n)\}$ is an isometry of $PM(\mathbf{T})$ onto ℓ^∞.

7. Let $\alpha, \beta \in \mathbf{T}$, let N be an integer, and let μ be the measure carried by the arithmetic progression $\{\alpha + j\beta\}_{j=-N}^{N}$, which places the mass zero at α and the mass j^{-1} at $\alpha + j\beta$, $1 \leqq |j| \leqq N$. Show that $\|\mu\|_{PM(\mathbf{T})} \leqq \pi + 2$. *Hint*: See exercise 3.4.

8. Let $f \in A(\mathbf{T})$ be real valued and monotonic in a neighborhood of $t_0 \in \mathbf{T}$. Show that $|f(t) - f(t_0)| = O[(\log|t - t_0|^{-1})^{-1}]$ as $t \to t_0$.

9. Let $\mu, \mu_n \in M(\mathbf{T})$, $n = 1, 2, \ldots$. Prove that $\mu_n \to \mu$ in the weak-star topology if, and only if, $\|\mu_n\|_{M(\mathbf{T})} = O(1)$ and $\hat{\mu}_n(j) \to \hat{\mu}(j)$ for all j.

10. Show that $\mu \in M(\mathbf{T})$ *is absolutely continuous* if, and only if, $\lim_{\tau \to 0} \|\mu_\tau - \mu\|_{M(\mathbf{T})} = 0$, where μ_τ is the translate of μ by τ (defined by $\mu_\tau(E) = \mu(E - \tau)$).

11. Let $\mu \in M(\mathbf{T})$. Show that $\sigma_n(\mu, t)$ converge to zero at every $t \notin$ support(μ), the convergence being uniform on every closed set disjoint from support(μ).

12. Let $\mu \in M(\mathbf{T})$ be singular with respect to dt (that is, there exists a Borel set E_0 of Lebesgue measure zero, such that $\mu(E) = \mu(E \cap E_0)$ for every Borel set E). Show that $\sigma_n(\mu, t) \to 0$ almost everywhere (dt).

13. Show that the conclusion of exercise 11 is false if instead of "$\mu \in M(\mathbf{T})$" we assume only $\mu \in \mathscr{D}^1$; however, if we replace Fejér's kernel by Poisson's, the conclusion is valid for every $\mu \in \mathscr{D}$. *Hint*: For every $\delta > 0$ and integer m, $\lim_{r \to 1} \dfrac{\partial^m \mathbf{P}(r, t)}{\partial t^m} = 0$ uniformly in $(\delta, 2\pi - \delta)$.

Chapter II

The Convergence of Fourier Series

We have mentioned already that the problems of convergence of Fourier series, that is, the convergence of the (symmetric) partial sums, $S_n(f)$, are far more delicate than the corresponding problems of summability with respect to "good" summability kernels such as Fejér's or Poisson's. As in the case of summability, problems of convergence "in norms" are usually easier than those of pointwise convergence. Many problems, concerning pointwise convergence for various spaces, are still unsolved and the convergence almost everywhere of the Fourier series of square summable functions was proved only recently (L. Carleson 1965). Convergence seems to be closely related to the existence and properties of the so-called conjugate function. In this chapter we give only a temporary incomplete definition of the conjugate function. A proper definition and the study of the basic properties of conjugation are to be found in chapter III.

1. CONVERGENCE IN NORM

1.1 Let B be a homogeneous Banach space on $\mathbf{T}$. As usual we write

$$S_n(f) = S_n(f, t) = \sum_{-n}^{n} \hat{f}(j) e^{ijt}. \tag{1.1}$$

We say that B *admits convergence in norm* if

$$\lim_{n \to \infty} \| S_n(f) - f \|_B = 0. \tag{1.2}$$

Our purpose in this section is to characterize the spaces B which have this property.

We have introduced the operators $S_n : f \to S_n(f)$ in chapter I. S_n is well defined in every homogeneous Banach space B; we denote its norm, as an operator on B, by $\| S_n \|^B$.

Theorem: *A homogeneous Banach space B admits convergence in norm if, and only if,* $\| S_n \|^B$ *are bounded (as* $n \to \infty$*), that is, if there exist a constant K such that*

$$\| S_n(f) \|_B \leqq K \| f \|_B \tag{1.3}$$

for all $f \in B$ *and* $n \geqq 0$,

PROOF: If $S_n(f)$ converge to f for all $f \in B$, then $S_n(f)$ are bounded for every $f \in B$. By the uniform boundedness theorem, it follows that $\| S_n \|^B = O(1)$. On the other hand, if we assume (1.3), let $f \in B$, $\varepsilon > 0$, and let P be a trigonometric polynomial satisfying $\| f - P \|_B \leqq \varepsilon / 2K$. For n greater than the degree of P, we have $S_n(P) = P$ and hence

$$\| S_n(f) - f \|_B = \| S_n(f) - S_n(P) + P - f \|_B$$

$$\leqq \| S_n(f - P) \|_B + \| P - f \|_B \leqq K \frac{\varepsilon}{2K} + \frac{\varepsilon}{2K} \leqq \varepsilon. \quad ◀$$

1.2 The fact that $S_n(f) = D_n * f$, where D_n is the Dirichlet kernel

$$D_n(t) = \sum_{-n}^{n} e^{ijt} = \frac{\sin (n + \frac{1}{2}) t}{\sin t/2} \tag{1.4}$$

yields a simple bound for $\| S_n \|^B$. In fact, since $\| D_n * f \|_B \leqq \| D_n \|_{L^1} \| f \|_B$, it follows that

$$\| S_n \|^B \leqq \| D_n \|_{L^1}. \tag{1.5}$$

The numbers $L_n = \| D_n \|_{L^1}$ are called the *Lebesgue constants*; they tend to infinity like a constant multiple of $\log n$ (see exercise 1 at the end of this section).

In the case $B = L^1(\mathbf{T})$ the inequality (1.5) becomes an equality. This can be seen as follows: as in chapter I we denote by $\mathbf{K}_n$ the Féjer kernel and remember that $\| \mathbf{K}_N \|_{L^1} = 1$. We clearly have $\| S_n \|^{L^1(\mathbf{T})} \geqq \| S_n(\mathbf{K}_N) \|_{L^1} = \| \sigma_N(D_n) \|_{L^1}$ and since $\sigma_N(D_n) \to D_n$ as $N \to \infty$, we obtain

$$\| S_n \|^{L^1(\mathbf{T})} \geqq \| D_n \|_{L^1}; \qquad \text{hence } \| S_n \|^{L^1(\mathbf{T})} = \| D_n \|_{L^1}.$$

It follows that $L^1(\mathbf{T})$ does not admit convergence in norm.

1.3 In the case $B = C(\mathbf{T})$, convergence in norm is simply uniform convergence. We show that Fourier series of continuous functions need not converge uniformly by showing that $\| S_n \|^{C(\mathbf{T})}$ are unbounded; more precisely we show that $\| S_n \|^{C(\mathbf{T})} = L_n$. For this, we consider continuous functions ψ_n satisfying

$$\| \psi_n \|_\infty = \sup_t |\psi_n(t)| \leqq 1$$

and such that $\psi_n(t) = \operatorname{sgn}(D_n(t))$ except in small intervals around the points of discontinuity of $\operatorname{sgn}(D_n(t))$. If the sum of the lengths of these intervals is smaller than $\varepsilon/2n$, we have

$$\| S_n \|^{C(\mathbf{T})} \geqq \| S_n(\psi_n) \|_\infty \geqq | S_n(\psi_n, 0) | = \frac{1}{2\pi} \left| \int D_n(t)\,\psi_n(t)\,dt \right| > L_n - \varepsilon$$

which, together with (1.5), proves our statement.

1.4 For a class of homogeneous Banach spaces on $\mathbf{T}$, the problem of convergence in norm can be related to invariance under conjugation. In chapter I we defined the conjugate series of a trigonometric series $\sum a_n e^{int}$ to be the series $-i \sum \operatorname{sgn}(n) a_n e^{int}$. If $f \in L^1(\mathbf{T})$ and if the series conjugate to $\sum \hat{f}(n) e^{int}$ is the Fourier series of some function $g \in L^1(\mathbf{T})$, we call g the *conjugate function* of f and denote it by $\tilde{f}$. This definition is adequate for the purposes of this section; however, it does not define $\tilde{f}$ for all $f \in L^1(\mathbf{T})$ and we shall extend it later.

DEFINITION: A space of functions $B \subset L^1(\mathbf{T})$ *admits conjugation* if for every $f \in B$, $\tilde{f}$ is defined and belongs to B.

If B is a homogeneous Banach space which admits conjugation, then the mapping $f \to \tilde{f}$ is a bounded linear operator on B. The linearity is evident from the definition and in order to prove the boundedness we apply the closed graph theorem. All that we have to do is show that the operator $f \to \tilde{f}$ is closed, that is, that if $\lim f_n = f$ and $\lim \tilde{f}_n = g$ in B, then $g = \tilde{f}$. This follows from the fact that for every integer j

$$\hat{g}(j) = \lim_{n\to\infty} \hat{\tilde{f}}_n(j) = \lim_{n\to\infty} -i \operatorname{sgn}(j) \hat{f}_n(j) = -i \operatorname{sgn}(j) \lim_{n\to\infty} \hat{f}_n(j) =$$

$$= -i \operatorname{sgn}(j) \hat{f}(j) = \hat{\tilde{f}}(j).$$

If B admits conjugation then the mapping

$$f \to f^\flat = \tfrac{1}{2}\hat{f}(0) + \tfrac{1}{2}(f + i\tilde{f}) \sim \sum_0^\infty \hat{f}(j) e^{ijt} \tag{1.6}$$

is a well-defined, bounded linear operator on B. Conversely, if the mapping $f \to f^{\flat}$ is well-defined in a space B, then B admits conjugation since $\tilde{f} = -i(2f^{\flat} - f - \hat{f}(0))$.

Theorem: *Let B be a homogeneous Banach space on $\mathbf{T}$ and assume that for $f \in B$ and for all n, $e^{int}f \in B$ and*

$$\| e^{int}f \|_B = \| f \|_B, \tag{1.7}$$

Then B admits conjugation if, and only if, B admits convergence in norm.

PROOF: By theorem 1.1 and the foregoing remarks, it is clearly sufficient to prove that the mapping $f \to f^{\flat}$ is well defined in B if, and only if, the operators S_n are uniformly bounded on B. Assume first that there exists a constant K such that $\| S_n \|^B \leqq K$. Define

$$S_n^{\flat}(f) = \sum_0^{2n} \hat{f}(j) e^{ijt} = e^{int} S_n(e^{-int}f); \tag{1.8}$$

by (1.7) we have $\| S_n^{\flat} \|^B \leqq K$.

Let $f \in B$ and $\varepsilon > 0$; let $P \in B$ be a trigonometric polynomial satisfying $\| f - P \|_B \leqq \varepsilon/2K$. We have

$$\| S_n^{\flat}(f) - S_n^{\flat}(P) \|_B = \| S_n^{\flat}(f - P) \|_B \leqq \frac{\varepsilon}{2}. \tag{1.9}$$

If n and m are both greater than the degree of P, $S_n^{\flat}(P) = S_m^{\flat}(P)$ and it follows from (1.9) that

$$\| S_n^{\flat}(f) - S_m^{\flat}(f) \|_B \leqq \varepsilon.$$

The sequence $S_n^{\flat}(f)$ is thus a Cauchy sequence in B; it converges and its limit has the Fourier series $\sum_0^{\infty} \hat{f}(j) e^{ijt}$. This means $f^{\flat} = \lim S_n^{\flat}(f) \in B$.

Assume conversely that $f \to f^{\flat}$ is well defined, hence bounded, in B. Then

$$S_n^{\flat}(f) = f^{\flat} - e^{i(2n+1)t}(e^{-i(2n+1)t}f)^{\flat}$$

which means that $\| S_n^{\flat} \|^B$ is bounded by twice the norm over B of the mapping $f \to f^{\flat}$. Since, by (1.7) and (1.8), $\| S_n \|^B = \| S_n^{\flat} \|^B$, the theorem follows. ◀

1.5 We shall see in chapter III that, for $1 < p < \infty$, $L^p(\mathbf{T})$ admits conjugation, hence:

***Theorem*:** *For $1 < p < \infty$, the Fourier series of every $f \in L^p(\mathbf{T})$ converges to f in the $L^p(\mathbf{T})$ norm.*

EXERCISES FOR SECTION 1

1. Show that the Lebesgue constants $L_n = \|D_n\|_{L^1}$ satisfy

$$L_n = 4/\pi^2 \log n + O(1).$$

Hint: †

$$L_n = \frac{1}{\pi}\int_0^\pi \left|\frac{\sin(n+\frac{1}{2})t}{\sin t/2}\right| dt = \frac{2}{\pi}\sum_{j=1}^{n-1}\int_{\frac{j\pi}{n+1/2}}^{\frac{(j+1)\pi}{n+1/2}} \frac{|\sin(n+\frac{1}{2})t|}{t}\, dt + O(1);$$

remember that

$$\int_{\frac{j\pi}{n+1/2}}^{\frac{(j+1)\pi}{n+1/2}} |\sin(n+\tfrac{1}{2})t|\, dt = \frac{2}{n+\frac{1}{2}}.$$

2. Show that if the sequence $\{N_j\}$ tends to infinity fast enough, then the Fourier series of the function

$$f(t) = \sum_1^\infty 2^{-j}\mathbf{K}_{N_j}(t)$$

does not converge in $L^1(\mathbf{T})$.

3. Let $\{a_n\}$ be an even sequence of positive numbers,convex on $(0,\infty)$ and vanishing at infinity (cf. I.4.1). Prove that the partial sums of the series $\sum a_n e^{int}$ are bounded in $L^1(\mathbf{T})$ if, and only if, $a_n \log n = O(1)$ and the series converges in $L^1(\mathbf{T})$ if, and only if, $\lim a_n \log n = 0$.

4. Show that $B = C^m(\mathbf{T})$ does not admit convergence in norm. *Hint*: $\mathbf{S}_n$ commute with derivation.

5. Let φ be a continuous, concave (i.e., $\varphi(h) + \varphi(h+2\delta) \leq 2\varphi(h+\delta)$), and increasing function on $[0,1]$, satisfying $\varphi(0) = 0$. Denote by Λ_φ the subspace of $C(\mathbf{T})$ consisting of all the functions f for which $\omega(f,h) = O(\varphi(h))$ as $h \to 0$ and by λ_φ the subspace of Λ_φ consisting of all the functions f for which $\omega(f,h) = o(\varphi(h))$ as $h \to 0$. ($\omega(f,h)$ is the modulus of continuity of f; see I.4.6.) Consider the following statements:
(a) $\varphi(h) = O(-(\log h)^{-1})$ as $h \to 0$.
(b) For every $f \in \lambda_\varphi$, $S[f]$ is uniformly convergent.
(c) $\varphi(h) = o(-(\log h)^{-1})$ as $h \to 0$.
(d) For every $f \in \Lambda_\varphi$, $S[f]$ is uniformly convergent.
Show that (a) is equivalent to (b) and that (c) is equivalent to (d).

† For another way, see [16].

2. CONVERGENCE AND DIVERGENCE AT A POINT

We have seen in the previous section that the Fourier series of a continuous function need not converge uniformly. In this section we show that it may even fail to converge pointwise, and then give two criteria for the convergence of Fourier series at a point.

2.1 Theorem: *There exists a continuous function whose Fourier series diverges at a point.*

We give two proofs which are in fact one; the first is "abstract" based on the uniform boundedness theorem, and is very short. The second is a construction of a concrete example in essentially the way one proves the uniform boundedness theorem.

PROOF A: The mappings $f \to S_n(f, 0)$ are continuous linear functionals on $C(\mathbf{T})$, We saw in the previous section that these functionals are not uniformly bounded and consequently, by the uniform boundedness theorem, there exists an $f \in C(\mathbf{T})$ such that $\{S_n(f, 0)\}$ is not bounded. In other words, the Fourier series of f diverges unboundedly at $t = 0$.

PROOF B: As we have seen in section 1, there exists a sequence of functions $\psi_n \in C(\mathbf{T})$ satisfying:

(2.1) $$\| \psi_n \|_\infty \leqq 1,$$

(2.2) $$|S_n(\psi_n, 0)| > \tfrac{1}{2} \| D_n \|_{L^1} > \tfrac{1}{10} \log n.$$

We put $\phi_n(t) = \sigma_{n^2}(\psi_n, t)$ and notice that φ_n is a trigonometric polynomial of degree n^2 satisfying

(2.1′) $$\| \varphi_n \|_\infty \leqq 1$$

and $$|S_n(\varphi_n, t) - S_n(\psi_n, t)| < 2,$$

hence

(2.2′) $$|S_n(\varphi_n, 0)| > \tfrac{1}{10} \log n - 2.$$

With $\lambda_n = 2^{3^n}$ we define

(2.3) $$f(t) = \sum_1^\infty \frac{1}{n^2} \varphi_{\lambda_n}(\lambda_n t)$$

and claim that f is a continuous function whose Fourier series diverges at $t = 0$. The continuity of f follows immediately from the uniform convergence of the series in (2.3); to show the divergence of the Fourier series of f at zero, we notice that $\varphi_{\lambda_j}(\lambda_j t) = \sum_m \widehat{\varphi_{\lambda_j}}(m) e^{i\lambda_j mt}$; hence

$$(2.4)\qquad \begin{aligned} |S_{\lambda_n^2}(f,0)| &= \left| S_{\lambda_n^2}\left(\sum_1^n \frac{1}{j^2}\varphi_{\lambda_j}(\lambda_j t), 0\right) + \sum_{n+1}^{\infty} \frac{1}{j^2}\widehat{\varphi_{\lambda_j}}(0) \right| = \\ &= \left| \sum_1^{n-1} \frac{1}{j^2}\varphi_{\lambda_j}(0) + \frac{1}{n^2} S_{\lambda_n}(\varphi_{\lambda_n}, 0) + \sum_{n+1}^{\infty} \frac{1}{j^2}\hat{\varphi}_{\lambda_j}(0) \right| \geqq \frac{K}{n^2}\log\lambda_n - 3, \end{aligned}$$

which tends to ∞, and the theorem follows.

Remark:

$$f(t) = \sum_1^{m-1} \frac{1}{n^2}\varphi_{\lambda_n}(\lambda_n t) + \sum_m^{\infty} \frac{1}{n^2}\varphi_{\lambda_n}(\lambda_n t).$$

The first sum is a trigonometric polynomial and therefore does not affect the convergence of the Fourier series of f. The second sum is periodic with period $2\pi/\lambda_m$ (since λ_m divides λ_k for $k \geqq m$); consequently the partial sums of the Fourier series of f are unbounded at every point of the form $2\pi j/\lambda_m$ for any positive integers j and m. If we want to obtain divergence at every rational multiple of 2π, all that we have to do is put $\lambda_n = n!2^{3^n}$.

2.2 Our first convergence criterion is really a simple Tauberian theorem due to Hardy.

Theorem: *Let $f \in L^1(\mathbf{T})$ and assume*

$$(2.5)\qquad \hat{f}(n) = O\left(\frac{1}{n}\right) \quad \text{as } |n| \to \infty.$$

Then $S_n(f,t)$ and $\sigma_n(f,t)$ converge for the same values of t and to the same limit. Also, if $\sigma_n(f,t)$ converges uniformly on some set, so does $S_n(f,t)$.

PROOF: The condition (2.5) implies the following weaker condition which is really all that we need: for every $\varepsilon > 0$ there exists a $\lambda > 1$ such that

$$(2.5')\qquad \limsup_{n\to\infty} \sum_{n<|j|\leqq\lambda n} |\hat{f}(j)| < \varepsilon.$$

Let $\varepsilon > 0$ and let $\lambda > 1$ be such that (2.5′) is valid. We have

$$
\begin{aligned}
S_n(f,t) &= \frac{[\lambda n]+1}{[\lambda n]-n}\sigma_{[\lambda n]}(f,t) - \frac{n+1}{[\lambda n]-n}\sigma_n(f,t) - \\
&\quad - \frac{[\lambda n]+1}{[\lambda n]-n} \sum_{n<|j|\leqq \lambda n} \left(1 - \frac{|j|}{[\lambda n]+1}\right)\hat{f}(j)e^{ijt},
\end{aligned}
\tag{2.6}
$$

(where $[\lambda n]$ denotes the integral part of λn). By (2.5′) there exists an n_0 such that if $n > n_0$, the last term in (2.6) is bounded by ε. If $\sigma_n(f,t_0)$ converge to a limit $\sigma(f,t_0)$, it follows from (2.6) that for n_1 sufficiently large, $n > n_1$ implies

$$
\left| S_n(f,t_0) - \sigma(f,t_0) \right| < 2\varepsilon, \tag{2.7}
$$

in other words,

$$
\lim S_n(f,t_0) = \sigma(f,t_0). \tag{2.8}
$$

The choice of n_1 depends only on the rate of convergence of $\sigma_n(f,t_0)$ to $\sigma(f,t_0)$ so that if this convergence is uniform on some set, so is (2.8). ◀

Corollary: *Let f be of bounded variation on* $\mathbf{T}$; *then the $S_n(f,t)$ converge to $\frac{1}{2}(f(t+0)+f(t-0))$ and in particular to $f(t)$ at every point of continuity. The convergence is uniform on closed intervals of continuity of f.*

PROOF: By Fejér's theorem the foregoing holds true for $\sigma_n(f,t)$, and the statement follows from the fact that for functions of bounded variation (2.5) is valid (cf. theorem I.4.5). ◀

2.3. ***Lemma***: *Let $f \in L^1(\mathbf{T})$ and assume* $\int_{-1}^{1} \left| \frac{f(t)}{t} \right| dt < \infty$. Then

$$
\lim_{n\to\infty} S_n(f,0) = 0.
$$

PROOF:

$$
\begin{aligned}
S_n(f,0) &= \frac{1}{2\pi}\int \frac{f(t)}{\sin\frac{t}{2}} \sin(n+\tfrac{1}{2})t\,dt = \\
&= \frac{1}{2\pi}\int f(t)\cos nt\,dt + \frac{1}{2\pi}\int \frac{f(t)\cos\frac{t}{2}}{\sin\frac{t}{2}} \sin nt\,dt.
\end{aligned}
\tag{2.9}
$$

By our assumption $\frac{f \cos t/2}{\sin t/2} \in L^1(\mathbf{T})$; hence, by the Riemann-Lebesgue lemma, all the integrals in (2.9) tend to zero. ◀

2.4 ***Theorem*** *(Principle of localization): Let $f \in L^1(\mathbf{T})$ and assume that f vanishes in an open interval I. Then $S_n(f, t)$ converge to zero for $t \in I$, and the convergence is uniform on closed subintervals of I.*

PROOF: The convergence to zero at every $t \in I$ is an immediate consequence of lemma 2.3. If I_0 is a closed subinterval of I, the functions $\varphi_{t_0}(t) = \frac{f(t - t_0)\cos t/2}{\sin t/2}$, $t_0 \in I_0$, form a compact family in $L^1(\mathbf{T})$, hence by remark I.2.8, the integrals in (2.9) corresponding to $f(t - t_0)$, $t_0 \in I_0$, tend to zero uniformly. ◀

The principle of localization is often stated as follows: let $f, g \in L^1(\mathbf{T})$ and assume that $f(t) = g(t)$ in some neighborhood of a point t_0. Then the Fourier series of f and g at t_0 are either both convergent and to the same limit or both divergent and in the same manner.

2.5 Another immediate application of lemma 2.3 yields

Theorem *(Dini's test): Let $f \in L^1(\mathbf{T})$. If*

$$\int_{-1}^{1} \left| \frac{f(t + t_0) - f(t_0)}{t} \right| dt < \infty$$

then

$$S_n(f, t_0) \to f(t_0).$$

EXERCISES FOR SECTION 2

1. Show that if a sequence of continuous functions on some interval is unbounded on a dense subset of the interval, then it is bounded only on a set of the first category. Use that to show that the Fourier series of f (defined in (2.3)) converges only on a set of the first category.

2. Show that for every given (denumerable) sequence $\{t_n\}$ there exists a continuous function whose Fourier series diverges at every t_n.

3. Let g be the 2π-periodic function defined by: $g(0) = 0$, $g(t) = t - \pi$ for $0 < t < 2\pi$.

(a) Discuss the convergence of the Fourier series of g.

(b) Show that $|S_n(g,t)| \leq \pi + 2$ for all n and t.

(c) Put $\varphi_n(t) = (\pi + 2)^{-1} e^{int} S_n(g,t)$; show that $\|\varphi_n\|_\infty \leq 1$ and $|S_n(\varphi_n, 0)| > K \log n$ for some constant $K > 0$.

(d) Show that for $|t| < \pi/2$, and all n and m,

$$|S_m(\varphi_n, t)| \leq \frac{K_1}{|t|}$$

where K_1 is a constant.

(e) Show that for a proper choice of the integers n_j and λ_j, the Fourier series of the continuous function

$$f(t) = \sum_{j=1}^{\infty} \frac{1}{j^2} e^{i\lambda_j t} \varphi_{n_j}(t)$$

diverges for $t = 0$ and converges for all other $t \in \mathbf{T}$.

*3. SETS OF DIVERGENCE

3.1 We consider a homogeneous Banach space B on $\mathbf{T}$.

DEFINITION: A set $E \subset \mathbf{T}$ is a *set of divergence* for B if there exists an $f \in B$ whose Fourier series diverges at every point of E.

3.2 DEFINITION: For $f \in L^1(\mathbf{T})$ we put

$$\text{(3.1)} \qquad \begin{aligned} S_n^*(f, t) &= \sup_{m \leq n} |S_m(f, t)| \\ S^*(f, t) &= \sup_n |S_n(f, t)|. \end{aligned}$$

Theorem: *E is a set of divergence for B if, and only if, there exists an element $f \in B$ such that*

$$\text{(3.2)} \qquad S^*(f, t) = \infty \qquad \text{for } t \in E.$$

The theorem is an easy consequence of the following:

Lemma: *Let $g \in B$; then there exist an element $f \in B$ and a positive even sequence $\{\Omega_j\}$ such that $\Omega_j \to \infty$ monotonically with j and $\hat{f}(j) = \Omega_j \hat{g}(j)$ for all $j \in \mathbf{Z}$.*

PROOF OF THE LEMMA: For each n let $\lambda(n)$ be such that $\|\sigma_{\lambda(n)}(g) - g\|_B < 2^{-n}$. We write $f = g + \sum_{n=1}^{\infty}(g - \sigma_{\lambda(n)}(g))$. The series defining f converges in norm; hence $f \in B$. Also $\hat{f}(j) = \Omega_j \hat{g}(j)$ where $\Omega_j = 1 + \sum_{n=1}^{\infty} \min(1, |j|/(\lambda(n) + 1))$. ◀

PROOF OF THE THEOREM: The condition (3.2) is clearly sufficient for the divergence of $\sum \hat{f}(j) e^{ijt}$ for all $t \in E$. Assume, on the other hand, that for some $g \in B$, $\sum \hat{g}(j) e^{ijt}$ diverges at every point of E. Let $f \in B$ and $\{\Omega_j\}$ be the function and the sequence corresponding to g by the lemma. We claim that (3.2) holds for f. This follows from: for $n > m$,

$$
\begin{aligned}
S_n(g,t) - S_m(g,t) &= \sum_{m+1}^{n} (S_j(f,t) - S_{j-1}(f,t))\Omega_j^{-1} \\
&= S_n(f,t)\Omega_n^{-1} - S_m(f,t)\Omega_{m+1}^{-1} + \qquad (3.3)\\
&\quad + \sum_{m+1}^{n-1} (\Omega_j^{-1} - \Omega_{j+1}^{-1}) S_j(f,t),
\end{aligned}
$$

hence

$$\left| S_n(g,t) - S_m(g,t) \right| \leqq 2S^*(f,t)\Omega_{m+1}^{-1}.$$

It follows that if $S^*(f,t) < \infty$, the Fourier series of g converges and $t \notin E$. ◀

Remark: Let $\{\omega_n\}$ be a sequence of positive numbers such that $\omega_j = O(\Omega_j)$, $\sum_1^\infty (\Omega_j^{-1} - \Omega_{j+1}^{-1})\omega_j < \infty$; then, for all $t \in E$, $S_j(f,t) \neq o(\omega_j)$. This follows immediately from (3.3).

3.3 For the sake of simplicity we assume throughout the rest of this section that

(3.4) *If* $f \in B$ *and* $n \in \mathbf{Z}$ *then* $e^{int} f \in B$ *and* $\| e^{int} f \|_B = \| f \|_B$.

Lemma: *Assume* (3.4); *then E is a set of divergence for B if, and only if, there exists a sequence of trigonometric polynomials* $P_j \in B$ *such that* $\sum \| P_j \|_B < \infty$ *and*

$$\sup_j S^*(P_j, t) = \infty \qquad \text{on } E. \tag{3.5}$$

PROOF: Assume the existence of a sequence $\{P_j\}$ satisfying $\sum \| P_j \|_B < \infty$ and (3.5). Denote by m_j the degree of P_j and let ν_j be integers satisfying

$$\nu_j > \nu_{j-1} + m_{j-1} + m_j.$$

Put $f(t) = \sum e^{i\nu_j t} P_j(t)$. For $n \leqq m_j$ we have

$$S_{\nu_j + n}(f,t) - S_{\nu_j - n - 1}(f,t) = e^{i\nu_j t} S_n(P_j, t);$$

hence $\sum \hat{f}(j) e^{ijt}$ diverges on E.

Conversely, assume that E is a set of divergence for B. By remark 3.2 there exists a monotonic sequence $\omega_n \to \infty$ and a function $f \in B$

such that $|S_n(f,t)| > \omega_n$ infinitely often for every $t \in E$. We now pick a sequence of integers $\{\lambda_j\}$ such that

(3.6) $$\| f - \sigma_{\lambda_j}(f) \|_B < 2^{-j}$$

and then integers μ_j such that

(3.7) $$\omega_{\mu_j} > 2 \sup_t S^*(\sigma_{\lambda_j}(f), t)$$

and write $P_j = \mathbf{V}_{\mu_{j+1}} * (f - \sigma_{\lambda_j}(f))$ where as usual $\mathbf{V}_\mu$ denotes de la Vallée Poussin's kernel (see I.2.13). It follows immediately from (3.6) that $\sum \| P_j \|_B < \infty$. If $t \in E$ and n is an integer such that $| S_n(f,t) | > \omega_n$, then for some j, $\mu_j < n \leqq \mu_{j+1}$ and

$$S_n(P_j, t) = S_n(f - \sigma_{\lambda_j}(f), t) = S_n(f,t) - S_n(\sigma_{\lambda_j}(f), t).$$

Hence, by (3.7), $| S_n(P_j, t) | > \frac{1}{2}\omega_n$, and (3.5) follows. ◀

Theorem: *Assume* (3.4). *Let* E_j *be sets of divergence for* B, $j = 1, 2, \cdots$; *then* $E = \bigcup_{j=1}^{\infty} E_j$ *is a set of divergence for* B.

PROOF: Let $\{P_n^j\}$ be the sequence of polynomials corresponding to E_j. Omitting a finite number of terms does not change (3.5), but permits us to assume $\sum_{j,n} \| P_n^j \|_B < \infty$ which shows, by the lemma, that E is a set of divergence for B. ◀

3.4 We turn now to examine the sets of divergence for $B = C(\mathbf{T})$.

Lemma: *Let* E *be a union of a finite number of intervals on* $\mathbf{T}$; *denote the measure of* E *by* δ. *There exists a trigonometric polynomial* φ *such that*

(3.8)
$$S^*(\varphi, t) > \frac{1}{2\pi} \log\left(\frac{1}{3\delta}\right) \qquad \text{on } E$$
$$\| \varphi \|_\infty \leqq 1.$$

PROOF: It is convenient to identify $\mathbf{T}$ with the unit circumference $\{z; |z| = 1\}$. Let I be a (small) interval on $\mathbf{T}$, $I = \{e^{it}, | t - t_0 | \leqq \varepsilon\}$; the function $\psi_I = (1 + \varepsilon - z e^{-it_0})^{-1}$ has a positive real part throughout the unit disc, its real part is larger than $1/3\varepsilon$ on I, and its value at the origin ($z = 0$) is $(1 + \varepsilon)^{-1}$. We now write $E \subseteq \bigcup_1^N I_j$, the I_j being small intervals of equal length 2ε such that $N\varepsilon < \delta$, and consider the function

$$\psi(z) = \frac{1+\varepsilon}{N} \sum \psi_{I_j}(z).$$

ψ has the following properties:

$$\text{(3.9)} \qquad \begin{aligned} &\mathrm{Re}(\psi(z)) > 0 \qquad \text{for } |z| \leqq 1 \\ &\psi(0) = 1 \\ &|\psi(z)| \geqq \mathrm{Re}(\psi(z)) > \frac{1}{3N\varepsilon} > \frac{1}{3\delta} \qquad \text{on } E. \end{aligned}$$

The function $\log \psi$ which takes the value zero at $z = 0$ is holomorphic in a neighborhood of $\{z; |z| \leq 1\}$ and has the properties

$$\text{(3.10)} \qquad \begin{aligned} &|\mathrm{Im}(\log \psi(z))| < \pi \qquad &\text{on } \mathbf{T} \\ &|\log \psi(z)| > \log(3\delta)^{-1} \qquad &\text{on } E. \end{aligned}$$

Since the Taylor series of $\log \psi$ converges uniformly on $\mathbf{T}$, we can take a partial sum $\Phi(z) = \sum_1^M a_n z^n$ of that series such that (3.10) is valid for Φ in place of $\log \psi$. We can now put

$$\varphi(t) = \frac{1}{\pi} e^{-iMt} \mathrm{Im}(\Phi(e^{it})) = \frac{1}{2\pi i} e^{-iMt} \left(\sum_1^M a_n e^{int} - \sum_1^M \bar{a}_n e^{-int} \right)$$

and notice that

$$|S_M(\varphi, t)| = \frac{1}{2\pi} |\Phi(e^{it})|. \qquad ◀$$

***Theorem**: Every set of measure zero is a set of divergence for $C(\mathbf{T})$.*

PROOF: If E is a set of measure zero, it can be covered by a union $\bigcup I_n$, the I_n being intervals of length $|I_n|$ such that $\sum |I_n| < 1$ and such that every $t \in E$ belongs to infinitely many I_n's. Grouping finite sets of intervals we can cover E infinitely often by $\bigcup E_n$ such that every E_n is a finite union of intervals and such that $|E_n| < e^{-2^n}$. Let φ_n be a polynomial satisfying (3.8) for $E = E_n$ and put $P_n = n^{-2} \varphi_n$, We clearly have $\sum \|P_n\|_\infty < \infty$ and $S^*(P_n, t) > 2^{n-1}/2\pi n^2$ on E_n. Since every $t \in E$ belongs to infinitely many E_n's, our theorem follows from lemma 3.3. ◀

3.5 ***Theorem**: Let B be a homogeneous Banach space on $\mathbf{T}$ satisfying the condition (3.4). Assume $B \supseteq C(\mathbf{T})$; then either $\mathbf{T}$ is a set of divergence for B or the sets of divergence for B are precisely the sets of measure zero.*

PROOF: By theorem 3.4 it is clear that every set of measure zero is a set of divergence for B. All that we have to show in order to complete the proof is that, if some set of positive measure is a set of divergence for B, then $\mathbf{T}$ is a set of divergence for B.

Assume that E is a set of divergence of positive measure. For $\alpha \in \mathbf{T}$ denote by E_α the translate of E by α; E_α is clearly a set of divergence for B. Let $\{\alpha_n\}$ be the sequence of all rational multiples of 2π and put $\tilde{E} = \bigcup E_{\alpha_n}$. By theorem 3.2 $\tilde{E}$ is a set of divergence, and we claim that $\mathbf{T} \setminus \tilde{E}$ is a set of measure zero. In order to prove that, we denote by χ the characteristic function of $\tilde{E}$ and notice that

$$\chi(t - \alpha_n) = \chi(t) \qquad \text{for all } t \text{ and } \alpha_n .$$

This means

$$\sum_j \hat{\chi}(j) e^{-i\alpha_n j} e^{ijt} = \sum_j \hat{\chi}(j) e^{ijt}$$

or

$$\hat{\chi}(j) e^{-i\alpha_n j} = \hat{\chi}(j) \qquad (\text{all } \alpha_n)$$

If $j \neq 0$, this implies $\hat{\chi}(j) = 0$; hence $\chi(t) = \text{constant}$ almost every where and, since χ is a characteristic function, this implies that the measure of $\tilde{E}$ is either zero or 2π. Since $\tilde{E} \supset E$, $\tilde{E}$ is almost all of $\mathbf{T}$. Now $\mathbf{T} \setminus \tilde{E}$ is a set of divergence (being of measure zero) and $\tilde{E}$ is a set of divergence, hence $\mathbf{T}$ is a set of divergence. ◀

3.6 Thus, for spaces B satisfying the conditions of theorem 3.5, and in particular for $B = L^p(\mathbf{T})$, $1 \leqq p < \infty$, or $B = C(\mathbf{T})$, either there exists a function $f \in B$ whose Fourier series diverges everywhere, or the Fourier series of every $f \in B$ converges almost everywhere. In the case $B = L^1(\mathbf{T})$ it was shown by Kolmogorov that the first possibility holds. The case of $B = L^2(\mathbf{T})$ was settled only recently by L. Carleson [4], who proved the famous "Lusin conjecture"; namely that the Fourier series of functions in $L^2(\mathbf{T})$ converge almost everywhere. This result was extended by Hunt [12] to all $L^p(\mathbf{T})$ with $p > 1$. The proof of these results is still rather complicated and we do not include it. We finish this section with Kolmogorov's theorem.

Theorem: *There exists a Fourier series diverging everywhere.*

PROOF: For arbitrary $\kappa > 0$ we shall describe a positive measure μ_κ of total mass one having the property that for almost all $t \in \mathbf{T}$

$$S^*(\mu_\kappa, t) = \sup_n \left| S_n(\mu_\kappa, t) \right| > \kappa . \tag{3.11}$$

Assume for the moment that such μ_κ exist; it follows from (3.11) that there exists an integer N_κ and a set E_κ of (normalized Lebesgue) measure greater than $1 - 1/\kappa$, such that for $t \in E_\kappa$

$$\sup_{n<N_\kappa} |S_n(\mu_\kappa, t)| > \kappa . \tag{3.12}$$

If we write now $\varphi_\kappa = \mu_\kappa * \mathbf{V}_{N_\kappa}$ ($\mathbf{V}_{N_\kappa}$ being de la Vallée Poussin's kernel), then φ_κ is a trigonometric polynomial, $\| \varphi_\kappa \|_{L^1} \leqq 3$ and

$$S^*(\varphi_\kappa, t) \geqq \sup_{n<N_\kappa} |S_n(\varphi_\kappa, t)| = \sup_{n<N_\kappa} |S_n(\mu_\kappa, t)| > \kappa$$

on E_k. Applying lemma 3.3 with $P_j = j^{-2}\varphi_{2^j}$ we obtain that $E = \bigcap_m \bigcup_{m \leq j} E_{2^j}$ is a set of divergence for $L^1(\mathbf{T})$. Since E is almost all $\mathbf{T}$, Kolmogorov's theorem would follow from theorem 3.5.

The description of the measures μ_κ is very simple; however, for the proof that (3.11) holds for almost all $t \in \mathbf{T}$, we shall need the following very important theorem of Kronecker (see VI.9).

Theorem (*Kronecker*): *Let $x_1, \ldots, x_N$ be real numbers such that $x_1, \ldots, x_N, \pi$ are linearly independent over the field of rational numbers. Let $\varepsilon > 0$ and $\alpha_1, \ldots, \alpha_N$ be real numbers, then there exists an integer n such that*

$$\left| e^{inx_j} - e^{i\alpha_j} \right| < \varepsilon, \qquad j = 1, \ldots, N .$$

We construct now the measures μ_κ as follows: let N be an integer, let x_j, $j = 1, \ldots, N$ be real numbers such that $x_1, \ldots, x_N, \pi$ are linearly independent over the rationals and such that $|x_j - (2\pi j/N)| < 1/N^2$, and let μ be the measure $1/N \sum \delta_{x_j}$.

For $t \in \mathbf{T}$ we have

$$S_n(\mu, t) = \int D_n(t-x)\, d\mu(x) = \frac{1}{N} \sum_1^N D_n(t - x_j) =$$

$$= \frac{1}{N} \sum_1^N \frac{\sin (n + \frac{1}{2})(t - x_j)}{\sin \frac{1}{2}(t - x_j)} .$$

For almost all $t \in \mathbf{T}$, the numbers $t - x_1, \ldots, t - x_N, \pi$ are linearly independent over the rationals. By Kronecker's theorem there exist, for each such t, integers n such that

$$\left| e^{i(n+\frac{1}{2})(t-x_j)} - i \operatorname{sgn}\left(\sin \frac{t-x_j}{2}\right) \right| < \frac{1}{2}, \qquad j = 1, \ldots, N;$$

hence

$$\frac{\sin(n+\frac{1}{2})(t-x_j)}{\sin\frac{1}{2}(t-x_j)} > \frac{1}{2}\left|\sin\frac{t-x_j}{2}\right|^{-1} \qquad \text{for all } j.$$

It follows that

$$S_n(\mu, t) > \frac{1}{2N}\sum_{j=1}^{N}\left|\sin\frac{t-x_j}{2}\right|^{-1} \tag{3.13}$$

and since the x_j's are so close to the roots of unity of order N, the sum in (3.13) is bounded below by $\frac{1}{2}\int_{1/N}^{\pi}|\sin t/2|^{-1}dt > \log N > \kappa$, provided we take N large enough. ◀

EXERCISE FOR SECTION 3

1. Let B be a homogeneous Banach space on $\mathbf{T}$. Show that for every $f \in B$ there exist $g \in B$ and $h \in L^1(\mathbf{T})$ such that $f = g * h$. *Hint*: Use lemma 3.2 and theorem I.4.1.

Chapter III

The Conjugate Function and Functions Analytic in the Unit Disc

We defined the conjugate function for some summable functions by means of their conjugate Fourier series. Our first purpose in this chapter is to extend the notion to all summable functions and to study the basic properties of the conjugate function for various classes of functions. This is done mainly in the first two sections. In section 1 we use the "complex variable" approach to define the conjugate function and obtain some basic results about the distribution functions of conjugates to functions belonging to various classes. In section 2 we introduce the Hardy-Littlewood maximal functions and use them to obtain results about the so-called maximal conjugate function. We show that the conjugate function can also be defined by a singular integral and use this to obtain some of its local properties. In section 3 we discuss the Hardy spaces H^p. As further reading we mention [11].

1. THE CONJUGATE FUNCTION

1.1 We identify $\mathbf{T}$ with the unit circumference $\{z; z = e^{it}\}$ in the complex plane. The unit disc $\{z; |z| < 1\}$ is denoted by D and the closed unit disc, $\{z; |z| \leqq 1\}$, by $\bar{D}$. For $f \in L^1(\mathbf{T})$ we denote by $f(re^{it})$, $r < 1$, the *Poisson integral of* f,

$$(1.1) \qquad f(re^{it}) = (\mathbf{P}(r,\cdot) * f)(t) = \sum_{-\infty}^{\infty} r^{|n|} \hat{f}(n) e^{int}.$$

In chapter I we have considered $\mathbf{P}(r, \cdot) * f$ as a family of functions on $\mathbf{T}$, depending on the parameter r, $0 \leqq r < 1$. The main idea in this section is to consider it as a function of the complex variable $z = re^{it}$ in D.

The functions $r^{|n|}e^{int}$, $-\infty < n < \infty$, are harmonic in D, and, since the series in (1.1) converges uniformly on compact subsets of D, it follows that $f(re^{it})$ is harmonic in D. We saw in I.3.3 that at every point t where f is the derivative of its integral (hence almost everywhere) $f(e^{it}) = \lim_{r\to 1} f(re^{it})$. Actually it is not very hard to see that for almost all t, $f(z) \to f(e^{it})$ as $z \to e^{it}$ nontangentially (i.e., if $z \to e^{it}$, remaining in a sector of the form $\{\zeta; |\arg(1 - \zeta e^{-it})| \leqq \alpha < \pi\}$. (See [28], Vol. 1, p. 101.)

The harmonic conjugate to (1.1) is the function

$$(1.2)\qquad \tilde{f}(re^{it}) = -i\sum_{-\infty}^{\infty} \operatorname{sgn}(n) r^{|n|}\hat{f}(n)e^{int} = (Q(r,\cdot) * f)(t)$$

where

$$(1.2')\qquad Q(r,t) = -i\sum_{-\infty}^{\infty} \operatorname{sgn}(n) r^{|n|} e^{int} = \frac{2r\sin t}{1 - 2r\cos t + r^2}$$

is the harmonic conjugate of Poisson's kernel $\mathbf{P}(r,t)$ (normalized by the condition $Q(0,t) = \operatorname{sgn}(0) = 0$). We shall show that $\tilde{f}(re^{it})$ has a radial limit for almost all t. Denoting this radial limit by $\tilde{f}(e^{it})$ we shall show that if f has a conjugate in the sense of section II.1, then this conjugate is $\tilde{f}(e^{it})$. We may therefore call $\tilde{f}$ the *conjugate function* of f.

1.2 ***Lemma:*** *Every function harmonic and bounded in D is the Poisson integral of some bounded function on* **T**.

PROOF: Let F be harmonic and bounded in D. Let $r_n \uparrow 1$ and write $f_n(e^{it}) = F(r_n e^{it})$. The sequence $\{f_n\}$ is a bounded sequence in $L^\infty(\mathbf{T})$; hence for some sequence $n_j \to \infty$, f_{n_j} converges in the weak-star topology ($L^\infty(\mathbf{T})$ being the dual of $L^1(\mathbf{T})$) to some function $F(e^{it})$. Let $\rho e^{i\tau} \in D$, then

$$\frac{1}{2\pi}\int \mathbf{P}(\rho, t-\tau)F(e^{it})\,dt = \lim_{j\to\infty}\frac{1}{2\pi}\int \mathbf{P}(\rho, t-\tau) f_{n_j}(e^{it})\,dt =$$

$$= \lim_{j\to\infty} F(r_{n_j}\rho e^{i\tau}) = F(\rho e^{i\tau}). \qquad ◂$$

1.3 ***Lemma***: *Assume* $f \in L^1(\mathbf{T})$ *and let* $\tilde{f}(re^{it})$ *be defined by* (1.2). *Then, for almost all* t, $\tilde{f}(re^{it})$ *tends to a limit as* $r \to 1$.

PROOF: Since the mapping $f \to \tilde{f}(re^{it})$ is clearly linear and since any $f \in L^1(\mathbf{T})$ can be written as $f_1 - f_2 + if_3 - if_4$ with $f_j \geqq 0$ in $L^1(\mathbf{T})$, there is no loss of generality in assuming $f \geqq 0$. The function $F(z) = e^{-f(z) - i\tilde{f}(z)}$ is holomorphic (hence harmonic) in D. Since the Poisson integral of a nonnegative function is nonnegative $f(z) \geqq 0$, and since $\tilde{f}$ is real valued (being the harmonic conjugate of the real-valued f), it follows that $|F(z)| \leqq 1$ in D. By lemma 1.2 (and I.3.3) F has a radial limit of modulus $e^{-f(e^{it})}$ almost everywhere. Since $f \in L^1(\mathbf{T}), f(e^{it}) < \infty$, hence $\lim_{r \to 1} F(re^{it}) \neq 0$ almost everywhere; and at every point where $F(e^{it})$ exists and is nonzero, $\tilde{f}(re^{it})$ has a finite radial limit. ◀

1.4 DEFINITION: The *conjugate function* of a function $f \in L^1(\mathbf{T})$ is the function $\tilde{f}(e^{it}) = \lim_{r \to 1} \tilde{f}(re^{it})$.

If the series conjugate to the Fourier series of f is the Fourier series of some $g \in L^1(\mathbf{T})$, then the Poisson integral of g is clearly $\tilde{f}(re^{it})$, which converges radially to $g(e^{it})$ for almost all t (theorem I.3.3). It follows that in this case $\tilde{f} = g$ and our new definition of the conjugate function extends that of II.1.

We have seen in I.4.2 that $\sum_{n=2}^{\infty} \frac{\cos nt}{\log n}$ is a Fourier series while the conjugate series, $\sum_{n=2}^{\infty} \frac{\sin nt}{\log n}$ is not. Since $\sum_{n=2}^{\infty} \frac{\sin nt}{\log n}$ converges everywhere, its sum is the conjugate function of $f = \sum \frac{\cos nt}{\log n}$ and we can check that $\sum \frac{\sin nt}{\log n} \notin L^1(\mathbf{T})$. Thus the conjugate function of a summable function need not be summable.

Remark: At this point we cannot deduce that $\sum \frac{\sin nt}{\log n} \notin L^1(\mathbf{T})$ from the mere fact that the series is not a Fourier series. However, we shall prove in section 3 that if $\tilde{f} \in L^1(\mathbf{T})$, for some $f \in L^1(\mathbf{T})$, then $\tilde{f}(re^{it})$ is the Poisson integral of $\tilde{f}$. From that we can deduce that if $\tilde{f} \in L^1(\mathbf{T})$ then its Fourier series is $\tilde{S}[f]$ so that if $\tilde{S}[f]$ is not a Fourier series then $\tilde{f} \notin L^1(\mathbf{T})$,

The difficulty in asserting immediately that $\tilde{f}(re^{it})$ is the Poisson integral of $\tilde{f}$ stems from the fact that we have only established pointwise convergence almost everywhere of $\tilde{f}(re^{it})$ to $\tilde{f}(e^{it})$ and this type of convergence is not sufficient to imply convergence of integrals.

1.5 We denote the (Lebesgue) measure of a measurable set $E \subset \mathbf{T}$ by $|E|$,

DEFINITION: The *distribution function* of a measurable, real-valued function f on $\mathbf{T}$ is the function

$$\mathfrak{m}(x) = \mathfrak{m}_f(x) = \left|\{t; f(t) \leqq x\}\right|, \qquad -\infty < x < \infty .$$

Distribution functions are clearly continuous to the right and monotone, increasing from zero at $x = -\infty$ to 2π as $x \to \infty$. The basic property of distribution functions is: for every continuous function F on $\mathbf{R}$

$$\int_T F(f(t))\,dt = \int F(x)\,d\,\mathfrak{m}_f(x) \tag{1.3}$$

DEFINITION: A measurable function f is *of weak L^p type*, $0 < p < \infty$, if there exists a constant C such that for all $\lambda > 0$

$$\mathfrak{m}_{|f|}(\lambda) \geqq 2\pi - C\lambda^{-p} \tag{1.4}$$

(or equivalently, $\left|\{t; |f(t)| \geqq \lambda\}\right| \leqq C\lambda^{-p}$).

Every $f \in L^p(\mathbf{T})$ is clearly of weak L^p type. In fact, for all $\lambda > 0$

$$\|f\|_{L^p}^p = \frac{1}{2\pi}\int_0^\infty x^p\,d\,\mathfrak{m}_{|f|}(x) \geqq \frac{1}{2\pi}\int_\lambda^\infty x^p\,d\,\mathfrak{m}_{|f|}(x)$$

$$\geqq \frac{1}{2\pi}\lambda^p \int_\lambda^\infty d\,\mathfrak{m}_{|f|}(x) = \frac{\lambda^p}{2\pi}(2\pi - \mathfrak{m}_{|f|}(\lambda))$$

hence (1.4) is satisfied with $C = 2\pi\|f\|_{L^p}^p$. It is equally clear that there are functions of weak L^p type which are not in $L^p(\mathbf{T})$; $|\sin t|^{-1/p}$ is a simple example.

Lemma: *If f is of weak L^p type then $f \in L^{p'}(\mathbf{T})$ for every $p' < p$.*

PROOF:

$$\int |f|^{p'}dt = \int_0^\infty x^{p'}d\,\mathfrak{m}_{|f|}(x) \leqq \mathfrak{m}_{|f|}(1) + \int_1^\infty x^{p'}\,d\,\mathfrak{m}_{|f|}(x) =$$

$$= \mathfrak{m}_{|f|}(1) - [x^{p'}(2\pi - \mathfrak{m}_{|f|}(x))]_1^\infty + \int_1^\infty (2\pi - \mathfrak{m}_{|f|}(x))\,d(x^{p'})$$

$$\leqq 2\pi + C\int_1^\infty x^{-p}d(x^{p'}) = 2\pi + Cp'\int_1^\infty x^{p'-p-1}dx < \infty . \blacktriangleleft$$

1.6 ***Theorem:*** *If $f \in L^1(\mathbf{T})$ then $\tilde{f}$ is of weak L^1 type.*

PROOF: We assume first that $f \geqq 0$; also, we normalize f by assuming $\|f\|_{L^1} = 1$. We want to evaluate the measure of the set of points where $|\tilde{f}| > \lambda$. The function $H_\lambda(z) = 1 + \frac{1}{\pi}\arg\frac{z - i\lambda}{z + i\lambda} = 1 + \frac{1}{\pi}\operatorname{Im}\left(\log\frac{z - i\lambda}{z + i\lambda}\right)$ is clearly harmonic and nonnegative in the half plane $\operatorname{Re}(z) > 0$, and its level lines are circular arcs passing through the points $i\lambda$ and $-i\lambda$. The level line $H_\lambda(z) = \frac{1}{2}$ is the half circle $z = \lambda e^{i\vartheta}$, $-\pi/2 < \vartheta < \pi/2$, hence if $|z| \geqq \lambda$, $H_\lambda(z) \geqq \frac{1}{2}$. Also it is clear that

$$H_\lambda(1) = 1 - (2/\pi)\arctan\lambda < 2/\pi\lambda.$$

Now $H_\lambda(f(z) + i\tilde{f}(z))$ is a well-defined positive harmonic function in D, hence

$$(1.5)\quad \frac{1}{2\pi}\int H_\lambda(f(re^{it}) + i\tilde{f}(re^{it}))\,dt = H_\lambda(f(0)) = H_\lambda(1) < \frac{2}{\pi\lambda},$$

and remembering that $H_\lambda(f + i\tilde{f}) \geqq \frac{1}{2}$ if $|f + i\tilde{f}| \geqq \lambda$, we obtain,

$$|\{t; |\tilde{f}(re^{it})| > \lambda\}| \leqq \frac{8}{\lambda}.$$

Since the mapping $f \to \tilde{f}$ is linear it is clear that if we omit the normalization $\|f\|_{L^1} = 1$ we obtain, letting $r \to 1$, that for $f \geqq 0$ in $L^1(\mathbf{T})$

$$|\{t; |\tilde{f}(e^{it})| > \lambda\}| \leqq 8\|f\|_{L^1}\lambda^{-1}.$$

Every $f \in L^1(\mathbf{T})$ can be written as $f = f_1 - f_2 + if_3 - if_4$ where $f_j \geqq 0$ and $\|f_j\|_{L^1} \leqq \|f\|_{L^1}$. We have $\tilde{f} = \tilde{f}_1 - \tilde{f}_2 + i\tilde{f}_3 - i\tilde{f}_4$ and consequently

$$\{t; |\tilde{f}(e^{it})| > \lambda\} \subseteq \bigcup_{j=1}^{4}\left\{t; |\tilde{f}_j(e^{it})| > \frac{\lambda}{4}\right\}.$$

It follows that for $c = 128$ and every $f \in L^1(\mathbf{T})$

$$(1.6)\quad |\{t; |\tilde{f}(e^{it})| > \lambda\}| \leqq c\|f\|_{L^1}\lambda^{-1}. \qquad ◀$$

Corollary: *If $f \in L^1(\mathbf{T})$ then $\tilde{f} \in L^\alpha(\mathbf{T})$ for all $\alpha < 1$.*

PROOF: Lemma 1.5. ◀

1.7** ***Theorem:† *If $f\log^+|f| \in L^1(\mathbf{T})$, then $\tilde{f} \in L^1(\mathbf{T})$.*

† $\operatorname{Log}^+ x = \sup(\log x, 0)$ for $x \geqq 0$.

PROOF: We shall use the fact that for $g \in L^2(\mathbf{T})$ we have $\tilde{g} \in L^2(\mathbf{T})$ and $\| \tilde{g} \|_{L^2} \leqq \| g \|_{L^2}$. This is an immediate corollary of theorem I.5.5. As we have seen in 1.5, this implies

$$\mathfrak{m}_{|\tilde{g}|}(\lambda) \geqq 2\pi(1 - \| g \|_{L^2}^2 \lambda^{-2}). \tag{1.7}$$

We have to prove that $\int_1^\infty \lambda \, d\,\mathfrak{m}_{|\tilde{f}|}(\lambda) < \infty$ which is the same thing as $\int_1^R \lambda \, d\,\mathfrak{m}_{|\tilde{f}|}(\lambda) = O(1)$ as $R \to \infty$. Integrating by parts and remembering (1.6) we see that the theorem is equivalent to

$$\int_1^R (2\pi - \mathfrak{m}_{|\tilde{f}|}(\lambda))\, d\lambda = O\,(1) \qquad \text{as } R \to \infty. \tag{1.8}$$

In order to estimate $2\pi - \mathfrak{m}_{|\tilde{f}|}(\lambda)$ we write $f = g + h$, where $g = f$ when $|f| \leqq \lambda$ and $h = f$ when $|f| > \lambda$. We have $\tilde{f} = \tilde{g} + \tilde{h}$ and consequently

$$\{t; |\tilde{f}(e^{it})| > \lambda\} \subset \left\{t; |\tilde{g}(e^{it})| > \frac{\lambda}{2}\right\} \cup \left\{t; |\tilde{h}(e^{it})| > \frac{\lambda}{2}\right\}. \tag{1.9}$$

By (1.7)

$$\left| \left\{t; |\tilde{g}(e^{it})| > \frac{\lambda}{2}\right\} \right| \leqq 8\pi\lambda^{-2} \| g \|_{L^2}^2 = 8\pi\lambda^{-2} \int_0^\lambda x^2 \, d\,\mathfrak{m}_{|f|} \tag{1.10}$$

and by (1.6)

$$\left| \left\{t; |\tilde{h}(e^{it})| > \frac{\lambda}{2}\right\} \right| \leqq 2c\lambda^{-1} \|h\|_{L^1} = 2c\lambda^{-1} \int_\lambda^\infty x \, d\,\mathfrak{m}_{|f|};$$

for $x \geqq \lambda$, $(\log x)^{1/2} > (\log \lambda)^{1/2}$ and we obtain

$$\left| \left\{t; |\tilde{h}(e^{it})| > \frac{\lambda}{2}\right\} \right| \leqq \frac{2c}{\lambda\sqrt{\log \lambda}} \int_\lambda^\infty x\sqrt{\log x}\, d\,\mathfrak{m}_{|f|}. \tag{1.11}$$

By (1.9), (1.10), and (1.11) we have

$$2\pi - \mathfrak{m}_{|\tilde{f}|}(\lambda) \leqq 8\pi\lambda^{-2} \int_0^\lambda x^2 \, d\,\mathfrak{m}_{|f|} + \frac{2c}{\lambda\sqrt{\log \lambda}} \int_\lambda^\infty x\sqrt{\log x}\, d\,\mathfrak{m}_{|f|}.$$

Thus (1.8), and hence the theorem, will follow if we show that, as $R \to \infty$,

$$\int_1^R \lambda^{-2}\left(\int_0^\lambda x^2 \, d\,\mathfrak{m}_{|f|}\right) d\lambda = O\,(1)$$

(1.12)

$$\int_1^R \frac{1}{\lambda\sqrt{\log \lambda}} \left(\int_\lambda^\infty x\sqrt{\log x}\, d\,\mathfrak{m}_{|f|}\right) d\lambda = O\,(1)$$

The information that we have concerning $\mathfrak{m}_{|f|}$ is that it is a monotonic function tending to 2π at infinity and such that

$$\int_1^\infty x \log x \, d\mathfrak{m}_{|f|}(x) < \infty \tag{1.13}$$

In order to derive (1.12) from (1.13) we apply Fubini's theorem. The domain for the first integral is the trapezoid $\{(x, \lambda); 1 < \lambda < R, 0 < x < \lambda\}$ and integrating first with respect to λ we obtain

$$\int_1^R \lambda^{-2}\left(\int_0^\lambda x^2 \, d\mathfrak{m}_{|f|}\right) d\lambda = \int_0^1 \left(1 - \frac{1}{R}\right)x^2 \, d\mathfrak{m}_{|f|} + \int_1^R \left(\frac{1}{x} - \frac{1}{R}\right)x^2 \, d\mathfrak{m}_{|f|}$$

$$\leqq 2\pi + \int_1^R x \, d\mathfrak{m}_{|f|} = O(1)$$

The domain for the second integral is the strip $\{(x, \lambda); 1 < \lambda < R, \lambda < x\}$. Integrating again first with respect to λ we obtain

$$\int_1^R \frac{1}{\lambda\sqrt{\log \lambda}}\left(\int_\lambda^\infty x\sqrt{\log x} \, d\mathfrak{m}_{|f|}\right) d\lambda =$$

$$2\int_1^R x \log x \, d\mathfrak{m}_{|f|} + 2\sqrt{\log R}\int_R^\infty x\sqrt{\log x} \, d\mathfrak{m}_{|f|} = O(1)$$

and the proof is complete. ◀

1.8 ***Theorem*** (*M. Riesz*): *For $1 < p < \infty$, the mapping $f \to \tilde{f}$ is a bounded linear operator on $L^p(\mathbf{T})$.*

PROOF: We have mentioned already that for $p = 2$ the theorem is obvious (I.5.5). From Parseval's formula (I.7.1) it follows that if p and q are conjugate exponents, the mappings $f \to \tilde{f}$ in $L^p(\mathbf{T})$ and in $L^q(\mathbf{T})$ are, except for a sign, each other's adjoints and consequently if one is bounded, so is the other and by the same bound. Thus it is enough to prove the theorem for $1 < p < 2$.

Let $f \in L^p(\mathbf{T})$ be nonnegative; denote by $f(re^{it})$ its Poisson integral, by $\tilde{f}(re^{it})$ the harmonic conjugate, and write $H(re^{it}) = f(re^{it}) + i\tilde{f}(re^{it})$. We may clearly assume that f does not vanish identically, and, since $f \geqq 0$, it follows that $f(re^{it}) > 0$, hence $H(re^{it}) \neq 0$ in D. Let $G(re^{it})$

be the branch of $(H(re^{it}))^p$ which is real at $r=0$. Let γ be a real number satisfying

(1.14) $$\gamma < \frac{\pi}{2}, \quad p\gamma > \frac{\pi}{2},$$

For $0<r<1$ we have

$$\frac{1}{2\pi}\int |G(re^{it})|\,dt = \frac{1}{2\pi}\int_I |G(re^{it})|\,dt + \frac{1}{2\pi}\int_{II} |G(re^{it})|\,dt,$$

where $\int_I$ is taken over the set where $|\arg(H(z))| < \gamma$ and $\int_{II}$ is taken over the complementary set (defined by the condition $\gamma \leqq |\arg(H(z))| < \pi/2$, where $z = re^{it}$). In $\int_I$ we have

$$|H(z)| < f(z)(\cos\gamma)^{-1},$$

hence

(1.15) $$\frac{1}{2\pi}\int_I |G(re^{it})|\,dt \leqq (\cos\gamma)^{-p}\|f\|_{L^p}^p,$$

and, in particular,

(1.16) $$\frac{1}{2\pi}\int_I \operatorname{Re}(G(re^{it}))\,dt \leqq (\cos\gamma)^{-p}\|f\|_{L^p}^p$$

On the other hand, we have in $\int_{II}$

(1.17) $$|G(z)| \leqq \operatorname{Re}(G(z))(\cos p\gamma)^{-1}$$

(both factors being negative). Now, since

$$\frac{1}{2\pi}\int \operatorname{Re}(G(re^{it}))\,dt = G(0) = (\hat{f}(0))^p,$$

it follows from (1.16) that

$$\frac{1}{2\pi}\int_{II} |\operatorname{Re}(G(re^{it}))|\,dt \leqq (\hat{f}(0))^p + (\cos\gamma)^{-p}\|f\|_{L^p}^p$$

and this, combined with (1.17) and (1,15), implies

(1.18) $$\frac{1}{2\pi}\int |G(re^{it})|\,dt \leqq c_p\,\|f\|_{L^p}^p$$

where c_p is a constant depending only on p. Since $|\tilde{f}(re^{it})|^p \leqq |H(re^{it})|^p = |G(re^{it})|$, it follows from (1.18), letting $r\to 1$, that $\tilde{f}\in L^p(\mathbf{T})$ and

$$\|\tilde{f}\|_{L^p} \leqq c_p^{1/p}\|f\|_{L^p}.$$

The theorem now follows from the case $f \geqq 0$ and the linearity of the mapping $f \to \tilde{f}$. ◀

Remark: Theorem 1.8 can be proved by the method that we used in 1.7; see exercise 9 at the end of this section.

1.9 *Theorem*: *If f is real valued and $|f| \leqq 1$, then for* $0 \leqq \alpha < \pi/2$

$$\frac{1}{2\pi}\int e^{\alpha|\tilde{f}(e^{it})|}dt \leqq \frac{2}{\cos\alpha}. \tag{1.19}$$

PROOF: Put $F(z) = \tilde{f}(z) - if(z)$, Since $\cos\alpha f(z) \geqq \cos\alpha$, we have

$$\operatorname{Re}(e^{\alpha F(z)}) \geqq \cos\alpha\left|e^{\alpha F(z)}\right| = \cos\alpha\, e^{\alpha\tilde{f}(z)},$$

and since $\frac{1}{2\pi}\int \operatorname{Re}(e^{\alpha F(re^{it})})\,dt = \operatorname{Re}(e^{\alpha F(0)}) = \cos\alpha f(0) \leqq 1$, it follows that $\frac{1}{2\pi}\int e^{\alpha\tilde{f}(re^{it})}dt \leqq \frac{1}{\cos\alpha}$. Similarly, $\frac{1}{2\pi}\int e^{-\alpha\tilde{f}(re^{it})}\,dt \leqq \frac{1}{\cos\alpha}$, Adding and letting $r \to 1$ we obtain (1.19). ◀

Corollary: *If* $|f| \leqq 1$

$$\mathfrak{m}_{|\tilde{f}|}(\lambda) > 2\pi\left(1 - \frac{4}{\cos\sqrt{2}}\,e^{-\lambda}\right) \tag{1.20}$$

PROOF: Write $f = f_1 + if_2$ where f_1, f_2 are real valued. We have $\tilde{f} = \tilde{f}_1 + i\tilde{f}_2$ and consequently $|\tilde{f}(e^{it})| > \lambda$ happens only if either

$$|\tilde{f}_1(e^{it})| > 2^{-\frac{1}{2}}\lambda \quad \text{or} \quad |\tilde{f}_2(e^{it})| > 2^{-\frac{1}{2}}\lambda$$

Now, by (1.19) with $\alpha = \sqrt{2}$,

$$\left|\left\{t; |\tilde{f}_j(e^{it})| > \frac{1}{\sqrt{2}}\lambda\right\}\right| < \frac{4\pi}{\cos\sqrt{2}}e^{-\lambda}, \qquad j = 1,2$$

and (1.20) follows. ◀

* **1.10** We shall see in chapter VI that a finite Borel measure on **R** is completely determined by its Fourier-Stieltjes transform (just as measures on **T** are determined by their Fourier-Stieltjes coefficients). This means that two distribution functions, $\mathfrak{m}_1(x)$ and $\mathfrak{m}_2(x)$, of real-valued functions on **T** are equal if $\int e^{i\xi x}d\mathfrak{m}_1(x) = \int e^{i\xi x}\,d\mathfrak{m}_2(x)$ for all $\xi \in \mathbf{R}$. Using this remark we shall show now that if f is the characteristic function of some set $U \subset \mathbf{T}$, then $\mathfrak{m}_{\tilde{f}}(\lambda)$ depends only on the measure of U and not on the particular structure of U. Thus

we can compute $\mathfrak{m}_{\tilde{f}}(\lambda)$ explicitly by replacing U by an interval of the same measure.

Theorem: *Let $U \subset \mathbf{T}$ be a set of measure 2α. Let f be the characteristic function of U and let χ_α be the characteristic function of $(-\alpha,\alpha)$. Write $\mathfrak{m}_\alpha(\lambda) = \mathfrak{m}_{\tilde{\chi}_\alpha}(\lambda)$. Then $\mathfrak{m}_{\tilde{f}}(\lambda) = \mathfrak{m}_\alpha(\lambda)$.*

PROOF: Applying Cauchy's formula to the analytic functions $F^\xi(z) = e^{i\xi(\tilde{f}(z)-if(z))}$ on $z = re^{it}$, letting $r \to 1$ and remembering that $f = 0$ on $\mathbf{T}\setminus U$ and $f = 1$ on U, we obtain

$$(1.21) \qquad \int_{\mathbf{T}\setminus U} e^{i\xi\tilde{f}(e^{it})}dt + e^{\xi}\int_U e^{i\xi\tilde{f}(e^{it})}dt = 2\pi F^\xi(0) = 2\pi e^{\xi\alpha/\pi}.$$

Rewriting (1.21) for $-\xi$ instead of ξ and then taking complex conjugates, we obtain

$$(1.22) \qquad \int_{\mathbf{T}\setminus U} e^{i\xi\tilde{f}(e^{it})}dt + e^{-\xi}\int_U e^{i\xi\tilde{f}(e^{it})}dt = 2\pi e^{-\xi\alpha/\pi}.$$

From (1.21) and (1.22) we obtain

$$(1.23) \qquad \begin{aligned} \int_U e^{i\xi\tilde{f}(e^{it})}dt &= 2\pi\frac{\sinh\frac{\xi\alpha}{\pi}}{\sinh\xi} \\ \int_{\mathbf{T}\setminus U} e^{i\xi\tilde{f}(e^{it})}dt &= 2\pi\frac{\sinh\xi\left(1-\frac{\alpha}{\pi}\right)}{\sinh\xi} \end{aligned}$$

We write now $\mathfrak{m}_{\tilde{f}}(\lambda) = \mathfrak{n}_1(\lambda) + \mathfrak{n}_2(\lambda)$ where

$$\mathfrak{n}_1(\lambda) = \left|U \cap \{t;\tilde{f}(e^{it}) \leqq \lambda\}\right|$$

and

$$\mathfrak{n}_2(\lambda) = \left|(\mathbf{T}\setminus U) \cap \{t;\tilde{f}(e^{it}) \leqq \lambda\}\right|$$

and we can rewrite (1.23) as

$$(1.24) \qquad \begin{aligned} \int e^{i\xi x}d\mathfrak{n}_1(x) &= 2\pi\frac{\sinh\frac{\xi\alpha}{\pi}}{\sinh\xi} \\ \int e^{i\xi x}d\mathfrak{n}_2(x) &= 2\pi\frac{\sinh\xi\left(1-\frac{\alpha}{\pi}\right)}{\sinh\xi} \end{aligned}$$

We see that $\mathfrak{n}_1(x)$ and $\mathfrak{n}_2(x)$ are uniquely determined by α and so they are the same for $\tilde{f}$ and $\tilde{\chi}_\alpha$. We thus obtain that $\tilde{f}$ and $\tilde{\chi}_\alpha$ have the same distribution of values not only on **T** but also on U for $\tilde{f}$ and $(-\alpha,\alpha)$ for $\tilde{\chi}_\alpha$.

The Fourier series of χ_α is

$$\sum_{-\infty}^{\infty} \frac{\sin n\alpha}{\pi n} e^{int} = \frac{\alpha}{\pi} + 2\sum_{1}^{\infty} \frac{\sin n\alpha}{\pi n} \cos nt$$

hence

$$\begin{aligned}\tilde{\chi}_\alpha(r e^{it}) &= 2\sum_{1}^{\infty} r^n \frac{\sin n\alpha}{\pi n} \sin nt = \sum_{1}^{\infty} r^n \frac{\cos n(t-\alpha) - \cos n(t+\alpha)}{\pi n} \\ &= \frac{1}{\pi} \operatorname{Re}\left(\sum_{1}^{\infty} \frac{1}{n} r^n e^{in(t-\alpha)} - \sum_{1}^{\infty} \frac{1}{n} r^n e^{in(t+\alpha)}\right) \\ &= \frac{1}{\pi} \log\left| \frac{r e^{it} - e^{i\alpha}}{r e^{it} - e^{-i\alpha}} \right|\end{aligned}$$

and finally

$$\tilde{\chi}_\alpha(e^{it}) = \frac{1}{\pi} \log\left| \frac{e^{it} - e^{i\alpha}}{e^{it} - e^{-i\alpha}} \right| = \frac{1}{2\pi} \log \frac{1-\cos(t-\alpha)}{1-\cos(t+\alpha)} . \tag{1.25}$$

It follows from (1.25) that for $\lambda > 1$ the set $\{t; \tilde{\chi}_\alpha(e^{it}) > \lambda\}$ is an interval containing $t = -\alpha$ and contained in $(-\alpha - \beta_1, -\alpha + \beta_2)$ where $\dfrac{\beta_1}{2\alpha + \beta_1} = \dfrac{\beta_2}{2\alpha - \beta_2} = e^{-\pi\lambda}$, hence

$$\mathfrak{m}_\alpha(\lambda) > 2\pi - 5\alpha\, e^{-\pi\lambda} . \tag{1.26}$$

Corollary: *Let f be the characteristic function of a set U of measure 2α on* **T**. *Then, for $\lambda > 1$*

$$\left|\{t; |\tilde{f}(e^{it})| > \lambda\}\right| < 10\alpha\, e^{-\lambda\pi}. \tag{1.27}$$

EXERCISES FOR SECTION 1

1. Show that there exists a constant A such that for all n, λ and every $f \in C(\mathbf{T})$, $\|f\|_\infty \leqq 1$,

$$|\{t; |S_n(f,t)| > \lambda\}| < A e^{-\lambda} .$$

2. Show that for $1 \leqq p < \infty$ there exist constants A_p such that for all n, λ and $f \in L^p(\mathbf{T})$, $\|f\|_{L^p} \leqq 1$

$$|\{t; |S_n(f,t)| > \lambda\}| < \frac{A_p}{\lambda^p} .$$

3. Prove that if $f \in L^p(\mathbf{T})$, $1 < p < \infty$, then

$$\lim_{r \to 1} \left\| \tilde{f}(r e^{it}) - \tilde{f}(e^{it}) \right\|_{L^p} = 0.$$

4. Prove that if $f \in L^p(\mathbf{T})$, $g \in L^q(\mathbf{T})$ where $p^{-1} + q^{-1} = 1$, $1 < p < \infty$, then $\sum_{-\infty}^{\infty} \hat{f}(n)\overline{\hat{g}(n)}$ converges.

5. Show that theorem 1.9 is sharp in the sense that there exist real-valued functions f such that $|f| \leq 1$ and $\int e^{\pi|\tilde{f}|/2}\, dt = \infty$. *Hint*: Take $f = 2\chi_\alpha - 1$.

6. Prove that if $f \in C(\mathbf{T})$ then $e^{\tilde{f}} \in L^1(\mathbf{T})$ no matter how large $\sup|f|$ is. *Hint*: Write $f = P + f_1$ where P is a polynomial and $\| f_1 \|_\infty < 1$.

7, Show that $\log \left| \dfrac{e^{it} - e^{i\alpha}}{e^{it} - e^{i\beta}} \right| = \operatorname{Re}\left(\log\left(\dfrac{e^{it} - e^{i\alpha}}{e^{it} - e^{i\beta}} \right)\right)$ is a constant multiple of the conjugate function of the characteristic function of (α, β) by examining $\operatorname{Im}\left(\log \left(\dfrac{e^{it} - e^{i\alpha}}{e^{it} - e^{i\beta}} \right)\right)$.

8. Let $1 < p < \infty$. Show that there exists a constant c_p such that for $f \in L^p(\mathbf{T})$, $\lambda > 0$,

$$|\{t; |\tilde{f}(e^{it})| > \lambda\}| \leq c_p \| f \|_{L^p}^p \lambda^{-p}.$$

Remark: This is an immediate consequence of 1.8; try, however, to prove it using 1.10 rather then 1.8 and then use it in the following exercise to prove 1.8. *Hint*: Assume that f is real valued, denote $U_\lambda = \{t; \tilde{f}(e^{it}) > \lambda\}$ and $V_\lambda = \{t; \tilde{f}(e^{it}) < -\lambda\}$. Denote by g_λ the characteristic function of U_λ; deduce from Parseval's formula that ($q = p/(p-1)$)

$$\lambda |U_\lambda| \leq \int \tilde{f}(e^{it}) g_\lambda(e^{it})\, dt = -\int f(e^{it}) \tilde{g}_\lambda(e^{it})\, dt \leq 2\pi \| f \|_{L^p} \| \tilde{g}_\lambda \|_{L^q}$$

and use (1.27) to evaluate $\| \tilde{g}_\lambda \|_{L^q}$. Repeat for V_λ.

*9. Prove theorem 1.8 by the method of 1.7.

* 2. THE MAXIMAL FUNCTION OF HARDY AND LITTLEWOOD

2.1 DEFINITION: The *maximal function* of a function $f \in L^1(\mathbf{T})$ is the function

$$M_f(t) = \sup_{0 < h \leq \pi} \left| \frac{1}{2h} \int_{t-h}^{t+h} f(\tau)\, d\tau \right|. \tag{2.1}$$

If we allow the value $+\infty$ then $M_f(t)$ is well defined for all $t \in \mathbf{T}$. We shall see presently that $M_f(t)$ is finite for almost all $t \in \mathbf{T}$ and that M_f is of weak L^1 type. This will follow from the following simple

Lemma: *From any family $\Omega = \{I_\alpha\}$ of intervals on $\mathbf{T}$ one can extract a sequence $\{I_n\}$ of pairwise disjoint intervals, such that*

$$\left| \bigcup_{n=1}^{\infty} I_n \right| > \frac{1}{4} \left| \bigcup_{\alpha} I_\alpha \right|. \tag{2.2}$$

PROOF: Denote $a_1 = \sup_{I \in \Omega} |I|$ and let I_1 be any interval of Ω satisfying $|I_1| > \frac{3}{4} a_1$; let Ω_2 be the subfamily of all the intervals in Ω which do not intersect I_1. Denote $a_2 = \sup_{I \in \Omega_2} |I|$ and let $I_2 \in \Omega_2$ such that $|I_2| > \frac{3}{4} a_2$. We continue by induction; having picked $I_1, \dots, I_k$ we consider the family Ω_{k+1} of the intervals of Ω which intersect none of I_j, $j \leqq k$, and pick $I_{k+1} \in \Omega_{k+1}$ such that $|I_{k+1}| > \frac{3}{4} a_{k+1}$ where $a_{k+1} = \sup_{I \in \Omega_{k+1}} |I|$. We claim that the sequence $\{I_n\}$ so obtained satisfies (2.2). In fact, denoting by J_n the interval of length $4|I_n|$ of which I_n is the center part, we claim that $\bigcup J_n \supseteq \bigcup I_\alpha$, which clearly implies (2.2).

We notice first that $a_k \to 0$ and consequently $\bigcap \Omega_k = \emptyset$. For $I \in \Omega$ let k denote the first index such that $I \notin \Omega_k$; then $I \cap I_{k-1} \neq \emptyset$ and, since $|I_{k-1}| \geqq \frac{3}{4}|I|$, $I \subseteq J_{k-1}$, and the lemma is proved. ◀

2.2 ***Theorem:*** *For $f \in L^1(\mathbf{T})$, M_f is of weak L^1 type,*

PROOF: Since $M_f(t) \leqq M_{|f|}(t)$, we may assume $f \geqq 0$. Let $\lambda > 0$; if $M_f(t) > \lambda$ let I_t be an interval centered at t such that

$$\int_{I_t} f(\tau)\, d\tau > \lambda |I_t|. \tag{2.3}$$

Thus we cover the set $\{t; M_f(t) > \lambda\}$ by a family of intervals $\{I_t\}$. Let $\{I_n\}$ be a pairwise disjoint subsequence of $\{I_t\}$ satisfying (2.2). Then, by (2.2) and (2.3),

$$|\{t; M_f(t) > \lambda\}| \leqq \left|\bigcup I_t\right| \leqq 4 \left|\bigcup I_n\right| \leqq \frac{4}{\lambda} \int_{\bigcup I_n} f(\tau)\, d\tau \leqq \frac{4}{\lambda} \int_{\mathbf{T}} f(\tau)\, d\tau. \tag{2.4}$$

◀

2.3 ***Lemma:*** *Let $f \in L^1(\mathbf{T})$ and let $\mathfrak{m}(\lambda) = \mathfrak{m}_{|f|}(\lambda)$ be the distribution function of $|f|$. Then*

$$|\{t; M_f(t) > 2\lambda\}| \leqq \frac{4}{\lambda} \int_{\lambda+}^{\infty} y\, d\mathfrak{m}(y). \tag{2.5}$$

PROOF: Writing $f = g + h$ where $g = f$ whenever $|f| \leqq \lambda$ and $h = f$ whenever $|f| > \lambda$, we have $M_f(t) \leqq M_g(t) + M_h(t) \leqq \lambda + M_h(t)$; hence, by (2.4),

$$|\{t; M_f(t) > 2\lambda\}| \leqq |\{t; M_h(t) > \lambda\}| \leqq \frac{4}{\lambda} \int |h(\tau)| d\tau = \frac{4}{\lambda} \int_{\lambda+}^{\infty} y\, d\mathfrak{m}(y).$$ ◀

Corollary:

(a) *For* $1 < p < \infty$ *there exists a constant* c_p *such that if* $f \in L^p(\mathbf{T})$, *then* $M_f \in L^p(\mathbf{T})$ *and* $\| M_f \|_{L^p} \leqq c_p \| f \|_{L^p}$.

(b) If $f \log^+ |f| \in L^1(\mathbf{T})$, *then* $M_f \in L^1(\mathbf{T})$ *and*

$$\| M_f \|_{L^1} \leqq 2 + 4 \int_{\mathbf{T}} |f| \log^+ |f| \, dt.$$

PROOF: Denote by $\mathfrak{m}(\lambda)$ and $\mathfrak{n}(\lambda)$ the distribution functions of $|f|$ and M_f respectively. We can rewrite (2.5) in the form

$$(2.6) \qquad 2\pi - \mathfrak{n}(2\lambda) \leqq \frac{4}{\lambda} \int_{\lambda}^{\infty} y\, d\mathfrak{m}(y) \leqq \frac{4}{\lambda^p} \int_{\lambda}^{\infty} y^p\, d\mathfrak{m}(y);$$

hence if $f \in L^p(\mathbf{T})$, $1 \leqq p < \infty$, we have $\lambda^p(2\pi - \mathfrak{n}(\lambda)) \to 0$ as $\lambda \to \infty$.

We have $\|M_f\|_{L^p}^p = \dfrac{1}{2\pi} \displaystyle\int_0^{\infty} \lambda^p d\mathfrak{n}(\lambda) = \dfrac{2^p}{2\pi} \int_0^{\infty} \lambda^p d\mathfrak{n}(2\lambda)$; integrating by parts we obtain ($1 \leqq p < \infty$)

$$\int_0^{\infty} \lambda^p\, d\mathfrak{n}(2\lambda) = [\lambda^p(2\pi - \mathfrak{n}(2\lambda))]_0^{\infty} + \int_0^{\infty} (2\pi - \mathfrak{n}(2\lambda)) p \lambda^{p-1}\, d\lambda \leqq$$

$$\begin{cases} 8p \displaystyle\int_0^{\infty} \lambda^{p-2} \int_{\lambda}^{\infty} y\, d\mathfrak{m}(y)\, d\lambda & \text{if } p > 1 \\[2ex] 2\pi + 8 \displaystyle\int_1^{\infty} \lambda^{-1} \int_{\lambda}^{\infty} y\, d\mathfrak{m}(y)\, d\lambda & \text{if } p = 1 \end{cases}$$

and integrating by parts again we finally obtain:

$$\int_0^{\infty} \lambda^p d\mathfrak{n}(2\lambda) \leqq \begin{cases} \dfrac{8p}{p-1} \displaystyle\int_0^{\infty} \lambda^p\, d\mathfrak{m}(\lambda) = \dfrac{8p}{p-1} 2\pi \| f \|_{L^p}^p; & p > 1 \\[2ex] 2\pi + 8 \displaystyle\int_1^{\infty} \lambda \log \lambda\, d\mathfrak{m}(\lambda) = 2\pi + 8 \int_T |f| \log^+ |f| dt; & p = 1. \end{cases}$$ ◀

2.4 ***Lemma:*** *Let k be a nonnegative even function on $(-\pi,\pi)$, monotone nonincreasing on $(0,\pi)$ and such that $\int_{-\pi}^{\pi} k(t)\,dt = 1$. Then for all $f \in L^1(\mathbf{T})$*

$$\left| \int k(t-\tau) f(\tau)\, d\tau \right| \leqq M_f(t). \tag{2.7}$$

PROOF: The definition (2.1) is equivalent to

$$M_f(t) = \sup_{0<h\leqq\pi} \left| \int \phi_h(t-\tau) f(\tau)\, d\tau \right|$$

where ϕ_h is the characteristic function of $(-h,h)$ multiplied by $1/2h$ (so that $\int \phi_h\, dt = 1$). A function k satisfying the conditions of the lemma can be uniformly approximated by convex combinations of ϕ_h, $0 < h \leqq \pi$, and (2.7) is then obvious. ◀

2.5 Let $f \in L^1(\mathbf{T})$, let $f(re^{it})$ be its Poisson integral and let $\tilde{f}(re^{it})$ be the harmonic conjugate,

DEFINITION: The *maximal conjugate function* of f is the function

$$\tilde{\tilde{f}}(e^{it}) = \sup_{0<r<1} \left| \tilde{f}(re^{it}) \right|.$$

Theorem: *Let $f \in L^p(\mathbf{T})$, $1 < p < \infty$; then $\tilde{\tilde{f}} \in L^p(\mathbf{T})$ and*

$$\| \tilde{\tilde{f}} \|_{L^p} \leqq C_p \| f \|_{L^p}.$$

PROOF; Remembering that $\tilde{f}(re^{it})$ is the Poisson integral of $\tilde{f}(e^{it})$ and that the Poisson kernel satisfies the condition of lemma 2.4, we obtain $\left| \tilde{f}(re^{it}) \right| \leqq M_{\tilde{f}}(t)$; hence $\tilde{\tilde{f}}(e^{it}) \leqq M_{\tilde{f}}(t)$ and the theorem follows from 2.3 and 1.8. ◀

2.6 We have defined the conjugate $\tilde{f}$ of a function $f \in L^1(\mathbf{T})$ as the boundary value of the harmonic function $\tilde{f}(re^{it}) = (Q(r,\cdot) * f)(t)$ where

$$Q(r,t) = \frac{2r\sin t}{1 - 2r\cos t + r^2} \tag{2.8}$$

is the conjugate Poisson kernel. Since the limit

$$Q(1,t) = \lim_{r\to 1} Q(r,t) = \frac{\sin t}{1-\cos t} = \frac{\cos\frac{t}{2}}{\sin\frac{t}{2}} = \cot\frac{t}{2} \tag{2.9}$$

is so obvious and so explicit, we are tempted to reverse the order of the operations and write

(2.10) $$\tilde{f} = Q(1,t) * f$$

The difficulty, however, is that $Q(1,t)$ is not Lebesgue integrable so that the convolution (2.10) is, as yet, undefined. We now aim at showing that, the convolution appearing in (2.10) can be defined as an improper integral and that, with this definition, (2.10) is valid almost everywhere.

Lemma: *For $f \in L^1(\mathbf{T})$ and $\vartheta = 1 - r$, we have*

$$E(r,t) = \left| \frac{1}{2\pi} \int_{\vartheta}^{2\pi-\vartheta} Q(1,\tau) f(t-\tau)\, d\tau - \tilde{f}(re^{it}) \right| \leqq 4M_{|f|}(t).$$

PROOF: Write

(2.11) $$E(r,t) \leqq \left| \frac{1}{2\pi} \int_{\vartheta}^{2\pi-\vartheta} (Q(1,\tau) - Q(r,\tau)) f(t-\tau)\, d\tau \right| + $$
$$+ \left| \frac{1}{2\pi} \int_{-\vartheta}^{\vartheta} Q(r,\tau) f(t-\tau)\, d\tau \right| = E_1(r,t) + E_2(r,t).$$

We notice that the function

(2.12) $$Q(1,t) - Q(r,t) = \frac{(1-r)^2 \sin t}{(1-\cos t)(1 - 2r\cos t + r^2)} = $$
$$= \frac{1-r}{1+r} Q(1,t) \mathbf{P}(r,t)$$

is odd and monotone decreasing on $(0,\pi)$. For $\vartheta < t < \pi$

$$\frac{1-r}{1+r} Q(1,t) \leqq \frac{1-r}{1+r} \left(\sin \frac{\vartheta}{2} \right)^{-1} < \pi$$

so that

(2.13) $$Q(1,t) - Q(r,t) < \pi \mathbf{P}(r,t).$$

It follows that

(2.14) $$E_1(r,t) \leqq \frac{1}{2\pi} \int_{\vartheta}^{2\pi-\vartheta} |Q(1,\tau) - Q(r,\tau)|\, |f(t-\tau| d\tau) \leqq$$
$$\leqq \frac{1}{2} \int \mathbf{P}(r,\tau) |f(t-\tau)|\, d\tau \leqq \pi M_{|f|}(t).$$

In order to estimate $E_2(r,t)$ it is sufficient to notice that in $(-\vartheta,\vartheta)$ we have $|Q(r,t)| < \dfrac{2}{1-r}$ and consequently

$$(2.15)\qquad E_2(r,t) \leqq \frac{1}{\pi(1-r)} \int_{-\vartheta}^{\vartheta} |f(t-\tau)|\,d\tau \leqq \frac{2}{\pi} M_{|f|}(t). \quad ◀$$

***Corollary*:**

$$\sup_{0<\vartheta<\pi} \left| \frac{1}{2\pi} \int_{\vartheta}^{2\pi-\vartheta} Q(1,\tau) f(t-\tau)\,d\tau \right| \leqq \tilde{\tilde{f}}(e^{it}) + 4M_{|f|}(t).$$

2.7 The estimates (2.14) and (2.15) are clearly very wasteful. They do not take into account the fact that $Q(r,t)$ is odd and we can improve them by writing

$$(2.16)\qquad E_1(r,t) = \left| \frac{1}{2\pi} \int_{\vartheta}^{\pi} (Q(1,\tau) - Q(r,\tau))(f(t-\tau) - f(t+\tau))\,d\tau \right|$$

and

$$(2.17)\qquad E_2(r,t) = \left| \frac{1}{2\pi} \int_{0}^{\vartheta} Q(r,t)(f(t-\tau) - f(t+\tau))\,d\tau \right|$$

$$\leqq \frac{1}{\pi(1-r)} \left| \int_{0}^{\vartheta_1} (f(t-\tau) - f(t+\tau))\,d\tau \right|$$

where $0 < \vartheta_1 < \vartheta$ (the mean value theorem).

At every point t of continuity of f, and more generally, at every point in which the primitive of f is differentiable, we have:

$$(2.18)\qquad \int_{0}^{\vartheta_1} (f(t-\tau) - f(t+\tau))\,d\tau = o(\vartheta_1).$$

By (2.17) it is clear that if (2.18) holds, $E_2(r,t) \to 0$.

***Theorem*:** *Let $f \in L^1(\mathbf{T})$; at every $t \in \mathbf{T}$ for which (2.18) is valid we have, $(\vartheta = 1 - r)$,*

$$E(r,t) = \left| \frac{1}{2\pi} \int_{\vartheta}^{2\pi-\vartheta} Q(1,\tau) f(t-\tau)\,d\tau - \tilde{f}(re^{it}) \right| \to 0$$

as $r \to 1$.

PROOF: As in (2.11), $E(r,t) \leqq E_1(r,t) + E_2(r,t)$. We have already remarked that under the assumption (2.18), $\lim_{r\to1} E_2(r,t) = 0$ so that we can confine our attention to $E_1(r,t)$. For $\varepsilon > 0$, let $\eta > 0$ be such that for $0 < \vartheta_1 \leqq \eta$

$$(2.19) \qquad \left| \int_0^{\vartheta_1} (f(t-\tau) - f(t+\tau))\, d\tau \right| < \varepsilon \vartheta_1$$

and write

$$(2.20) \quad 2\pi E_1(r,t) = \left| \left(\int_\vartheta^\eta + \int_\eta^\pi \right) (Q(1,\tau) - Q(r,\tau))(f(t-\tau) - f(t+\tau))\, d\tau \right|.$$

The second integral tends to zero by virtue of the fact that on (η, π), $Q(r,\tau) \to Q(1,\tau)$ uniformly. Put

$$\Phi(\vartheta_1) = \int_0^{\vartheta_1} (f(t-\tau) - f(t+\tau))\, d\tau ,$$

then, integrating the first integral in (2.20) by parts, we see that it is bounded by

$$[\Phi(\vartheta_1)(Q(1,\vartheta_1) - Q(r,\vartheta_1))]_\vartheta^\eta + \varepsilon \left| \int_\vartheta^\eta \vartheta_1\, d(Q(1,\vartheta_1) - Q(r,\vartheta_1)) \right| ;$$

integrating by parts once more and remembering that $Q(1,\vartheta_1) < \pi/\vartheta_1$, it follows from (2.13) and (2.19) that $E_1(r,t) < 10\varepsilon + o(1)$ and the theorem is proved. ◀

2.8 Let F be defined on $\mathbf{T}$ and assume that for all $\vartheta > 0$ F is integrable on $\mathbf{T} \setminus (-\vartheta, \vartheta)$.

DEFINITION: The *principal value of* $\int_{\mathbf{T}} F(t)\, dt$ is

$$PV \int_{\mathbf{T}} F(t)\, dt = \lim_{\vartheta \to 0} \int_\vartheta^{2\pi - \vartheta} F(t)\, dt .$$

For $f \in L^1(\mathbf{T})$ condition (2.18) is satisfied for almost all $t \in \mathbf{T}$; since $\tilde{f}(re^{it}) \to \tilde{f}(e^{it})$ almost everywhere, we obtain,

Theorem: *Let* $f \in L^1(\mathbf{T})$. *The principal value of* $\frac{1}{2\pi} \int f(t-\tau) \cot \frac{\tau}{2}\, d\tau$ *exists for almost all* $t \in \mathbf{T}$, *and, almost everywhere,*

$$\tilde{f}(e^{it}) = PV \frac{1}{2\pi} \int f(t-\tau) \cot \frac{\tau}{2}\, d\tau .$$

2.9 Theorem 2.7 can be used both ways. We can use it to show the existence of the principal value of $PV \int f(t-\tau) \cot \frac{\tau}{2}\, d\tau$ if we know that $\tilde{f}(e^{it})$ exists or to obtain the existence of $\tilde{f}(e^{it})$ at points where

$$PV\int f(t-\tau)\cot\frac{\tau}{2}\,d\tau$$

clearly exists. For instance, if f satisfies a Lipschitz condition at t, that is, if $|f(t+h)-f(t)| \leqq K|h|^\alpha$ for some $K>0$ and $\alpha>0$, then $\int_0^\pi |f(t-\tau)-f(t+\tau)|\cot\frac{\tau}{2}\,d\tau<\infty$ and it follows that $\tilde{f}(e^{it})$ exists and

$$\tilde{f}(e^{it}) = \frac{1}{2\pi}\int_0^\pi (f(t-\tau)-f(t+\tau))\cot\frac{\tau}{2}\,d\tau. \tag{2.21}$$

If f satisfies a Lipschitz condition uniformly on a set $E\subset\mathbf{T}$, that is, if for some $K>0$ and $\alpha>0$

$$|f(t+h)-f(t)| \leqq K|h|^\alpha \qquad \text{for all } t\in E,$$

then the integrals (2.21) are uniformly bounded and

$$\frac{1}{2\pi}\int_\varepsilon^\pi (f(t-\tau)-f(t+\tau))\cot\frac{\tau}{2}\,d\tau\to\tilde{f}(e^{it})$$

uniformly in $t\in E$ as $\varepsilon\to 0$. It follows, rexamining the proof of 2.7, that $\tilde{f}(re^{it})\to\tilde{f}(e^{it})$ uniformly for $t\in E$ as $r\to 1$. In particular, if E is an interval, it follows that $\tilde{f}(e^{it})$ is continuous on E.

2.10 Conjugation is not a local operation; that is, it is *not* true that if $f(t)=g(t)$ in some interval I, then $\tilde{f}(t)=\tilde{g}(t)$ on I, or equivalently, that if $f(t)=0$ on I, then $\tilde{f}(t)=0$ on I. However,

***Theorem*:** *If $f(t)=0$ on an interval I, then $\tilde{f}(t)$ is analytic on I.*

PROOF: By the previous remarks $\tilde{f}$ is continuous on I. Thus the function $F=f+i\tilde{f}$ is analytic in D and is continuous and purely imaginary on I. By Schwarz's reflection principle F admits an analytic extension through I, and since $F(e^{it})=i\tilde{f}(e^{it})$ on I, the theorem follows. ◀

Remark: Using (2.21) we can estimate the successive derivatives of $\tilde{f}$ at points $t\in I$ and show that $\tilde{f}$ is analytic on I without the use of the "complex" reflection principle.

EXERCISES FOR SECTION 2

1. Assume $f\in \mathrm{Lip}_\alpha(\mathbf{T})$, $0<\alpha\leqq 1$. Show that $\tilde{f}\in \mathrm{Lip}_{\alpha'}(\mathbf{T})$ for all $\alpha'<\alpha$.

*2. Assume $f\in \mathrm{Lip}_\alpha(\mathbf{T})$, $0<\alpha<1$. Show that $\tilde{f}\in \mathrm{Lip}_\alpha(\mathbf{T})$. *Hint*:

$$\tilde{f}(t+h)-\tilde{f}(t)=\frac{1}{2\pi}\int (f(t+h-\tau)-f(t+h))\cot\frac{\tau}{2}d\tau -$$

$$-\frac{1}{2\pi}\int (f(t-\tau)-f(t))\cot\frac{\tau}{2}\,d\tau$$

$$= O(h^{\alpha})+\int_{2h}^{2\pi-2h}(f(t-\tau)-f(t+h))\cot\frac{\tau+h}{2}\,d\tau -$$

$$-\int_{2h}^{2\pi-2h}(f(t-\tau)-f(t))\cot\frac{\tau}{2}\,d\tau$$

$$= O(h^{\alpha})+\int_{2h}^{2\pi-2h}(f(t-\tau)-f(t))\left(\cot\frac{\tau+h}{2}-\cot\frac{\tau}{2}\right)d\tau -$$

$$-(f(t+h)-f(t))\int_{2h}^{2\pi-2h}\cot\frac{\tau+h}{2}\,d\tau.$$

3. Assume $f\in C^{n}(\mathbf{T})$, $n\geqq 1$. Show that $\tilde{f}\in C^{n-1}(\mathbf{T})$ and $\tilde{f}^{(n-1)}\in \mathrm{Lip}_{\alpha}(\mathbf{T})$ for all $\alpha<1$.

4. Localize exercises 1–3, that is, assume that f satisfies the respective conditions on an open interval $I\subset \mathbf{T}$ and show that the conclusions hold in I.

*3. THE HARDY SPACES

In this section we study some spaces of functions holomorphic in the unit disc D. These spaces are closely related to spaces of functions on $\mathbf{T}$ and we obtain, for example, a characterization of L^p functions and of measures whose Fourier coefficients vanish for negative values of n. We also prove that if for some $f\in L^1(\mathbf{T})$, $\tilde{f}(e^{it})$ is summable then $S[\tilde{f}]=\tilde{S}[f]$, and, finally, we obtain results concerning the absolute convergence of some classes of Fourier series. We start with some preliminary remarks about products of Moebius functions.

3.1 Let $0<|\zeta|<1$; the function $b(z,\zeta)=\dfrac{\bar{\zeta}(\zeta-z)}{|\zeta|(1-z\bar{\zeta})}$ defines, as is well known, a conformal representation of $\bar{D}$ onto itself, taking ζ into zero and zero into $|\zeta|$. The important thing for us now is that $b(z,\zeta)$ vanishes only at $z=\zeta$ and $|b(z,\zeta)|=1$ on $|z|=1$. If $0<|\zeta|<r$, then $b\left(\dfrac{z}{r},\dfrac{\zeta}{r}\right)=r\dfrac{\bar{\zeta}(\zeta-z)}{|\zeta|(r^2-z\bar{\zeta})}$ is holomorphic in $|z|\leqq r$, vanishes only at $z=\zeta$, and $\left|b\left(\dfrac{z}{r},\dfrac{\zeta}{r}\right)\right|=1$ on $|z|=r$. For $\zeta=0$ we define $b(z,0)=z$.

Let f be holomorphic in $|z| \leqq r$ and denote its zeros there by $\zeta_1, \ldots, \zeta_k$ (counting each zero as many times as its multiplicity). The function $f_1(z) = f(z)\left(\prod_1^k b\left(\frac{z}{r}, \frac{\zeta_n}{r}\right)\right)^{-1}$ is holomorphic in $|z| \leqq r$, is zero-free and satisfies $|f_1(z)| = |f(z)|$ for $|z| = r$. Since $\log|f_1(z)|$ is harmonic in $|z| \leqq r$ we have

$$\log|f_1(0)| = \frac{1}{2\pi}\int \log|f_1(re^{it})|\,dt$$

and if we assume, for simplicity, that $f(0) \neq 0$, the above formula is equivalent to *Poisson-Jensen's formula*:

$$\text{(3.1)} \qquad \log|f(0)| + \log\prod_{n=1}^{k} r|\zeta_n|^{-1} = \frac{1}{2\pi}\int \log|f(re^{it})|\,dt.$$

We implicitly assumed that f has no zeros of modulus r; however, since both sides of the formula depend continuously on r, the above is valid even if f vanishes on $|z| = r$. The reader should check the form that Poisson-Jensen's formula takes when f vanishes at $z = 0$.

The term $\log\prod_1^k r|\zeta_n|^{-1} = \sum_1^k \log(r|\zeta_n|^{-1})$ is positive, and removing it from (3.1) we obtain *Jensen's inequality*

$$\text{(3.2)} \qquad \log|f(0)| \leqq \frac{1}{2\pi}\int \log|f(re^{it})|\,dt;$$

or, if f has a zero of order s at $z = 0$,

$$\log\left|\lim_{z\to 0} z^{-s}f(z)\right| + \log(r^s) \leqq \frac{1}{2\pi}\int \log|f(re^{it})|\,dt.$$

Another form of Jensen's inequality is: let f be holomorphic in $|z| \leqq r$ and let $\zeta'_1, \ldots, \zeta'_m$ be (some) zeros of f in $|z| \leqq r$, counted each one at most as many times as its multiplicity. Then

$$\text{(3.3)} \qquad \log|f(0)| + \sum_1^m \log(r|\zeta'_n|^{-1}) \leqq \frac{1}{2\pi}\int \log|f(re^{it})|\,dt.$$

Inequality (3.3) is obtained from (3.1) by deleting some (positive) terms of the form $\log(r|\zeta_n|^{-1})$ from the left-hand side.

3.2 Let $p > 0$ and let f be holomorphic in D. We introduce the notation

$$\text{(3.4)} \qquad h_p(f, r) = \frac{1}{2\pi}\int |f(re^{it})|^p\,dt.$$

If $0<r<1$ and $\rho<1$ we have $f(r\rho e^{it})=f(re^{it})*\mathbf{P}(\rho,t)$ and consequently for $p\geqq 1$ we have

$$h_p(f,r\rho) = \|f(r\rho e^{it})\|_{L^p}^p \leqq \|f(re^{it})\|_{L^p}^p = h_p(f,r)$$

or, in other words, $h_p(f,r)$ is a monotone nondecreasing function of r. The case $p=2$ is particularly obvious since for $f(z)=\sum_0^\infty a_n z^n$, we have $h_2(f,r)=\sum_0^\infty |a_n|^2 r^{2n}$. We show now that the same is true for all $p>0$.

Lemma: *Let f be holomorphic in D and $p>0$. Then $h_p(f,r)$ is a monotone nondecreasing function of r.*

PROOF: We reduce the case of an arbitrary positive p to the case $p=2$. Let $r_1<r<1$. Assume first that f has no zeros on $|z|\leqq r$ and consider the function† $g(z)=(f(z))^{p/2}$; then

$$\frac{1}{2\pi}\int |f(r_1,e^{it})|^p\,dt = \frac{1}{2\pi}\int |g(r_1,e^{it})|^2\,dt \leqq \frac{1}{2\pi}\int |g(re^{it})|^2\,dt$$

$$= \frac{1}{2\pi}\int |f(re^{it})|^p\,dt$$

or $h_p(f,r_1)\leqq h_p(f,r)$.

If f has zeros inside $|z|<r$ but not on $|z|=r$, we denote the zeros, repeating each according to its multiplicity, by $\zeta_1,\dots,\zeta_k$, and write

$$f_1(z) = f(z)\left(\prod_1^k b\left(\frac{z}{r},\frac{\zeta_n}{r}\right)\right)^{-1}.$$

For $|z|<r$ we have $|f(z)|<|f_1(z)|$, for $|z|=r$ we have $|f(z)|=|f_1(z)|$, and f_1 is zero-free in $|z|\leqq r$. It follows that

$$h_p(f,r_1) < h_p(f_1,r_1) \leqq h_p(f_1,r) = h_p(f,r).$$

Since $h_p(f,r)$ is a continuous function of r, the same is true even if f does have zeros on $|z|=r$, and the lemma is proved. ◀

3.3 ***Lemma***: *Let $\{\zeta_n\}$ be a sequence of complex numbers satisfying $|\zeta_n|<1$ and $\sum(1-|\zeta_n|)<\infty$. Then the product*

$$B(z) = \prod_1^\infty b(z,\zeta_n) = z^m \prod_{\zeta_n\neq 0} \frac{\bar{\zeta}_n(\zeta_n - z)}{|\zeta_n|(1-z\bar{\zeta}_n)} \tag{3.5}$$

converges absolutely and uniformly in every disc of the form $|z|\leqq r<1$.

† Any branch of $(f(z))^{p/2}$.

PROOF: It is sufficient to show that $\sum \left| 1 - \dfrac{\bar{\zeta}_n(\zeta_n - z)}{|\zeta_n|\,(1 - z\bar{\zeta}_n)} \right|$ converges uniformly in $|z| \leqq r < 1$. But

$$\left| 1 - \frac{\bar{\zeta}_n(\zeta_n - z)}{|\zeta_n|(1 - z\bar{\zeta}_n)} \right| = \left| \frac{(|\zeta_n| + z\bar{\zeta}_n)(1 - |\zeta_n|)}{|\zeta_n|(1 - z\bar{\zeta}_n)} \right| \leqq \frac{1 + r}{1 - r}(1 - |\zeta_n|)$$

and the series converges since $\sum (1 - |\zeta_n|) < \infty$. ◀

The product (3.5), often called the *Blaschke product* corresponding to $\{\zeta_n\}$, is clearly holomorphic in D and it vanishes precisely at the points ζ_n. Nothing prevents, of course, repeating the same complex number a (finite) number of times in $\{\zeta_n\}$, so that we can prescribe not only the zeros but their multiplicities as well. Since all the terms in (3.5) are bounded by 1 in modulus, we have $|B(z)| < 1$ in D.

3.4 We now introduce the spaces H^p (H for Hardy) and N (N for Nevanlinna).

DEFINITION: *The space* H^p, $p > 0$, is the (linear) space of all functions f holomorphic in D, such that

$$(3.6) \qquad \|f\|_{H^p}^p = \lim_{r \to 1} h_p(f, r) = \sup_{0<r<1} h_p(f, r) < \infty .$$

The space N is the space of all functions f holomorphic in D, such that

$$(3.7) \qquad \|f\|_N = \sup_{0<r<1} \frac{1}{2\pi} \int \log^+ |f(re^{it})|\, dt < \infty .$$

Remarks: (a) For $p \geqq 1$, $\|\ \|_{H^p}$ as defined in (3.6) is a norm and we shall see later that H^p endowed with this norm, can be identified with a closed subspace of $L^p(\mathbf{T})$. For $p < 1$, $\|\ \|_{H^p}^p$ satisfies the triangle inequality and is homogeneous of degree p. It can be used as a metric for H^p; $\|\ \|_{H^p}$ is homogeneous of degree one but does not satisfy the triangle inequality. $\|\ \|_N$ is not homogeneous and does not satisfy the triangle inequality.

(b) If $p' < p$ we have $N \supset H^{p'} \supset H^p$.

The space H^2 has a simple characterization:

Lemma: *Let* $f(z) = \sum_0^\infty a_n z^n$; *then* $f \in H^2$ *if, and only if,* $\sum_0^\infty |a_n|^2$ *is finite.*

PROOF: $h_2(f, r) = \sum_0^\infty |a_n|^2 r^{2n}$. It follows that $\|f\|_{H^2}^2 = \sum_0^\infty |a_n|^2$. ◀

An immediate consequence is that in this case f is the Poisson integral of $f(e^{it}) \sim \sum_0^\infty a_n e^{int}$.

3.5 ***Lemma:*** *Let $f \in N$, and denote its zeros in D by $\zeta_1, \zeta_2, \dots$, counting each one according to its multiplicity. Then $\sum (1 - |\zeta_n|) < \infty$.*

Remark: The convergence of the series $\sum (1 - |\zeta_n|)$ is equivalent to the convergence of the product $\prod |\zeta_n|$ hence to the boundedness (below) of the series $\sum \log |\zeta_n|$ (not counting the zeros at the origin, if any).

PROOF: We may assume $f(0) \neq 0$. By Jensen's inequality (3.3), if M is fixed and r sufficiently close to 1

$$\log |f(0)| - \| f \|_N \leqq \sum_1^M \log |\xi_n| - M \log r .$$

Letting $r \to 1$ we obtain

$$\log |f(0)| - \| f \|_N \leqq \sum_1^M \log |\zeta_n|$$

and since M is arbitrary the lemma follows. ◀

3.6 If we combine lemma 3.3 with 3.5 we see that if $f \in N$, the Blaschke product corresponding to the sequence of zeros of f is a well-defined holomorphic function in D, having the same zeros (with the same multiplicities) as f and satisfying $|B(z)| < 1$ in D. If we write $F(z) = f(z)(B(z))^{-1}$ then F is holomorphic and satisfies $|F(z)| > |f(z)|$ in D. We shall refer to $f = BF$ as *the canonical factorization* of f.

Theorem: *Let $f \in H^p$, $p > 0$, and let $f = BF$ be its canonical factorization. Then $F \in H^p$ and $\| F \|_{H^p} = \| f \|_{H^p}$.*

PROOF: The Blaschke product B has the form

$$B(z) = \lim_{N \to \infty} z^m \prod_1^N b(z, \zeta_n) .$$

If we write $F_N(z) = f(z)(z^m \prod_1^N b(z, \zeta_n))^{-1}$, then F_N converges to F uniformly on every disc of the form $|z| \leqq r < 1$. Since the absolute value of the finite product appearing in the definition of F_N tends to one uniformly as $|z| \to 1$, it follows from lemma 3.2 that $F_N \in H^p$ and $\| F_N \|_{H^p} = \| f \|_{H^p}$. Let $r < 1$, then

$$h_p(F, r) = \lim_{N \to \infty} h_p(F_N, r) \leqq \lim_{N \to \infty} \| F_N \|_{H^p}^p = \| f \|_{H^p}^p .$$

Now $h_p(F,r) \leqq \|f\|_{H^p}^p$ for all $r<1$ is equivalent to $\|F\|_{H^p} \leqq \|f\|_{H^p}$, and since the reverse inequality is obvious, the theorem is proved. ◀

Theorem 3.6 is a key theorem in the theory of H^p spaces. It allows us to operate mainly with zero-free functions which, by the fact of being zero-free, can be raised to arbitrary powers and thereby move from one H^p to a more convenient one. This idea was already used in the proof of lemma 3.2.

Our first corollary to theorem 3.6 deals with Blaschke products.

3.7 ***Corollary***: *Let B be a Blaschke product. Then* $|B(e^{it})| = 1$ *almost everywhere.*

PROOF: Since $|B(z)| < 1$ in D it follows from lemma 1.2 that $B(e^{it})$ exists as a radial (actually: nontangential) limit for almost all $t \in \mathbf{T}$. The canonical factorization of $f = B$ is trivial, the function F is identically one, and consequently

$$\|B\|_{H^2}^2 = \frac{1}{2\pi}\int |B(e^{it})|^2\,dt = 1 .$$

Since $|B(e^{it})| \leqq 1$, the above equality can hold only if $|B(e^{it})| = 1$ almost everywhere. ◀

3.8 ***Theorem***: *Assume $f \in H^p$, $p > 0$. Then the radial limit $\lim_{r\to 1} f(re^{it})$ exists for almost all $t \in \mathbf{T}$ and, denoting it by $f(e^{it})$, we have*

$$\|f\|_{H^p}^p = \frac{1}{2\pi}\int |f(e^{it})|^p\,dt .$$

PROOF: The case $p = 2$ follows from 3.4.

For arbitrary $p > 0$, let $f = BF$ be the canonical factorization of f, and write $G(z) = (F(z))^{p/2}$. Then G belongs to H^2 and consequently $G(re^{it}) \to G(e^{it})$ for almost all $t \in \mathbf{T}$; at every such t, $F(re^{it})$ converges to some $F(e^{it})$ such that $|F(e^{it})|^{p/2} = |G(e^{it})|$. Since B has a radial limit of absolute value one almost everywhere we see that $f(e^{it}) = \lim f(re^{it})$ exists and $|f(e^{it})|^{p/2} = |G(e^{it})|$ almost everywhere. Now $\|f\|_{H^p}^p = \|F\|_{H^p}^p = \|G\|_{H^2}^2 = \frac{1}{2\pi}\int |G(e^{it})|^2\,dt = \frac{1}{2\pi}\int |f(e^{it})|^p\,dt$ and the proof is complete. ◀

3.9 The convergence assured by theorem 3.8 is pointwise convergence almost everywhere. For $p \geqq 2$ we know that f is the Poisson integral of $f(e^{it})$ and consequently $f(re^{it})$ converges to $f(e^{it})$ in the $L^p(\mathbf{T})$ norm. We shall show that the same holds for $p = 1$ (hence for $p \geqq 1$); first, however, we use the case $p \geqq 2$ to prove:

Theorem: *Let* $0 < p < p'$ *and suppose* $f \in H^p$ *and* $f(e^{it}) \in L^{p'}(\mathbf{T})$. *Then* $f \in H^{p'}$.

PROOF: As before, if we write $f = BF$, $G(z) = (F(z))^{p/2}$, then $G \in H^2$ and $G(e^{it}) \in L^{2p'/p}(\mathbf{T})$. G is the Poisson integral of $G(e^{it})$ and consequently $G \in H^{2p'/p}$ which means $F \in H^{p'}$, hence $f \in H^{p'}$. ◀

Corollary: *Let* $f \in L^1(\mathbf{T})$ *and assume* $\tilde{f} \in L^1(\mathbf{T})$; *then* $(f + i\tilde{f}) \in H^1$.

PROOF: We know (corollary 1.6) that $(f + i\tilde{f}) \in H^\alpha$ for all $\alpha < 1$ and by the assumption $(f + i\tilde{f})(e^{it}) \in L^1(\mathbf{T})$. ◀

3.10 ***Theorem:*** *Every function* f *in* H^1 *can be written as a product* $f = f_1 f_2$ *with* $f_1, f_2 \in H^2$.

PROOF: Let $f = BF$ be the canonical factorization of f. Write $f_1 = F^{1/2}$, $f_2 = BF^{1/2}$. ◀

3.11 We can now prove:

Theorem: *Let* $f \in H^1$ *and let* $f(e^{it})$ *be its boundary value. Then* f *is the Poisson integral of* $f(e^{it})$.

PROOF: We prove the theorem by showing that $f(re^{it})$ converges to $f(e^{it})$ in the L^1 norm. This implies that if $f(z) = \sum a_n z^n$, then a_n is the Fourier coefficient of $f(e^{it})$ which is clearly equivalent to f being the Poisson integral of $f(e^{it})$.

Write $f = f_1 f_2$ with $f_j \in H^2$, $j = 1, 2$; we know that as $r \to 1$, $\|f_j(re^{it}) - f_j(e^{it})\|_{L^2} \to 0$. Now,

$$f(re^{it}) - f(e^{it}) = f_1(re^{it}) f_2(re^{it}) - f_1(e^{it}) f_2(e^{it});$$

adding and subtracting $f_1(e^{it}) f_2(re^{it})$ and using the Cauchy-Schwarz inequality, we obtain

$$\|f(re^{it}) - f(e^{it})\|_{L^1} \leqq \|f_2\|_{L^2} \|f_1(re^{it}) - f_1(e^{it})\|_{L^2} + \|f_1\|_{L^2} \|f_2(re^{it}) - f_2(e^{it})\|_{L^2}$$

and the proof is complete. ◀

Remark: See exercise 2 at the end of the section for an extension of the theorem to the case $0 < p < 1$.

Corollary: *Let $f \in L^1(\mathbf{T})$ and $\tilde{f} \in L^1(\mathbf{T})$. Then $S[\tilde{f}] = \tilde{S}[f]$.*

PROOF: From 3.9 and the theorem above follows that $\tilde{f}(re^{it})$ is the Poisson integral of $\tilde{f}(e^{it})$ (see remark 1.4). ◀

3.12 Theorem: *Assume $p \geqq 1$. A function f belongs to H^p if, and only if, it is the Poisson integral of some $f(e^{it}) \in L^p(\mathbf{T})$ satisfying*

$$(3.8) \qquad \hat{f}(n) = 0 \qquad \text{for all } n < 0.$$

PROOF; Let $f \in H^p$; by theorem 3.11, f is the Poisson integral of $f(e^{it})$ and (3.8) is clearly satisfied. On the other hand, let $f \in L^p(\mathbf{T})$ and assume (3.8); the Poisson integral of f,

$$f(re^{it}) = \sum_0^\infty \hat{f}(n) r^n e^{int} = \sum_0^\infty \hat{f}(n) z^n$$

is holomorphic in D, and since $\| f(re^{it}) \|_{L^p} \leqq \| f(e^{it}) \|_{L^p}$, it follows that $f(z) \in H^p$. ◀

3.13 For $p > 1$, we can prove that every $f \in H^p$ is the Poisson integral of $f(e^{it})$ without appeal to theorem 3.11 or any other result obtained in this section. We just repeat the proof of lemma 1.2 (which is the case $p = \infty$ of 3.12): if $f \in H^p$, $\| f(re^{it}) \|_{L^p}$ is bounded as $r \to 1$; we can pick a sequence $r_n \to 1$ such that $f_n(e^{it}) = f(r_n e^{it})$ converge weakly in $L^p(\mathbf{T})$ to some $f(e^{it})$. Since weak convergence in L^p implies convergence of Fourier coefficients, it is clear that (3.8) is satisfied and that the function f with which we started is the Poisson integral of $f(e^{it})$.

For $p = 1$ the proof as given above is insufficient. $L^1(\mathbf{T})$ is a subspace of $M(\mathbf{T})$, the space of Borel measures on $\mathbf{T}$, which is the dual of $C(\mathbf{T})$, and the argument above can be used to show that every $f \in H^1$ is the Poisson integral of some measure μ on $\mathbf{T}$. This measure has the property

$$(3.9) \qquad \hat{\mu}(n) = 0, \qquad \text{for all } n < 0.$$

All that we have to do in order to complete the (alternative) proof of theorem 3.12 in the case $p = 1$ is to prove that the measures satisfying (3.9), often called *analytic measures*, are absolutely continuous with respect to the Lebesgue measure on $\mathbf{T}$.

Theorem: (*F. and M. Riesz*): *Let μ be a Borel measure on* $\mathbf{T}$ *satisfying*

$$\hat{\mu}(n) = 0 \qquad \textit{for all } n < 0. \tag{3.9}$$

Then μ is absolutely continuous with respect to Lebesgue measure.

We first prove:

Lemma: *Let $E \subset \mathbf{T}$ be a closed set of measure zero. There exists a function ϕ holomorphic in D and continuous in $\bar{D}$ such that*:

$$\begin{aligned} &\text{(i)} \quad \phi(e^{it}) = 1 &&\text{on } E \\ &\text{(ii)} \quad |\phi(z)| < 1 &&\text{on } \bar{D} \setminus E. \end{aligned} \tag{3.10}$$

PROOF: Since E is closed and of measure zero we can construct a function $\psi(e^{it})$ on $\mathbf{T}$ such that $\psi(e^{it}) > 0$ everywhere, $\psi(e^{it})$ is continuously differentiable in each component of $\mathbf{T} \setminus E$, $\psi(e^{it}) \to \infty$ as t approaches E, and $\psi \in L^2(\mathbf{T})$. The Poisson integral $\psi(z)$ of $\psi(e^{it})$ is positive on D and $\psi(z) \to \infty$ as z approaches E. The conjugate function is continuous in $\bar{D} \setminus E$ (see the end of section 2) and consequently, if we put $\varphi(z) = \dfrac{\psi(z) + i\tilde{\psi}(z)}{\psi(z) + i\tilde{\psi}(z) + 1}$, then φ is holomorphic in D and continuous in $\bar{D} \setminus E$. At every point where $\psi(z) < \infty$ we have $|\varphi(z)| < 1$, and as $\psi(z) \to \infty$, $\varphi(z) \to 1$. If we define $\varphi(z) = 1$ on E then φ satisfies (3.10). ◀

PROOF OF THE THEOREM: Assume that μ satisfies the condition (3.9). We can assume $\hat{\mu}(0) = 0$ as well (otherwise consider $\mu - \hat{\mu}(0)\,dt$) and it then follows from Parseval's formula that

$$\langle \bar{f}, \mu \rangle = \overline{\int f \, d\mu} = 0 \tag{3.11}$$

for every $f \in C(\mathbf{T})$ which is the boundary value of a holomorphic function in D or, equivalently, such that $\hat{f}(n) = 0$ for all negative n. Let $E \subset \mathbf{T}$ be closed and of (Lebesgue) measure zero. Let φ be a function satisfying (3.10). Then, by (3.11)

$$\int \varphi^m \, d\mu = 0 \qquad \text{for all } m > 0$$

and by (3.10)

$$\lim_{m \to \infty} \int \varphi^m \, d\mu = \mu(E).$$

Thus $\mu(E) = 0$ for every closed set E of Lebesgue measure zero and, since μ is regular, the theorem follows. ◀

3.14 Theorem 3.13 can be given a more complete form in view of the following important

Theorem: *Let $f \in H^p$, $p > 0, f \not\equiv 0$; then*

$$\log |f(e^{it})| \in L^1(\mathbf{T}).$$

Remarks: The same conclusion holds under the weaker assumption $f \in N$. We state it for H^p since we did not prove the existence of $f(e^{it})$ for $f \in N$ (cf. [28], Vol. 1, p. 276). Since $p \log^+ |f| \leqq |f|^p$, we already know that $\log^+ |f(e^{it})| \in L^1(\mathbf{T})$. Thus the content of the theorem is that $f(e^{it})$ cannot be too small on a large set.

PROOF: Replacing f by $z^{-m}f$, if f has a zero of order m at $z = 0$, we may assume $f(0) \neq 0$. Let $r < 1$; then, by Jensen's inequality

$$\log |f(0)| - \|f\|_N \leqq -\frac{1}{2\pi} \int \log^- |f(re^{it})| \, dt \leqq 0$$

(where $\log^- x = -\log x$ if $x < 1$ and zero otherwise). It follows that $\int |\log |f(re^{it})|| \, dt$ is bounded as $r \to 1$ and the theorem follows from Fatou's lemma. ◀

Corollary: *If $f \neq 0$ is in H^p, $f(e^{it})$ can vanish only on a set of measure zero.*

Combining theorem 3.13 with our last corollary we obtain that analytic measures are equivalent to Lebesgue's measure (i.e., they all have the same null sets).

3.15 ***Theorem***: *Let E be a closed proper subset of* $\mathbf{T}$. *Any continuous function on E can be approximated uniformly by Taylor polynomials.*†

PROOF: We denote by $C(E)$ the algebra of all continuous functions on E endowed with the supremum norm. The theorem claims that the restrictions to E of Taylor polynomials are dense in $C(E)$. Consider a measure μ carried by E orthogonal to all z^n, $n = 0, 1, \ldots$.

† We use the term "Taylor polynomial" to designate trigonometric polynomials of the form $\sum_0^N a_n e^{int}$.

Now $\langle \overline{z^n, \mu} \rangle = \hat{\mu}(n) = 0$ so that μ is analytic; hence $\mu = f\,dt$ with $f \in H^1$, f carried by E. By theorem 3.14 $f = 0$; hence there is no nontrivial functional on $C(E)$, which is orthogonal to all Taylor polynomials, and the theorem follows from the Hahn-Banach theorem. ◀

3.16 We finish this section with another application of 3.10.

Theorem: (*Hardy*): *Let $f(z) = \sum_0^\infty a_n z^n \in H^1$. Then $\sum_1^\infty |a_n| n^{-1} < \infty$.*

Remark: The theorem can also be stated: "Let $f \in L^1(\mathbf{T})$ satisfy (3.8), then $\sum_1^\infty |\hat{f}(n)| n^{-1} \le \|f\|_{L^1}$."

PROOF: If $F(e^{it})$ is a primitive of $f(e^{it})$, then F is continuous on $\mathbf{T}$ and consequently its Fourier series is Abel summable to F at every $t \in \mathbf{T}$. In particular $\sum_1^\infty (a_n/n) r^n$ tends to a finite limit as $r \to 1$. If we assume $a_n \geqq 0$ for all n then $\sum a_n/n$ is clearly convergent (compare with the proof of I.4.2).

In the general case write $f = f_1 f_2$ with $f_1 \in H^2$ and $f_2 \in H^2$. Writing $f_j(z) = \sum A_{j,n} z^n$ we have $a_n = \sum_{k=0}^n A_{1,k} A_{2,n-k}$ and consequently $|a_n| \leqq a_n^* = \sum_{k=0}^n |A_{1,k}|\,|A_{2,n-k}|$. The functions $f_j^*(z) = \sum |A_{j,n}| z^n$ are clearly in H^2 so that, by the Cauchy-Schwarz inequality, $f^*(z) = f_1^*(z) f_2^*(z) = \sum a_n^* z^n \in H^1$. Since $a_n^* \geqq 0$ it follows from the first part of the proof that $\sum_1^\infty (a_n^*/n) < \infty$ and the theorem follows from $|a_n| \leqq a_n^*$. ◀

3.17 Let $f \in H^1$ and assume that $f(e^{it})$ is of bounded variation on $\mathbf{T}$. If $f \sim \sum_0^\infty a_n e^{int}$ then $\sum_1^\infty i n a_n e^{int}$ is the Fourier-Stieltjes series of df. Thus the measure df satisfies the condition of theorem 3.13 and consequently $df = f'\,dt$ and $f'(z)$ is in H^1. Combining this with 3.16 we obtain:

Theorem: *Let $f \in H^1$ and assume that $f(e^{it})$is of bounded variation on $\mathbf{T}$. Then $f(e^{it})$ is absolutely continuous and $\sum |\hat{f}(n)| < \infty$.*

An equivalent form of the theorem is (see 3.9):

Theorem: *Let $f, \tilde{f} \in L^1(\mathbf{T})$ and assume that both f and $\tilde{f}$ are of bounded variation. Then they are both absolutely continuous and $\sum_{-\infty}^\infty |\hat{f}(n)| < \infty$.*

EXERCISES FOR SECTION 3

1. Deduce theorem 3.13 (F. and M. Riesz) from theorem 3.11.
2. Show that for all $p > 0$, if $f \in H^p$, then $\int |f(e^{it}) - f(re^{it})|^p dt \to 0$ as

$r \to 1$. *Hint*: Reduce the general case to the case in which f is zero-free. In the case that f is zero-free, write $f_1 = f^{\frac{1}{2}}$; then $f_1 \in H^{2p}$ and

$$f(e^{it}) - f(re^{it}) = (f_1(e^{it}) - f_1(re^{it}))(f_1(e^{it}) + f_1(re^{it}));$$

hence show that if the statement is valid for $2p$ is also valid for p. Use the fact that it is valid for $p \geqq 2$.

3. Let E be a closed set of measure zero on $\mathbf{T}$. Let φ be a continuous function on E.

(a) Show that there exists a function Φ, holomorphic in D and continuous on D such that $\Phi(e^{it}) = \varphi(e^{it})$ on E,

(b) Show that Φ can be chosen satisfying the additional condition

$$\sup_{z \in \bar{D}} |\Phi(z)| = \sup_{e^{it} \in E} |\varphi(e^{it})|.$$

Hint: Construct Φ by successive approximation using 3.15 and lemma 3.13.

4. Let $f \in L^1(\mathbf{T})$ be absolutely continuous and assume $f' \log^+ |f'| \in L^1(\mathbf{T})$. Prove that $\sum |\hat{f}(n)| < \infty$.

Chapter IV

Interpolation of Linear Operators and the Theorem of Hausdorff-Young

Interpolation of norms and of linear operators is really a topic in functional analysis rather than harmonic analysis proper; but, though less so than ten years ago, it still seems esoteric among authors in functional analysis and we include a brief account. The interpolation theorems that are the most useful in Fourier analysis are the Riesz-Thorin theorem and the Marcinkiewicz theorem. We give a general description of the complex interpolation method and prove the Riesz-Thorin theorem in section 1. In the second section we use Riesz-Thorin to prove the Hausdorff-Young theorem. We do not discuss the Marcinkiewicz theorem although it appeared implicitly in the proof of theorem III.1.7. We refer the reader to Zygmund ([28] chap. XII) for a complete account of Marcinkiewicz's theorem.

1. INTERPOLATION OF NORMS AND OF LINEAR OPERATORS

1.1 Let B be a normed linear space and let F be defined in some domain Ω in the complex plane, taking values in B. We say that F is holomorphic in Ω if, for every continuous linear functional μ on B, the numerical function $h(z) = \langle F(z), \mu \rangle$ is holomorphic in Ω.

Assume now that B is a linear space with two norms $\| \ \|_0$ and $\| \ \|_1$ defined on it. We consider the family $\mathscr{B}$ of all B-valued functions which are holomorphic and bounded, with respect to both norms, in a neighborhood of the strip $\Omega = \{z; 0 \leqq \operatorname{Re}(z) \leqq 1\}$. $\mathscr{B}$ is a linear space which we norm as follows: for $F \in \mathscr{B}$ put

(1.1) $$\|F\| = \sup_y \{\|F(iy)\|_0, \|F(1+iy)\|_1\}.$$

For $0 < \alpha < 1$, the set $\mathscr{B}_\alpha = \{F \in \mathscr{B}; F(\alpha) = 0\}$ is a linear subspace of $\mathscr{B}$. We shall say that $\| \ \|_0$ and $\| \ \|_1$ are *consistent* if $\mathscr{B}_\alpha$ is closed in $\mathscr{B}$ for all $0 < \alpha < 1$. A convenient criterion for consistency is the following:

Lemma: *Assume that for every $f \in B$, $f \neq 0$, there exists a functional μ continuous with respect to both $\| \ \|_0$ and $\| \ \|_1$, such that $\langle f, \mu \rangle \neq 0$. Then $\| \ \|_0$ and $\| \ \|_1$ are consistent.*

PROOF: Let $0 < \alpha < 1$ and let $F_n \in \mathscr{B}_\alpha$, $F_n \to F$ in $\mathscr{B}$. Let μ be an arbitrary linear functional continuous with respect to both norms. The functions $\langle F_n(z), \mu \rangle$ are bounded on the strip Ω and tend to $\langle F(z), \mu \rangle$ uniformly on the lines $z = iy$ and $z = 1 + iy$. By the theorem of Phragmèn-Lindelöf the convergence is uniform throughout Ω and in particular $\langle F(\alpha), \mu \rangle = \lim_{n \to \infty} \langle F_n(\alpha), \mu \rangle = 0$. Since this is true for every functional μ it follows that $F(\alpha) = 0$, that is, $F \in \mathscr{B}_\alpha$ and the lemma is proved. ◀

Remark: The condition of the lemma is satisfied if $\| \ \|_0$ and $\| \ \|_1$ both majorize a third norm $\| \ \|_2$. This follows from the Hahn-Banach theorem: if $f \neq 0$, there exists a functional μ continuous with respect to $\| \ \|_2$ such that $\langle f, \mu \rangle \neq 0$. It is clear that if $\| \ \|_j > \| \ \|_2$, μ is continuous with respect to $\| \ \|_j$, $j = 0, 1$.

1.2 We interpolate consistent norms on B as follows: for $0 < \alpha < 1$, the quotient space $\mathscr{B}/\mathscr{B}_\alpha$ is algebraically isomorphic to B (through the mapping $F \to F(\alpha)$). Since $\mathscr{B}_\alpha$ is closed in $\mathscr{B}$, $\mathscr{B}/\mathscr{B}_\alpha$ has a canonical quotient norm which we can transfer to B through the forementioned isomorphism; we denote this new norm on B by $\| \ \|_\alpha$.

The usefulness of this method of interpolating norms comes from the fact that it permits us to interpolate linear operators in the following sense:

Theorem: *Let B (resp. B') be a normed linear space with two consistent norms $\| \ \|_0$ and $\| \ \|_1$ (resp. $\| \ \|'_0$ and $\| \ \|'_1$). Denote the interpolating norms by $\| \ \|_\alpha$ (resp. $\| \ \|'_\alpha$), $0 < \alpha < 1$. Let $\mathbf{S}$ be a linear transformation from B to B' which is bounded as*

(1.2) $$(B, \| \ \|_j) \xrightarrow{\mathbf{S}} (B', \| \ \|'_j), \qquad j = 0, 1.$$

Then **S** *is bounded as*

$$(1.3) \qquad (B, \| \ \|_\alpha) \xrightarrow{\mathbf{S}} (B', \| \ \|'_\alpha),$$

and its norm $\| \mathbf{S} \|_\alpha$ *satisfies*

$$(1.4) \qquad \| \mathbf{S} \|_\alpha \leqq \| \mathbf{S} \|_0^{1-\alpha} \| \mathbf{S} \|_1^\alpha .$$

PROOF: We denote by $\mathscr{B}'$ the space of holomorphic B'-valued functions which is used in defining $\| \ \|'_\alpha$. The map $B \xrightarrow{\mathbf{S}} B'$ can be extended to a map $\mathscr{B} \xrightarrow{\mathbf{S}} \mathscr{B}'$ by writing $\mathbf{S}F(z) = \mathbf{S}(F(z))$. To show that $\mathbf{S}F$ so defined is holomorphic, we consider an arbitrary functional μ, continuous with respect to $\| \ \|'_0$ or $\| \ \|'_1$, and notice that $\langle \mathbf{S}F(z), \mu \rangle = \langle F(z), \mathbf{S}^* \mu \rangle$. Since $\mathbf{S}F(z)$ is clearly bounded it follows that $\mathbf{S}F \in \mathscr{B}'$.

Let $f \in B$, $\| f \|_\alpha = 1$; then there exists an $F \in \mathscr{B}$ such that $F(\alpha) = f$ and such that $\| F \| < 1 + \varepsilon$. Applying $\mathbf{S}$ to F, we obtain

$$\| \mathbf{S}f \|'_\alpha \leqq \| \mathbf{S}F \|' \leqq (1 + \varepsilon) \max (\| \mathbf{S} \|_0, \| \mathbf{S} \|_1);$$

hence

$$\| \mathbf{S} \|_\alpha \leqq \max (\| \mathbf{S} \|_0, \| \mathbf{S} \|_1)$$

which proves the continuity of (1.3). To prove the better estimate (1.4), we consider the function $e^{a(z-\alpha)} F(z)$, where $e^a = \| \mathbf{S} \|_0 \| \mathbf{S} \|_1^{-1}$. We have

$$\begin{aligned} \| \mathbf{S}f \|'_\alpha &\leqq \| \mathbf{S}(e^{a(z-\alpha)} F(z)) \|' \\ &= \sup_t \{ e^{-a\alpha} \| \mathbf{S}F(it) \|'_0, \ e^{a(1-\alpha)} \| \mathbf{S}F(1 + it) \|'_1 \} \\ &\leqq (1 + \varepsilon) \sup \{ e^{-a\alpha} \| \mathbf{S} \|_0, \ e^{a(1-\alpha)} \| \mathbf{S} \|_1 \} \\ &= (1 + \varepsilon) \| \mathbf{S} \|_0^{1-\alpha} \| \mathbf{S} \|_1^\alpha . \end{aligned}$$

◀

Remark: The idea of using the function $e^{a(z-\alpha)}$ goes back to Hadamard (the "three-circles theorem"); it can be used to show that, for every $f \in B$,

$$(1.5) \qquad \| f \|_\alpha \leqq \| f \|_0^{1-\alpha} \| f \|_1^\alpha .$$

1.3 A very important example of interpolation of norms is the following: let $(\mathfrak{X}, d\mathfrak{x})$ be a measure space, let $1 \leqq p_0 < p_1 \leqq \infty$, and let B be a subspace of $L^{p_0} \cap L^{p_1}(d\mathfrak{x})$. We claim that the norms $\| \ \|_0$ and $\| \ \|_1$ induced on B by $L^{p_0}(d\mathfrak{x})$ and $L^{p_1}(d\mathfrak{x})$, respectively, are consistent. By lemma 1.1, all we have to show is that, given $f \in B$, $f \neq 0$, there exists a linear functional μ, continuous with respect to both norms, such that $\langle f,\mu\rangle \neq 0$; we can take as μ the functional defined by $\langle f,\mu\rangle = \int f\bar{g}\,d\mathfrak{x}$ where $g \in L^1 \cap L^\infty(d\mathfrak{x})$ has the property† that $f\bar{g} > 0$ whenever $|f| > 0$.

Theorem: *Let $(\mathfrak{X}, d\mathfrak{x})$ be a measure space, $B = L^{p_0} \cap L^{p_1}(d\mathfrak{x})$ (with $1 \leqq p_0 < p_1 \leqq \infty$). Denote by $\| \ \|_j$ the norms induced by $L^{p_j}(d\mathfrak{x})$, and by $\| \ \|_\alpha$ the interpolating norms. Then $\| \ \|_\alpha$ coincides with the norm induced on B by $L^{p_\alpha}(d\mathfrak{x})$ where*

$$(1.6) \qquad p_\alpha = \frac{p_0 p_1}{p_0\alpha + p_1(1-\alpha)} \qquad \left(= \frac{p_0}{1-\alpha} \ \text{ if } p_1 = \infty \right).$$

PROOF: Let $f \in B$ and $\|f\|_{L^{p_\alpha}} \leqq 1$. Consider $F(z) = |f|^{a(z-\alpha)+1} e^{i\varphi}$ where $f = |f| e^{i\varphi}$ and

$$a = \frac{p_0 - p_1}{p_0\alpha + p_1(1-\alpha)} \qquad \left(= \frac{-1}{1-\alpha} \ \text{ if } p_1 = \infty \right).$$

We have $F(\alpha) = f$ and consequently $\|f\|_\alpha \leqq \|F\|$; now $|F(iy)| = |f|^{1-a\alpha} = |f|^{p_\alpha/p_0}$ so that

$$\|F(iy)\|_0 = \left(\int |f|^{p_\alpha} d\mathfrak{x} \right)^{1/p_0} \leqq 1;$$

similarly $\|F(1+iy)\|_1 \leqq 1$ (use the same argument if $p_1 < \infty$ and check directly if $p_1 = \infty$); hence $\|f\|_\alpha \leqq 1$. This proves $\| \ \|_\alpha \leqq \| \ \|_{L^{p_\alpha}}$. In order to prove the reverse inequality, we denote by q_0, q_1 the conjugate exponents of p_0 and p_1 and notice that the exponent conjugate to p_α is

$$(1.6') \qquad q_\alpha = \frac{q_0 q_1}{q_0\alpha + q_1(1-\alpha)} \qquad \left(= \frac{q_1}{\alpha} \ \text{ if } q_0 = \infty \right).$$

We now set $B' = L^{q_0} \cap L^{q_1}(d\mathfrak{x})$ and denote by $\mathscr{B}'$ the corresponding space of holomorphic B'-valued functions.

Let $f \in B$ and assume $\|f\|_{L^{p_\alpha}} > 1$; then, since B' is dense in L^{q_α}, there exists a $g \in B'$ such that $\|g\|_{L^{q_\alpha}} \leqq 1$ and such that $\int fg\,d\mathfrak{x} > 1$. As in the first part of this proof, there exists a function $G \in \mathscr{B}'$ such that $G(\alpha) = g$ and $\|G\|$ (with respect to q_0, q_1) is bounded by 1. Let $F \in \mathscr{B}$ such that $F(\alpha) = f$. The function $h(z) = \int F(z)G(z)\,d\mathfrak{x}$ (remember that for each $z \in \Omega$, $F(z) \in B$ and $G(z) \in B'$) is holomorphic and bounded in Ω (see Appendix). Now $h(\alpha) > 1$, hence, by the Phragmèn-Lindelöf

† If we write $f = |f| e^{i\varphi}$ with real-valued φ, we may take $g = \min(1, |f|^{p_0}) e^{i\varphi}$.

theorem, $|h(z)|$ must exceed 1 on the boundary. However, on the boundary $|h(z)| \leqq \|F\| \|G\| \leqq \|F\|$ so that $\|F\| > 1$. This proves $\|f\|_\alpha \geqq 1$ and it follows that $\| \quad \|_\alpha$ and $\| \quad \|_{L^{p_\alpha}}$ are identical. ◀

1.4 As a corollary to theorems 1.2 and 1.3, we obtain the Riesz-Thorin theorem.

Theorem: *Let* $(\mathfrak{X}, d\mathfrak{x})$ *and* $(\mathfrak{Y}, d\mathfrak{y})$ *be two measure spaces. Let* $B = L^{p_0} \cap L^{p_1}(d\mathfrak{x})$ *and* $B' = L^{p'_0} \cap L^{p'_1}(d\mathfrak{y})$ *and let* $\mathbf{S}$ *be a linear transformation from* B *to* B', *continuous as* $(B, \| \quad \|_j) \xrightarrow{\mathbf{S}} (B', \| \quad \|'_j)$, $j = 0, 1$, *where* $\| \quad \|_j$ *(resp.* $\| \quad \|'_j$*) is the norm induced by* $L^{p_j}(d\mathfrak{x})$ *(resp.* $L^{p'_j}(d\mathfrak{y})$*). Then* $\mathbf{S}$ *is continuous as*

$$(B, \| \quad \|_\alpha) \xrightarrow{\mathbf{S}} (B', \| \quad \|'_\alpha),$$

where $\| \quad \|_\alpha$ *(resp.* $\| \quad \|'_\alpha$*) is the norm induced by* $L^{p_\alpha}(d\mathfrak{x})$ *(resp.* $L^{p'_\alpha}(d\mathfrak{y})$, p_α *and* p'_α *are defined in* (1.6)).

A bounded linear transformation $\mathbf{S}$ from one normed space B to another can be completed in one and only one way, to a transformation having the same norm, from the completion of B into the completion of the range space of $\mathbf{S}$. Thus, under the assumption of 1.4, $\mathbf{S}$ can be extended as a transformation from $L^{p_\alpha}(d\mathfrak{x})$ into $L^{p'_\alpha}(d\mathfrak{y})$ with norm satisfying (1.4). The same remark is clearly valid for theorem 1.2.

***1.5** As a first application of the Riesz-Thorin theorem we give Bochner's proof of M. Riesz' theorem III.1.8. We show that $L^p(\mathbf{T})$ admits conjugation in the case that p is an even integer. It then follows by interpolation that the same is true for all $p \geqq 2$. Using duality we can then obtain the result for all $p > 1$.

Let f be a real-valued trigonometric polynomial and assume, for simplicity, $\hat{f}(0) = 0$. As usual we denote the conjugate by $\tilde{f}$ and put $f^\flat = \frac{1}{2}(f + i\tilde{f})$. $f^\flat$ is a Taylor polynomial† and its constant term is zero; the same is clearly true for $(f^\flat)^p$, p being any positive integer. Consequently

$$\frac{1}{2\pi} \int (f^\flat(t))^p \, dt = 0.$$

Assume now that p is even, $p = 2k$, and consider the real part of the identity above; we obtain:

† We use the term "Taylor polynomial" to designate trigonometric polynomials of the form $\sum_0^N a_n e^{int}$.

$$\frac{1}{2\pi}\int (\tilde{f})^{2k}dt - \binom{2k}{2}\frac{1}{2\pi}\int (\tilde{f})^{2k-2}f^2\,dt + \binom{2k}{4}\frac{1}{2\pi}\int (\tilde{f})^{2k-4}f^4\,dt - \cdots = 0.$$

By Hölder's inequality

$$\left|\frac{1}{2\pi}\int (\tilde{f})^{2k-2m}f^{2m}\,dt\right| \leqq \|\tilde{f}\|_{L^{2k}}^{2k-2m}\|f\|_{L^{2k}}^{2m};$$

hence

$$\|\tilde{f}\|_{L^{2k}}^{2k} \leqq \binom{2k}{2}\|\tilde{f}\|_{L^{2k}}^{2k-2}\|f\|_{L^{2k}}^{2} + \binom{2k}{4}\|\tilde{f}\|_{L^{2k}}^{2k-4}\|f\|_{L^{2k}}^{4} + \cdots$$

or, denoting

$$Y = \|\tilde{f}\|_{L^{2k}}\|f\|_{L^{2k}}^{-1},$$

we have

$$Y^{2k} \leqq \binom{2k}{2}Y^{2k-2} + \binom{2k}{4}Y^{2k-4} + \cdots + 1$$

which implies that Y is bounded by a constant depending on k (i.e., on p). Thus the mapping $f \to \tilde{f}$ is bounded in the $L^p(\mathbf{T})$ norm for all polynomials f, and, since polynomials are dense in $L^p(\mathbf{T})$, the theorem follows.

EXERCISES FOR SECTION 1

1. Prove inequality (1.5).
2. Let $\{a_n\}$ be a sequence of numbers. Find $\min(\sum |a_n|)$ under the conditions $\sum |a_n|^2 = 1$, $\sum |a_n|^4 = a$.

2. THE THEOREM OF HAUSDORFF-YOUNG

The theorem of Riesz-Thorin enables us to prove now a theorem that we stated without proof at the end of I.4 (theorem I.4.7); it is known as the Hausdorff-Young theorem:

2.1 ***Theorem***: *Let $1 \leqq p \leqq 2$ and let q be the conjugate exponent, that is, $q = p/(p-1)$. If $f \in L^p(\mathbf{T})$ then $\sum |\hat{f}(n)|^q < \infty$. More precisely $(\sum |\hat{f}(n)|^q)^{1/q} \leqq \|f\|_{L^p}$.*

PROOF: The mapping $f \overset{\mathcal{F}}{\to} \{\hat{f}(n)\}$ is a transformation of functions on the measure space $(\mathbf{T}, dt)$ into functions on $(\mathbf{Z}, dn)$, $\mathbf{Z}$ being the group of integers and dn the so-called counting measure, that is, the measure that places a unit mass at each integer. We know that the norm of the mapping as $L^1(\mathbf{T}) \xrightarrow{\mathcal{F}} L^\infty(\mathbf{Z}) = \ell^\infty$ is 1 (I.1.4) and we know that it is an isometry of $L^2(\mathbf{T})$ onto $L^2(\mathbf{Z}) = \ell^2$ (I.5.5). It follows from the Riesz-Thorin theorem that $\mathcal{F}$ is a transformation of norm $\leqq 1$ from $L^p(\mathbf{T})$ into $L^q(\mathbf{Z}) = \ell^q$, which is precisely the statement of our theorem. We can add that since the exponentials are mapped with no loss in norm, the norm of $f \xrightarrow{\mathcal{F}} \hat{f}$ on $L^p(\mathbf{T})$ into ℓ^q is exactly 1. ◀

2.2 ***Theorem:*** *Let $1 \leqq p \leqq 2$ and let q be the conjugate exponent. If $\{a_n\} \in \ell^p$ then there exists a function $f \in L^q(\mathbf{T})$ such that $a_n = \hat{f}(n)$. Moreover, $\|f\|_{L^q} \leqq (\sum |a_n|^p)^{1/p}$.*

PROOF: Theorem 2.2 is the exact analog to 2.1 with the roles of the groups $\mathbf{T}$ and $\mathbf{Z}$ reversed. The proof is identical: if $\{a_n\} \in \ell^1$ then $f(t) = \sum a_n e^{int}$ is continuous on $\mathbf{T}$ and $\hat{f}(n) = a_n$. The case $p = 2$ is again given by theorem I.5.5 and the case $1 < p < 2$ is obtained by interpolation. ◀

***2.3** We have already made the remark (end of I.4) that theorem 2.1 cannot be extended to the case $p > 2$ since there exist continuous functions f such that $\sum |\hat{f}(n)|^{2-\varepsilon} = \infty$ for all $\varepsilon > 0$. An example of such a function is $f(t) = \sum_{n=2}^{\infty} \frac{e^{in \log n}}{n^{1/2}(\log n)^2} e^{int}$ (see [28], vol. I, p. 199); another example is $g(t) = \sum m^{-2} 2^{-m/2} f_m(t)$ where f_m are the Rudin-Shapiro polynomials (see exercise I.6.6, part c.) We can try to explain the phenomenon by a less explicit but more elementary construction.

The first remark is that this, like many problems in analysis, is a problem of comparison of norms. It is sufficient, we claim, to show that, given $p < 2$, there exist functions g such that $\|g\|_\infty \leqq 1$ and $\sum |\hat{g}(n)|^p$ is arbitrarily big. If we assume that, we may assume that our functions g are polynomials (replace g by $\sigma_n(g)$ with sufficiently big n) and then, taking a sequence $p_j \to 2$, g_j satisfying $\|g_j\|_\infty \leqq 1$ and $\sum_{n=-\infty}^{\infty} |\hat{g}_j(n)|^{p_j} > 2^j$, we can write $f = \sum_{j=1}^{\infty} j^{-2} e^{im_j t} g_j(t)$ where the integers m_j increase fast enough to ensure that $e^{im_j t} g_j(t)$ and $e^{im_k t} g_k(t)$ have no frequencies in common if $j \neq k$. The series defining f converges uniformly and for any $p < 2$ we have

$$\sum |\hat{f}(n)|^p = \sum_j \sum_n \frac{1}{j^2} |\hat{g}_j(n)|^p \geqq \sum_{p_j > p} \frac{1}{j^2} \sum |\hat{g}_j(n)|^{p_j} = \infty .$$

One way to show the existence of the functions g above is to show that, given $\varepsilon > 0$, there exist functions g satisfying

$$(2.1) \qquad \|g\|_\infty \leqq 1, \quad \|g\|_{L^2} \geqq \tfrac{1}{2}, \quad \sup_n |\hat{g}(n)| < \varepsilon.$$

In fact, if (2.1) is valid then

$$\sum |\hat{g}(n)|^p \geqq \varepsilon^{p-2} \sum |\hat{g}(n)|^2 > \tfrac{1}{4}\varepsilon^{p-2},$$

and if ε can be chosen arbitrarily small, the corresponding g will have $\sum |\hat{g}(n)|^p$ arbitrarily large.

Functions satisfying (2.1) are not hard to find; however, it is important to realize that when we need a function satisfying certain conditions, it may be easier to construct an example rather than look for one in our inventory. We therefore include a construction of functions satisfying (2.1). The key remark in the construction is simple yet very useful: if P is a trigonometric polynomial of degree N, $f \in L^1(\mathbf{T})$ and $\lambda > 2N$ is an integer, then the Fourier coefficients of $\varphi(t) = f(\lambda t)P(t)$ are either zero or have the form $\hat{f}(m)\hat{P}(k)$. This follows from the identity $\hat{\varphi}(n) = \sum_{\lambda m + k = n} \hat{f}(m)\hat{P}(k)$ and the fact that there is at most one way to write $n = \lambda m + k$ with integers m, k such that $|k| \leqq N < \lambda/2$.

Consider now any continuous function of modulus 1 on $\mathbf{T}$, which is not an exponential (of the form e^{int}); for example $\psi(t) = e^{i\cos t}$. Since $\sum |\hat{\psi}(n)|^2 = 1$ and the sum contains more than one term, it follows that $\sup_n |\hat{\psi}(n)| = \rho < 1$. Let M be an integer such that $\rho^M < \varepsilon$. Let $\eta < 1$ be such that $\eta^M > \frac{1}{2}$. Let $\varphi = \sigma_N(\psi)$, where the order N is high enough to ensure $\eta < |\varphi(t)| < 1$. It follows from the preceding remark that if we set $\lambda = 3N$ and $g(t) = \prod_{j=1}^{M} \varphi(\lambda^j t)$, the Fourier coefficients of g are products of M Fourier coefficients of φ; hence $|\hat{g}(n)| \leqq \rho^M < \varepsilon$. On the other hand $\frac{1}{2} < \eta^M < |g(t)| < 1$ and (2.1) is valid.

***2.4** We can use the polynomials satisfying (2.1) to show also that theorem 2.2 does not admit an extension to the case $p > 2$. In fact, we can construct a trigonometric series $\sum a_n e^{int}$ which is *not a Fourier-Stieltjes* series, and such that $\sum |a_n|^p < \infty$ for all $p > 2$.

Let g_j be a trigonometric polynomial satisfying (2.1) with $\varepsilon = 2^{-j}$. Since now $p > 2$ we have

$$\sum |\hat{g}_j(n)|^p \leqq \varepsilon^{p-2} \sum |\hat{g}_j(n)|^2 \leqq 2^{-j(p-2)}$$

and consequently, for any choice of the integers m_j, $\sum j e^{im_j t} g_j(t) = \sum a_n e^{int}$ does satisfy $\sum |a_n|^p < \infty$ for all $p > 2$. We now choose the integers

m_j increasing very rapidly in order to well separate the blocks corresponding to $j\,e^{im_jt}g_j$ in the series above. If we denote by N_j the degree of the polynomial g_j, we can take m_j so that $m_j - 3N_j > m_{j-1} + 3N_{j-1}$. If $\sum a_n e^{int}$ is the Fourier-Stieltjes series of a measure μ then

$$\mu * e^{im_jt}\mathbf{V}_{N_j} = j\,e^{im_jt}g_j \qquad (\mathbf{V}_{N_j} \text{ being de la Vallée Poussin's kernel})$$

and consequently

$$3\,\|\mu\|_{M(\mathbf{T})} > j\,\|g_j\|_{L^1} > \frac{j}{4}$$

which is impossible. We have thus proved

Theorem: (a) *There exists a continuous function f such that for all $p < 2$, $\sum |\hat{f}(n)|^p = \infty$.*

(b) *There exists a trigonometric series, $\sum a_n e^{int}$, which is not a Fourier-Stieltjes series, such that $\sum |a_n|^p < \infty$ for all $p > 2$.*

***2.5** We finish this section with another construction: that of a set E of positive measure on $\mathbf{T}$ which carries no function with Fourier coefficients in ℓ^p for any $p < 2$. Such a set clearly must be totally disconnected and therefore carries no continuous functions. Its characteristic function, however, is a bounded function the Fourier coefficients of which belong to no ℓ^p, $p < 2$.

Theorem: *There exists a compact set E on $\mathbf{T}$ such that E has positive measure and such that the only function f carried by E with $\sum |\hat{f}(j)|^p < \infty$ for some $p < 2$, is $f \equiv 0$.*

First, we introduce the notation†

$$(2.2) \qquad \|f\|_{\mathcal{F}\ell^\infty} = \sup |\hat{f}(j)|, \qquad \|f\|_{\mathcal{F}\ell^p} = (\sum |\hat{f}(j)|^p)^{1/p};$$

and prove:

Lemma: *Let $\varepsilon > 0$, $1 \leqq p < 2$. There exists a closed set $E_{\varepsilon,p} \subset \mathbf{T}$ having the following properties:*

(1) *The measure of $E_{\varepsilon,p}$ is $> 2\pi - \varepsilon$.*

(2) *If f is carried by $E_{\varepsilon,p}$ then*

$$\|f\|_{\mathcal{F}\ell^\infty} \leqq \varepsilon \|f\|_{\mathcal{F}\ell^p}.$$

PROOF: Let $\gamma > 0$. Put

$$\varphi_\gamma(t) = \begin{cases} \dfrac{\gamma - 2\pi}{\gamma} & 0 < t < \gamma \mod 2\pi \\ 1 & \gamma \leqq t \leqq 2\pi \mod 2\pi. \end{cases}$$

† Notice that $\|\quad\|_{\mathcal{F}\ell^1}$ is the same as $\|\quad\|_{A(\mathbf{T})}$.

Then, by theorem 2.1

(2.3) $$\|\varphi_\gamma\|_{\mathscr{F}\ell^q} \leqq \|\varphi_\gamma\|_{L^p} \leqq 2\pi\gamma^{\frac{1}{p}-1}, \qquad \text{where } \frac{1}{p}+\frac{1}{q}=1.$$

We notice that $\hat{\varphi}_\gamma(0)=0$ so that, if we choose the integers $\lambda_1, \lambda_2, \ldots, \lambda_N$ increasing fast enough, every Fourier coefficient of $\sum_1^N \varphi_\gamma(\lambda_j t)$ is essentially a Fourier coefficient of one of the summands. It then follows that

(2.4) $$\left\| \frac{1}{N}\sum_1^N \varphi_\gamma(\lambda_j t) \right\|_{\mathscr{F}\ell^q} \leqq 2N^{\frac{1}{q}-1} \|\varphi_\gamma\|_{\mathscr{F}\ell^q}.$$

We take a large value for N and put $\gamma = \varepsilon/N$ and

$$\Phi(t) = \frac{1}{N}\sum_1^N \varphi_\gamma(\lambda_j t).$$

Then, by (2.3) and (2.4), it follows that

$$\|\Phi\|_{\mathscr{F}\ell^q} \leqq 4\pi\gamma^{\frac{1}{p}-1} N^{\frac{1}{q}-1} = 4\pi\varepsilon^{\frac{1}{p}-1} N^{\frac{1}{q}-\frac{1}{p}}$$

so that if N is large enough $\|\Phi\|_{\mathscr{F}\ell^q} < \varepsilon$. We can take

$$E_{\varepsilon,p} = \{t; \Phi(t)=1\} = \bigcap_1^N \{t; \varphi_\gamma(\lambda_j t)=1\}.$$

Since $\varphi_\gamma(\lambda_j t) \neq 1$ on a set of measure γ, it follows that

$$|E_{\varepsilon,p}| \geqq 2\pi - N\gamma = 2\pi - \varepsilon.$$

Now if f is carried by $E_{\varepsilon,p}$, then for arbitrary n,

$$\hat{f}(n) = \frac{1}{2\pi}\int e^{-int} f(t)\,dt = \frac{1}{2\pi}\int e^{-int} f(t)\,\Phi(t)\,dt.$$

It follows from Parseval's formula that

$$|\hat{f}(n)| = \left|\sum \hat{f}(n-m)\hat{\Phi}(m)\right| \leqq \|\Phi\|_{\mathscr{F}\ell^q}\|f\|_{\mathscr{F}\ell^p} \leqq \varepsilon\|f\|_{\mathscr{F}\ell^p};$$

and the proof of the lemma is complete. ◀

PROOF OF THE THEOREM: Take $\varepsilon_n = 3^{-n}$, $p_n = 2-\varepsilon_n$ and $E = \bigcap_{n=1}^\infty E_{\varepsilon_n, p_n}$. The measure of E is clearly positive, and if f is carried by E and $\|f\|_{\mathscr{F}\ell^p} < \infty$ for some $p<2$, it follows that

$$\|f\|_{\mathscr{F}\ell^\infty} \leqq \varepsilon_n \|f\|_{\mathscr{F}\ell^{p_n}} \leqq \varepsilon_n \|f\|_{\mathscr{F}\ell^p}$$

or all n large enough; hence $\hat{f}=0$ and so $f=0$. ◀

EXERCISES FOR SECTION 2

1. Verify that $2^{-(m+1)/2}f_m$ (f_m as defined in exercise I.6.6, part c) satisfy (2.1) when $2^{-(m+1)/2} < \varepsilon$.

2. Show that if $N > \varepsilon^{-1}$ and that if m_n increases fast enough, then g, defined by: $g(t) = e^{iNm_nt}$ for $2\pi n/N \leqq t \leqq 2\pi(n+1)/N$, $n = 1, \ldots, N$ satisfies (2.1).

3. Let $\{a_n\}$ be an even sequence of positive numbers. A closed set $E \subset \mathbf{T}$ is a set *of type* $U(a_n)$ if the only distribution μ carried by E and satisfying $\hat{\mu}(n) = o(a_n)$ as $|n| \to \infty$, is $\mu = 0$. Show that if $a_n \to 0$ there exist sets E of positive measure which are of type $U(a_n)$. *Hint*: For $0 < a < \pi$ we write (see exercise I.6.3):

$$\Delta_a(t) = \begin{cases} 1 - a^{-1}|t| & |t| \leqq a \\ 0 & a \leqq |t| \leqq \pi. \end{cases}$$

We have $\Delta_a \in A(\mathbf{T})$, $\|\Delta_a\|_{A(\mathbf{T})} = 1$, and $\hat{\Delta}_a(0) = a/2\pi$. Choose n_j so that $|n| > n_j$ implies $a_j < 10^{-j}$; put $E_j = \{t;\ \Delta_{2^{-j}}(2n_jt) = 0\}$ and $E = \bigcap_{j=1}^{\infty} E_j$. Notice that $|E| \geqq 2\pi - \sum 2^{1-j} > 0$. If μ is carried by E we have, for all m and j, $\langle e^{imt}\Delta_{3^{-j}}(2n_jt), \mu\rangle = 0$ since $\Delta_{3^{-j}}(2n_jt)$ vanishes in a neighborhood of E. By Parseval's formula

$$0 = \langle e^{imt}\Delta_{3^{-j}}(2n_jt), \mu\rangle = \frac{3^{-j}}{2\pi}\overline{\hat{\mu}(m)} + \sum_{k\neq 0} \hat{\Delta}_{3^{-j}}(k)\overline{\hat{\mu}(m+2n_jk)}$$

if $n_j > |m|$ and if $|\hat{\mu}(n)| \leqq a_n$ we have for $k \neq 0$, $|\hat{\mu}(m+2n_jk)| < 10^{-j}$, hence $(3^{-j}/2\pi)|\hat{\mu}(m)| < 10^{-j}$. Letting $j \to \infty$ we obtain $\hat{\mu}(m) = 0$, and, m being arbitrary, $\mu = 0$.

Chapter V

Lacunary Series and Quasi-analytic Classes

The theme of this chapter is that of I.4, namely, the study of the ways in which properties of functions or of classes of functions are reflected by the Fourier series. We consider important special cases of the following general problem: let Λ be a sequence of integers and B a homogeneous Banach space on $\mathbf{T}$; denote by B_Λ the closed subspace of B spanned by $\{e^{i\lambda t}\}_{\lambda \in \Lambda}$ or, equivalently, the space of all $f \in B$ with Fourier series of the form $\sum_{\lambda \in \Lambda} a_\lambda e^{i\lambda t}$. Describe the properties of functions in B_Λ in terms of their Fourier series (and Λ). An obvious example of the above is the case of a finite Λ in which all the functions in B_Λ are polynomials. If Λ is the sequence of nonnegative integers and $B = L^p(\mathbf{T})$, $1 \leqq p < \infty$, then B_Λ is the space of boundary values of functions in the corresponding H^p. In the first section we consider lacunary sequences Λ and show, for instance, that if Λ is lacunary à la Hadamard then $(L^1(\mathbf{T}))_\Lambda = (L^2(\mathbf{T}))_\Lambda$ and every bounded function in $(L^1(\mathbf{T}))_\Lambda$ has an absolutely convergent Fourier series.

In the second section we prove the Denjoy-Carleman theorem on the quasi-analyticity of classes of infinitely differentiable functions and discuss briefly some related problems.

1. LACUNARY SERIES

1.1 A sequence of positive integers $\{\lambda_n\}$ is said to be *Hadamard-lacunary*, or simply *lacunary*, if there exists a constant $q > 1$ such that $\lambda_{n+1} > q\lambda_n$ for all n. A power series $\sum a_n z^{\lambda_n}$ is lacunary if the sequence $\{\lambda_n\}$ is, and a trigonometric series is lacunary if all the frequencies appearing in it have the form $\pm \lambda_n$ where $\{\lambda_n\}$ is lacunary.

The reason for mentioning Hadamard's name is his classical theorem stating that the circle of convergence of a lacunary power series is a natural boundary for the function given by the sum of the series within its domain of convergence. The general idea behind Hadamard's theorem and behind most of the results concerning lacunary series is that the sparsity of the exponents appearing in the series forces on it a certain homogeneity of behavior.

1.2 Lacunarity can be used technically in a number of ways. Our first example is "local"; it illustrates how a Fourier coefficient that stands apart from the others is affected by the behavior of the function in a neighborhood of a point.

***Lemma*:** *Let $f \in L^1(\mathbf{T})$ and assume that $\hat{f}(j) = 0$ for all j satisfying $1 \leqq |n_0 - j| \leqq 2N$. Assume that $f(t) = O(t)$ as $t \to 0$. Then*

$$(1.1) \qquad |\hat{f}(n_0)| \leqq 2\pi^4 \Big(N^{-1} \sup_{|t| < N^{-1/4}} |t^{-1} f(t)| + N^{-2} \| f \|_{L^1}\Big).$$

PROOF: We use the condition $\hat{f}(j) = 0$ for j satisfying $1 \leqq |n_0 - j| \leqq 2N$ as follows: let g_N be any polynomial of degree $2N$ satisfying $\hat{g}_N(0) = 1$, then

$$\hat{f}(n_0) = \frac{1}{2\pi} \int e^{-in_0 t} f(t) g_N(t)\, dt.$$

We take $g_N = \| \mathbf{K}_N \|_{L^2}^{-2} \mathbf{K}_N^2$, $\mathbf{K}_N$ being Féjer's kernel of order N; remembering inequality (3.10) of chapter I and noticing that $\| \mathbf{K}_N \|_{L^2}^2 = \sum_{-N}^{N} \left(1 - \frac{|j|}{N+1}\right)^2 > \frac{N}{2}$ we obtain

$$g_N(t) \leqq 2\pi^4 N^{-3} t^{-4}.$$

We now write

$$\hat{f}(n_0) \leqq \frac{1}{2\pi} \int |f(t)|\, g_N(t)\, dt$$

$$= \left(\int_{|t| < N^{-1}} + \int_{N^{-1} < |t| < N^{-1/4}} + \int_{N^{-1/4} < |t| < \pi} \right) \frac{1}{2\pi} |f(t)|\, g_N(t)\, dt.$$

The first integral is bounded by

$$N^{-1} \sup_{|t|<N^{-1}} |t^{-1}f(t)| \frac{1}{2\pi} \int |g_N(t)|\,dt = N^{-1} \sup_{|t|<N^{-1}} |t^{-1}f(t)|.$$

The second integral is bounded by

$$\pi^3 N^{-3} \sup_{|t|<N^{-1/4}} |t^{-1}f(t)| \int_{N^{-1}}^{N^{-1/4}} t^{-3}\,dt \leqq \pi^3 N^{-1} \sup_{|t|<N^{-1/4}} |t^{-1}f(t)|.$$

The third integral is bounded by

$$\pi^3 N^{-2} \int |f(t)|\,dt = 2\pi^4 N^{-2} \|f\|_{L^1}.$$

Adding up the three estimates we obtain (1.1). ◀

Corollary: *Let* $\{\lambda_n\}$ *be a lacunary sequence and* $f \sim \sum a_n \cos \lambda_n t$ *be in* $L^1(\mathbf{T})$. *Assume that* f *is differentiable at one point. Then* $a_n = o(\lambda_n^{-1})$.

PROOF: Assume that f is differentiable at $t = 0$. Replacing it, if necessary, by $f - f(0)\cos t - f'(0)\sin t$ we can assume $f(0) = f'(0) = 0$. It follows that $f(t) = o(t)$ as $t \to 0$. The lacunarity condition on $\{\lambda_n\}$ is equivalent to saying that there exists a positive constant c such that, for all n, none of the numbers j satisfying $1 \leqq |\lambda_n - j| \leqq c\lambda_n$ is in the sequence; hence $\hat{f}(j) = 0$ for all such j. Applying the lemma with $n_0 = \lambda_n$ and $2N = c\lambda_n$ we obtain $a_n = 2\hat{f}(\lambda_n) = o(\lambda_n^{-1})$. ◀

Corollary: *The Weierstrass function* $\sum 2^{-n}\cos 2^n t$ *is* nowhere *differentiable.*

The condition $a_n = o(\lambda_n^{-1})$ clearly implies that $\sum |a_n| < \infty$. It is not hard to see (see Zygmund [28], chap. 2, §3, 4) that $f(t) = \sum a_n \cos \lambda_n t$ is then in $\text{Lip}_\alpha(\mathbf{T})$ for all $\alpha < 1$ and that it is differentiable on a set having the power of the continuum in every interval. Thus, for a lacunary series, differentiability at one point implies differentiability on an everywhere dense set. This is one example of the "certain homogeneity of behavior" mentioned earlier. We can obtain a more striking result if instead of differentiability we consider Lipschitz conditions. For instance, if $0 < \alpha < 1$, a lacunary series that satisfies a Lip_α condition at a point satisfies the same condition *everywhere* (see exercise 1 at the end of the section).

1.3 Another typical use of the condition of lacunarity is through its arithmetical consequences. A useful remark is that if $\lambda_{j+1} \geqq q\lambda_j$

with $q \geqq 3$, then every integer n has at most one representation of the form $n = \sum \eta_j\lambda_j$ where $\eta_j = -1, 0, 1$. With this remark in mind we consider products of the form

$$(1.2) \qquad P_N(t) = \prod_1^N (1 + a_j \cos(\lambda_j t + \varphi_j))$$

the a_j's being arbitrary complex numbers and $\varphi_j \in \mathbf{T}$.

The Fourier coefficients of a factor $1 + a_j \cos(\lambda_j t + \varphi_j)$ are: 1 for $n = 0$, $\frac{1}{2}a_j e^{i\varphi_j}$ for $n = \lambda_j$, $\frac{1}{2}a_j e^{-i\varphi_j}$ for $n = -\lambda_j$, and zero elsewhere. If we assume the lacunarity condition with $q \geqq 3$, it follows that $\hat{P}_N(n) = 0$ unless $n = \sum \eta_j \lambda_j$, with $\eta_j = -1, 0, 1$, in which case $\hat{P}_N(n) = \prod_{\eta_j \neq 0} \frac{1}{2}a_j e^{i\eta_j\varphi_j}$; in particular $\hat{P}_N(0) = 1$. If we compare the Fourier series of P_{N+1} to that of P_N we see that P_{N+1} contains P_N as a partial sum, and contains two more blocks: $\frac{1}{2}a_{N+1}e^{i\varphi_{N+1}}e^{i\lambda_{N+1}t}P_N$ and $\frac{1}{2}a_{N+1}\, e^{-i\varphi_{N+1}}e^{-i\lambda_{N+1}t}P_N$. The frequencies appearing in the first block lie within the interval $(\lambda_{N+1} - \sum_1^N \lambda_j,\ \lambda_{N+1} + \sum_1^N \lambda_j) \subset \subset (\lambda_{N+1}(q-2)/(q-1),\ \lambda_{N+1}q/(q-1))$ and the second block is symmetric to the first with respect to the origin. No matter what coefficients a_j we take, the (formal) infinite product

$$P = \prod_1^\infty (1 + a_j \cos(\lambda_j t + \varphi_j))$$

can be expanded as a well-defined trigonometric series, and if the product converges in the weak-star topology of $M(\mathbf{T})$ to a function f or a measure μ, then the corresponding trigonometric series is the Fourier series of f (resp. μ).

We shall refer to the finite or infinite products described above as Riesz products. Two classes of Riesz products will be of special interest.

1. The coefficients a_j are all real and $|a_j| \leqq 1$. In this case $1 + a_j\cos(\lambda_j t + \varphi_j) \geqq 0$ hence $P_N(t) \geqq 0$ for all N. It follows that $\| P_N \|_{L^1} = \hat{P}_N(0) = 1$ and, taking a weak-star limit point, it follows that P is a positive measure of total mass 1 (i.e., that the trigonometric series formally corresponding to P is the Fourier-Stieltjes series of a positive measure of mass 1).

2. The coefficients a_j are purely imaginary (in which case we shall write $P = \prod(1 + ia_j\cos(\lambda_j t + \varphi_j))$ with a_j real) and satisfy $\sum |a_j|^2 < \infty$. In this case $1 \leqq |1 + ia_j\cos(\lambda_j t + \varphi_j)| \leqq 1 + a_j^2$ and $1 \leqq |P_N(t)| \leqq \prod_1^\infty (1 + a_j^2) < \infty$. Since the P_N are uniformly bounded

we can pick a sequence N_j such that P_{N_j} converge weakly to a bounded function P whose Fourier series is the formal expansion of P.

1.4 The usefulness of the Riesz products can be seen in the proof of

Lemma: *Let* $f(t) = \sum_{-N}^{N} c_j e^{i\lambda_j t}$ *with* $\lambda_{-j} = -\lambda_j$, $\lambda_1 > 0$ *and, for some* $q > 1$, $\lambda_{j+1} > q\lambda_j$, $j = 1, 2, \ldots, N$. *Then*

$$\sum |c_j| \leqq A_q \|f\|_\infty \tag{1.3}$$

and

$$\|f\|_{L^2} \leqq B_q \|f\|_{L^1} \tag{1.4}$$

where A_q *and* B_q *are constants depending only on* q.

PROOF: We remark first that it is sufficient to prove (1.3) and (1.4) in the case that f is real valued (i.e., $c_{-j} = \bar{c}_j$) since we can then apply them separately to the real and imaginary parts of arbitrary f, thereby at most doubling the constants A_q and B_q.

Assume first that $q \geqq 3$. In order to prove (1.3) we consider the Riesz product $P(t) = \prod_1^N (1 + \cos(\lambda_j t + \varphi_j))$ where φ_j is defined by the condition $\overline{c_j} e^{i\varphi_j} = |c_j|$. We have $\|P\|_{L^1} = 1$ and consequently $|1/2\pi \int P(t)\overline{f(t)}\,dt| \leqq \|f\|_\infty$. Since $\hat{P}(\lambda_j) = \frac{1}{2} e^{i\varphi_j}$ we obtain from Parseval's formula: $\frac{1}{2} \sum \overline{c_j} e^{i\varphi_j} = \frac{1}{2} \sum |c_j| \leqq \|f\|_\infty$, and (1.3) follows with $A_q = 2$ for real-valued f and $A_q = 4$ in the general case.

For the proof of (1.4) we consider a Riesz product of the second type. We remark that $\|f\|_{L^2}^2 = \sum |c_j|^2$ and if we take $a_j = |c_j| \cdot \|f\|_{L^2}^{-1}$ and φ_j such that $i\overline{c_j} e^{i\varphi_j} = |c_j|$ then $P(t) = \prod (1 + i a_j \cos(\lambda_j t + \varphi_j))$ is uniformly bounded by $\prod (1 + a_j^2)^{\frac{1}{2}} \leqq e^{\frac{1}{2}\Sigma a_j^2} = e^{\frac{1}{2}}$. By Parseval's formula

$$\frac{1}{2}\|f\|_{L^2} = \frac{1}{2}\sum |c_j| a_j = \frac{1}{2\pi} \int P(t)\overline{f(t)}\,dt \leqq e^{\frac{1}{2}} \|f\|_{L^1}$$

which is (1.4) with $B_q = 2e^{\frac{1}{2}}$. Again if we put $B_q = 4e^{\frac{1}{2}}$ then (1.4) is valid for complex-valued functions as well.

If $1 < q < 3$ and we try to repeat the proofs above, we face the difficulty that, having set the product P the way we did, we cannot assert that $\hat{P}(\lambda_j)$ is $\frac{1}{2} e^{i\varphi_j}$ (or $\frac{1}{2} i a_j e^{i\varphi_j}$, in case 2) since λ_j may happen to satisfy nontrivial relations of the form $\lambda_j = \sum \eta_k \lambda_k$ with $\eta_k = 0, \pm 1$. We can, however, construct the Riesz products for subsequences of $\{\lambda_j\}$. Let $M = M_q$ be an integer large enough so that

(1.5) $$q^M > 3, \; 1 - \frac{1}{q^M - 1} > \frac{1}{q}, \text{ and } 1 + \frac{1}{q^M - 1} < q.$$

For $k \leqq m < M$ write $\lambda_j^{(m)} = \lambda_{m+jM}$ and notice that $\lambda_{j+1}^{(m)} > q^M \lambda_j^{(m)}$. By the remark concerning the frequencies appearing in a Riesz product it follows that all the frequencies n appearing in any product corresponding to $\{\lambda_j^{(m)}\}$ satisfy $\left| |n| - \lambda_j^{(m)} \right| < \dfrac{\lambda_j^{(m)}}{q^M - 1}$ for some j, hence by (1.5) if $k > 0$, $k \not\equiv m \pmod M$, $\pm \lambda_k$ does not appear as a frequency in a Riesz product constructed on $\{\lambda_j^{(m)}\}$. It follows that if

$$P(t) = \prod (1 + a_{m+jM} \cos(\lambda_{m+jM} t + \varphi_{m+jM}))$$

then

$$\frac{1}{2\pi} \int P(t) \overline{f(t)} \, dt = \sum \frac{1}{2} a_{m+jM} (e^{i\varphi_{m+jM}} c_{m+jM} + e^{-i\varphi_{m+jM}} c_{-m-jM})$$

and repeating the two constructions used above we obtain

(1.3′) $$\sum |c_{m+jM}| \leqq 4 \|f\|_\infty$$

(1.4′) $$\left(\sum |c_{m+jM}|^2\right)^{\frac{1}{2}} \leqq 4 e^{\frac{1}{2}} \|f\|_{L^1}.$$

Adding (1.3′) and (1.4′) for $m = 1, \ldots, M$ we obtain (1.3) and (1.4) with $A_q = 4M_q$ and $B_q = 4e^{\frac{1}{2}} M_q$. ◀

Theorem: *Let $\{\lambda_j\}$ be lacunary.* (a) *If $\sum c_j e^{i\lambda_j t}$ is the Fourier series of a bounded function, then $\sum |c_j| < \infty$.*

(b) *If $\sum c_j e^{i\lambda_j t}$ is a Fourier series, then $\sum |c_j|^2 < \infty$.*

PROOF: Write $f(t) \sim \sum c_j e^{i\lambda_j t}$ and apply (1.3) resp. (1.4) to $\sigma_n(f, t)$. ◀

1.5 The role of the Riesz products in the proof of lemma 1.4 may become clearer if we consider the statements obtained from 1.4 by duality. For an arbitrary sequence of integers Λ, we denote by C_Λ the space of all continuous functions f on $\mathbf{T}$ such that $\hat{f}(n) = 0$ if $n \notin \Lambda$. C_Λ is clearly a closed subspace of $C(\mathbf{T})$.

DEFINITION: A set of integers Λ is a *Sidon set* if every $f \in C_\Lambda$ has an absolutely convergent Fourier series.

It follows from the closed-graph theorem that Λ is a Sidon set if, and only if, there exists a constant K such that

(1.6) $$\sum |\hat{f}(n)| < K \|f\|_\infty$$

for every polynomial $f \in C_\Lambda$.

***Lemma*:** *A set (of integers) Λ is a Sidon set if, and only if, for every bounded sequence $\{d_\lambda\}_{\lambda\in\Lambda}$ there exists a measure $\mu \in M(\mathbf{T})$ such that $\hat{\mu}(\lambda) = d_\lambda$ for $\lambda \in \Lambda$.*

PROOF: Let Λ be a Sidon set and $\{d_\lambda\}$ a bounded sequence on Λ. The mapping $f \to \sum_{\lambda\in\Lambda} \hat{f}(\lambda)\,\bar{d}_\lambda$ is a well-defined linear functional on C_Λ. By the Hahn-Banach theorem it can be extended to a functional on $C(\mathbf{T})$, that is, a measure μ. For this measure μ we have

$$\hat{\mu}(\lambda) = \overline{\int e^{i\lambda t}\,\overline{d\mu}} = d_\lambda \qquad \text{for all } \lambda \in \Lambda. \tag{1.7}$$

Assume, on the other hand, that the interpolation (1.7) is always possible. Let $f \in C_\Lambda$ and write $d_\lambda = \operatorname{sgn} \hat{f}(\lambda)$. Then, by Parseval's formula $\sum |\hat{f}(\lambda)| = \sum \hat{f}(\lambda)\,\bar{d}_\lambda$ is summable to $\langle f, \mu\rangle$ where μ is a measure which satisfies (1.7). Since for series with positive terms summability is equivalent to convergence, $\sum |\hat{f}(\lambda)| < \infty$ and the proof is complete. ◀

The statement of part (a) of theorem 1.4 is that lacunary sequences are Sidon sets, and the Riesz product is simply an explicit construction of corresponding interpolating measures.

1.6 The statement of part (b) of theorem 1.4 is that for lacunary Λ, $(L^1(\mathbf{T}))_\Lambda = (L^2(\mathbf{T}))_\Lambda$. Every sequence $\{d_\lambda\}$ such that $\sum_{\lambda\in\Lambda} |d_\lambda|^2 < \infty$ defines, as above, a linear functional on $(L^2(\mathbf{T}))_\Lambda$ which, by 1.4, is a closed subspace of $L^1(\mathbf{T})$. Remembering that the dual space of $L^1(\mathbf{T})$ is $L^\infty(\mathbf{T})$, we obtain, using the Hahn-Banach theorem, that there exists a bounded measurable function g such that

$$\hat{g}(\lambda) = d_\lambda \qquad \lambda \in \Lambda \tag{1.8}$$

Here, again, Riesz products (of type 2) provide explicit construction of such functions g. One can actually prove the somewhat finer result:

***Theorem*:** *Let Λ be lacunary and assume that $\sum_{\lambda\in\Lambda} |d_\lambda|^2 < \infty$. Then there exists a continuous function g such that (1.8) is valid.*

We refer the reader to exercise 6 for the proof.

EXERCISES FOR SECTION 1

1. Let $\{\lambda_n\}$ be lacunary and let $f \sim \sum a_n \cos \lambda_n t$. Assume that f satisfies a Lip_α condition with $0 < \alpha < 1$ at $t = t_0$. Show that $a_n = O(\lambda_n^{-\alpha})$ as $n \to \infty$. Deduce that $f \in \mathrm{Lip}_\alpha(\mathbf{T})$.

2. Let $\{\lambda_n\}$ be a sequence of integers and assume that for some $0 < \alpha < 1$ the following statement is true: if $f(t) = \sum a_n e^{i\lambda_n t}$ satisfies a Lip_α condition at one point, then $f \in \mathrm{Lip}_\alpha(\mathbf{T})$. Show that $\{\lambda_n\}$ is lacunary. *Hint*: If $\{\lambda_n\}$ is not lacunary it contains a subsequence $\{\mu_k\}$ such that $\lim \mu_{2k-1}/\mu_{2k} = 1$ and $\lim \mu_{2k+1}/\mu_{2k} = \infty$. For an appropriate sequence a_k, $f(t) = \sum a_k(\cos \mu_{2k}t - \cos \mu_{2k-1}t)$ satisfies a Lip_α condition at $t = 0$ but $f \notin \mathrm{Lip}_\alpha(\mathbf{T})$.

3. Let $f \in L^1(\mathbf{T})$, $f \sim \sum a_n \cos \lambda_n t$ with $\{\lambda_n\}$ lacunary. Assume $f(t) = 0$ for $|t| \leqq \eta$, η being a positive number. Show that f is infinitely differentiable.

4. Show directly, without the use of Riesz products, that if $\lambda_{n+1} > 4\lambda_n$ and $f(t) = \sum_1^N a_n \cos \lambda_n t$, is real-valued, then

$$\sup |f(t)| > \tfrac{1}{2} \sum |a_n|.$$

Hint: Consider the sets $\{t; a_n \cos \lambda_n t > |a_n|/2\}$.

5. If $d_n \to 0$ as $n \to \infty$ we can write $d_n = \delta_n \psi_n$ where $\{\delta_n\}$ is bounded and $\psi_n = \hat{\psi}(n)$ for some $\psi \in L^1(\mathbf{T})$. (See theorem I.4.1 and exercise I.4.1.) Deduce that if Λ is a lacunary sequence and $d_\lambda \to 0$ as $|\lambda| \to \infty$, there exists a function $g \in L^1(\mathbf{T})$ such that

$$d_\lambda = \hat{g}(\lambda) \qquad \text{for } \lambda \in \Lambda.$$

6. If $\sum |d_n|^2 < \infty$ there exist sequences $\{\delta_n\}$ and $\{\psi_n\}$ such that $d_n = \delta_n \psi_n$, $\sum |\delta_n|^2 < \infty$, and $\psi_n = \hat{\psi}(n)$ for some $\psi \in L^1(\mathbf{T})$. Remembering that the convolution of a summable function with a bounded function is continuous, prove theorem 1.6.

7. Assume $\lambda_{j+1}/\lambda_j \geqq q > 1$. Show that there exists a number $M = M_q$ such that every integer n has at most one representation of the form $n = \sum \eta_j \lambda_{m_j + jM}$ where $\eta_j = -1, 0, 1$, and $1 \leqq m_j \leqq M$. Use this to show that the product $\prod_{j=1}^{\infty} (1 + \sum_{m=1}^{M} d_{m+jM} \cos(\lambda_{m+jM}t + \varphi_{m+jM}))$ has (formally) the Fourier coefficient $\frac{1}{2} d_k e^{i\varphi_k}$ at the point λ_k. Show that if $0 < d_k \leqq 1/M_q$ for all k, then the product above is the Fourier Stieltjes series of a positive measure which interpolates $\{\frac{1}{2} d_k e^{i\varphi_k}\}$ on $\{\lambda_k\}$.

8. Assume $\lambda_{j+1}/\lambda_j \geqq q > 1$ and $\sum |d_j|^2 < \infty$. Find a product analogous to that of exercise 7), which is the Fourier series of a bounded function, and which interpolates $\{d_j\}$ on $\{\lambda_j\}$.

*9. Show that the following condition is sufficient to imply that the sequence Λ is a Sidon set: to every sequence $\{d_\lambda\}$ such that $|d_\lambda| = 1$ there exists a measure $\mu \in M(\mathbf{T})$ such that

$$|\hat{\mu}(\lambda) - d_\lambda| < \tfrac{1}{2} \qquad \lambda \in \Lambda.$$

*10. Show that a finite union of lacunary sequences is a Sidon set.

*2. QUASI-ANALYTIC CLASSES

2.1 We consider classes of infinitely differentiable functions on $\mathbf{T}$. Let $\{M_n\}$ be a sequence of positive numbers; we denote by $C^*\{M_n\}$ the class of all infinitely differentiable functions f on $\mathbf{T}$ such that for an appropriate $R > 0$

$$\|f^{(n)}\|_\infty \leqq R^n M_n \qquad n = 1, 2, \ldots. \tag{2.1}$$

We shall denote by $C^{\#}\{M_n\}$ the class of infinitely differentiable functions on $\mathbf{T}$ satisfying:

$$\|f^{(n)}\|_{L^2} \leqq R^n M_n \qquad n = 1, 2, \ldots \tag{2.2}$$

for some R (depending on f).

The inclusion $C^*\{M_n\} \subseteq C^{\#}\{M_n\}$ is obvious; on the other hand, since the mean value of derivatives on $\mathbf{T}$ is zero, we obtain $\sup_{t \in \mathbf{T}} |f^{(n)}(t)| \leqq \|f^{(n+1)}\|_{L^2}$ and consequently $C^{\#}\{M_n\} \subseteq C^*\{M_{n+1}.\}$ Thus the two classes are fairly close to each other.

Examples: If $M_n = 1$ for all n, then $C^{\#}\{M_n\}$ is precisely the class of all trigonometric polynomials on $\mathbf{T}$. If $M_n = n!$, $C^{\#}\{M_n\}$ is precisely the class of all functions analytic on $\mathbf{T}$. (See exercise I.4.3.)

We recall that a sequence $\{c_n\}$, $c_n > 0$, is *log convex* if the sequence $\{\log c_n\}$ is a convex function of n. This amounts to saying that, given $k < l < m$ in the range of n, we have

$$\log c_l \leqq \frac{m-l}{m-k} \log c_k + \frac{l-k}{m-k} \log c_m \tag{2.3}$$

or equivalently

$$c_l \leqq c_k^{(m-l)/(m-k)} c_m^{(l-k)/(m-k)}. \tag{2.4}$$

2.2 The identity $\|f^{(n)}\|_{L^2} = (\sum |\hat{f}(j)|^2 j^{2n})^{\frac{1}{2}}$ allows an expression of condition (2.2) directly in terms of the Fourier coefficients of f. Also it implies

Lemma: *Let f be N times differentiable on $\mathbf{T}$. Then the sequence $\{\|f^{(n)}\|_{L^2}\}$ is monotone increasing and log convex for $1 \leqq n \leqq N$.*

PROOF: The fact that $\|f^{(n)}\|_{L^2} = (\sum |\hat{f}(j)|^2 j^{2n})^{\frac{1}{2}}$ is monotone increasing is obvious. In order to prove (2.4) we write $p = (m-k)/(m-l)$, $q = (m-k)/(l-k)$; then $1/p + 1/q = 1$ and by Hölder's inequality

$$\sum |a_j|^2 j^{2l} = \sum (|a_j|^{2/p} j^{2k/p})(|a_j|^{2/q} j^{2m/q}) \leqq \|f^{(k)}\|_{L^2}^{2/p} \|f^{(m)}\|_{L^2}^{2/q}$$

which is exactly (2.4).

It follows from lemma 2.2 (cf. exercise 2 at the end of this section) that for every sequence $\{M_n'\}$ there exists a sequence $\{M_n\}$ which is monotone increasing and log convex such that $C^{\#}\{M_n\} = C^{\#}\{M_n'\}$. Thus, when studying classes $C^{\#}\{M_n\}$ we may assume without loss of generality that $\{M_n\}$ is monotone increasing and log convex; throughout the rest of this section we always assume that, for $k < l < m$,

$$M_l \leqq M_k^{(m-l)/(m-k)} M_m^{(l-k)/(m-k)}. \tag{2.5}$$

2.3 For a (monotone increasing and log convex) sequence $\{M_n\}$ we define the *associated function* $\tau(r)$ by

$$\tau(r) = \inf_{n \geqq 0} M_n r^{-n} \tag{2.6}$$

We consider sequences M_n which increase faster that R^n for all $R > 0$; hence the infimum in (2.6) is attained and we can write $\tau(r) = \min_{n \geqq 0} M_n r^{-n}$. If we write $\mu_1 = M_1^{-1}$, and $\mu_n = M_{n-1}/M_n$ for $n > 1$; then μ_n is a monotone-decreasing sequence since by (2.5), $\mu_{n+1}/\mu_n = M_n^2/M_{n-1}M_{n+1} < 1$; we have $M_n r^{-n} = \prod_1^n (\mu_j r)^{-1}$ and consequently

$$\tau(r) = \prod_{\mu_j r > 1} (\mu_j r)^{-1}. \tag{2.6'}$$

The function $\tau(r)$ was implicitly introduced in I.4; thus it follows from I.4.4 that if $f \in C^{\#}\{M_n\}$ then, for the appropriate $R > 0$

$$|\hat{f}(j)| \leqq \tau(jR^{-1}),$$

and exercise I.4.6 is essentially an estimate for $\tau(r)$ in the case $M_n = n^{\alpha n}$.

2.4 An analytic function on $\mathbf{T}$ is completely determined by its Taylor expansion around any point $t_0 \in \mathbf{T}$, that is, by the sequence $\{f^{(n)}(t_0)\}_{n=0}^{\infty}$. In particular if $f^{(n)}(t_0) = 0$, $n = 0, 1, 2, \ldots$, it follows that $f = 0$ identically.

DEFINITION: A class of infinitely differentiable functions on $\mathbf{T}$ is *quasi-analytic* if the only function in the class, which vanishes with all its derivatives at some $t_0 \in \mathbf{T}$, is the function which vanishes identically.

The main result of this section is the so-called Denjoy-Carleman theorem which gives a necessary and sufficient conditions for the quasi-analyticity of classes $C^{\#}\{M_n\}$.

Theorem: *Let* $\{M_n\}$ *be monotone increasing and log convex. Let* $\tau(r)$ *be the associated function* (2.6). *The following three conditions are equivalent*:

(i) $C^{\#}\{M_n\}$ *is quasi-analytic*

(ii) $$\int_1^\infty \frac{\log \tau(r)}{1+r^2}\,dr = -\infty$$

(iii) $$\sum \frac{M_n}{M_{n+1}} = \infty .$$

The proof will consist in establishing the three implications (ii) ⇒ (i) (theorem 2.6 below), (i) ⇒ (iii) (theorem 2.8), and (iii) ⇒ (ii) (lemma 2.9).

We begin with:

2.5 ***Lemma:*** *Let* $\varphi(z) \not\equiv 0$ *be holomorphic and bounded in the half plane* $\operatorname{Re}(z) > 0$ *and continuous on* $\operatorname{Re}(z) \geqq 0$. *Then*

$$\int_0^\infty \log|\varphi(\pm iy)|\,\frac{dy}{1+y^2} > -\infty .$$

PROOF: The function $F(\zeta) = \varphi\left(\frac{1+\zeta}{1-\zeta}\right)$ is holomorphic and bounded in the unit disc D (and is continuous on $\bar{D}$ except possibly at $\zeta = 1$). By III.3.14 we have $\int_0^\pi \log|F(e^{it})|\,dt > -\infty$. The change of variables that we have introduced gives for the boundaries $e^{it} = (iy-1)/(iy+1)$ or $t = 2 \operatorname{arc\,cot} y$.

Consequently $dt = \frac{-2}{1+y^2}\,dy$ and

$$\int_0^\infty \log|\varphi(iy)|\,\frac{dy}{1+y^2} = \frac{1}{2}\int_0^\pi \log|F(e^{it})|\,dt > -\infty;$$

similarly $\int_0^\infty \log|\varphi(-iy)|\,\frac{dy}{1+y^2} > -\infty$ and the lemma is proved. ◀

2.6 ***Theorem:*** *A sufficient condition for the quasi-analyticity of the class* $C^{\#}\{M_n\}$ *is that* $\int_1^\infty \frac{\log\tau(r)}{1+r^2}\,dr = -\infty$ *where* $\tau(r)$ *is defined by* (2.6).

PROOF: Let $f \in C^{\#}\{M_n\}$ and assume that $f^{(n)}(0) = 0 \quad n = 0, 1, \ldots$. Define

$$\varphi(z) = \frac{1}{2\pi} \int_0^{2\pi} e^{-zt} f(t)\, dt.$$

Integrating by parts we obtain, $z \neq 0$

$$\varphi(z) = \frac{1}{2\pi} \left[\frac{-1}{z} e^{-zt} f(t) \right]_0^{2\pi} + \frac{1}{2\pi z} \int_0^{2\pi} e^{-zt} f'(t)\, dt$$

and since $f(0) = f(2\pi) = 0$ the first term vanishes for all $z \neq 0$ (we have used the same integration by parts in I.4; there we did not assume $f(0) = 0$ but considered only the case $z = im$, that is, $e^{-zt} f$ is 2π-periodic). Repeating the integration by parts n times (using $f^{(j)}(0) = f^{(j)}(2\pi) = 0$ for $j \leqq n$), we obtain

$$\varphi(z) = \frac{1}{2\pi z^n} \int_0^{2\pi} e^{-zt} f^{(n)}(t)\, dt.$$

For $\operatorname{Re}(z) \geqq 0$, $|e^{-zt}| \leqq 1$ on $(0, 2\pi)$ and consequently

$$|\varphi(z)| \leqq \frac{M_n}{|z|^n} \qquad \text{for } n = 0, 1, \ldots$$

hence

$$|\varphi(z)| \leqq \tau(|z|)$$

or

$$\log|\varphi(z)| < \log|\tau(|z|)|.$$

It follows that $\int_1^\infty \log|\varphi(iy)| \dfrac{dy}{1 + y^2} = -\infty$ and by lemma 2.5 $\varphi(z) = 0$. Since $\varphi(in) = \hat{f}(n)$ it follows that $f = 0$. ◀

2.7 Lemma: *Assume $\mu_j > 0$, $\sum_0^\infty \mu_j \leqq 1$. Write $\varphi(k) = \prod_0^\infty \dfrac{\sin \mu_j k}{\mu_j k}$. Then $f(t) = \sum_{-\infty}^\infty \varphi(k) e^{ikt}$ is carried by $[-1, 1]$ $(\mathrm{mod}\, 2\pi)$, it is infinitely differentiable and $\|f^{(n)}\|_{L^2} \leqq 2 \prod_0^n \mu_j^{-1}$.*

PROOF: All the factors in the product defining $\varphi(k)$ are bounded by 1 so that the product either converges or diverges to zero (actually it converges for all k) and $\varphi(k)$ is well defined. $\varphi(k)$ clearly tends to zero faster than any power of k so that the series defining f converges uniformly and f is infinitely differentiable. We have $\varphi(0) = 1$ so that $f \not\equiv 0$. The sequence $\left\{\frac{\sin \mu_j k}{\mu_j k}\right\}_{k=-\infty}^{\infty}$ is the sequence of Fourier coefficients of the function $\Gamma_j(t) = \begin{cases} \pi\mu_j^{-1} & |t| < \mu_j \\ 0 & \text{elsewhere} \end{cases}$; if we write $\varphi_N(k) = \prod_0^N \frac{\sin \mu_j k}{\mu_j k}$, we have $f_N(t) = \sum_{-\infty}^{\infty} \varphi_N(k) e^{ikt} = \Gamma_0 * \Gamma_1 * \cdots * \Gamma_N$ and the support of f_N is equal to $[-\sum_0^N \mu_j, \sum_0^N \mu_j]$ (mod 2π). Since f_N converges uniformly to f, the support of f is contained in $[-\sum_0^\infty \mu_j, \sum_0^\infty \mu_j]$ (mod 2π). Finally, since $\|f^{(n)}\|_{L^2}^2 = \sum |\varphi(k)|^2 k^{2n}$ and $|\varphi(k)| \leqq \prod_0^n (\mu_j k)^{-1} = (\prod_0^n \mu_j)^{-1} k^{-n-1}$, we obtain $\|f^{(n)}\|_{L_2} \leqq (\prod_0^n \mu_j)^{-1}(\sum_{k \neq 0} k^{-2})^{\frac{1}{2}}$ and the proof is complete. ◀

2.8 *Theorem*: *A necessary condition for the quasi-analyticity of of the class* $C^\#\{M_n\}$ *is that* $\sum \frac{M_n}{M_{n+1}} = \infty$.

PROOF: Assume that $\sum \frac{M_n}{M_{n+1}} < \infty$. Without loss of generality we may assume $\sum \frac{M_n}{M_{n+1}} < \frac{1}{2}$ (replacing M_n by $M_n' = M_n R^n$ does not change the class $C^\#\{M_n\}$ while $\sum \frac{M_n'}{M_{n+1}'} = R^{-1} \sum \frac{M_n}{M_{n+1}}$). Write $\mu_0 = \mu_1 = \frac{1}{4}$, $\mu_j = \frac{M_{j-1}}{M_j}$, $j \geqq 2$. Then $\prod_0^n \mu_j^{-1} = 16 \frac{M_n}{M_1}$ and the function f defined by lemma 2.7 has a zero of infinite order (actually vanishes outside of $[-1, 1]$), is not identically zero, and $f \in C^\#\{M_n\}$. ◀

2.9 *Lemma*: *Under the assumption of theorem* 2.4 *we have*

$$\sum \frac{M_n}{M_{n+1}} \leqq 2e^4 \int_{e^2}^{\infty} \frac{-\log \tau(r)}{1 + r^2} dr.$$

PROOF: As before we write $\mu_n = \frac{M_{n-1}}{M_n}$. We define the counting function $\mathscr{M}(r)$ of $\{\mu_n\}$ by:

$$\mathscr{M}(r) = \text{the number of elements } \mu_j \text{ such that } \mu_j r \geqq e,$$

and recall that $\tau(r) = \prod_{\mu_j r > 1} (\mu_j r)^{-1}$; hence

$$-\log \tau(r) = \sum_{\mu_j r > 1} \log(\mu_j r) \geqq \sum_{\mu_j r \geqq e} \log(\mu_j r) \geqq \mathscr{M}(r).$$

Thus for $k = 2, 3, \ldots$

$$\text{(2.7)} \qquad \int_{e^k}^{e^{k+1}} \frac{-\log \tau(r)}{1+r^2}\, dr \geqq \frac{\mathcal{M}(e^k)}{2\, e^{2k+2}} \int_{e^k}^{e^{k+1}} dr > \frac{1}{2\, e^2} \frac{\mathcal{M}(e^k)}{e^k};$$

on the other hand,

$$\text{(2.8)} \qquad \sum_{e^{1-k} < \mu_j < e^{2-k}} \mu_j \leqq (\mathcal{M}(e^k) - \mathcal{M}(e^{k-1}))\, e^{2-k} \leqq e^2 \frac{\mathcal{M}(e^k)}{e^k}$$

and the theorem follows by summing (2.7) and (2.8) with respect to k, $k = 2, 3, \ldots$. ◀

Remark: Theorems 2.6, 2.8, and lemma 2.9 together prove theorem 2.4. We see in particular that if $C^{\#}\{M_n\}$ is not quasi-analytic, it contains functions (which are not identically zero) having arbitrarily small supports.

For further reading, generalizations, and related topics we mention [17].

EXERCISES FOR SECTION 2

1. Show that $\{c_n\}$ is log convex if, and only if, $c_n^2 \leqq c_{n-1}c_{n+1}$ for all n.

2. (a) Let $\{c_n^\alpha\}_{n=1}^\infty$ be a log-convex sequence for all α belonging to some index set I. Assume that $M_n = \sup_{\alpha \in I} c_n^\alpha < \infty$ for all n. Prove that $\{M_n\}$ is log convex.

(b) Let $\{M_n'\}$ be a sequence of positive numbers. Let $\{c_n^\alpha\}$ be the family of all log-convex sequences satisfying $c_n^\alpha \leqq M_n'$ for all n. Put $M_n = \sup_\alpha c_n^\alpha$. Then $C^{\#}\{M_n\} = C^{\#}\{M_n'\}$.

3. Let $M_j \leqq j!$ for infinitely many values of j. Show that $C^*\{M_n\}$ and $C^{\#}\{M_n\}$ are quasi-analytic. *Hint*: Assuming $f \in C^*\{M_n\}$ and $f^{(k)}(0) = 0$ for all k, use Taylor's expansion with remainder to show $f \equiv 0$.

4. We say that *a function* $\varphi \in C^\infty(\mathbf{T})$ *is quasi-analytic if* $C^{\#}\{\| \varphi^{(n)} \|_{L^2}\}$ is quasi-analytic. Let $f \in C^\infty(\mathbf{T})$; show that if the sequence $\{\lambda_j\}$ increases fast enough and if we set

$$f_1(t) = \sum_j \sum_{\lambda_{2j} < |k| \leqq \lambda_{2j+1}} \hat{f}(k)\, e^{ikt},$$

then both f_1 and $f_2 = f - f_1$ are quasi-analytic. Thus, every infinitely differentiable function is the sum of two quasi-analytic functions.

5. Show that $C^{\#}\{n!(\log n)^{\alpha n}\}$ is quasi-analytic if $0 \leqq \alpha \leqq 1$, and is nonquasi-analytic if $\alpha > 1$.

6. Let $\tau(r)$ be the function associated with a sequence $\{M_n\}$. (a) Show that $(\tau(r))^{-1}$ is log-convex function of r. (b) Show that $M_n = \max_r r^n \tau(r)$.

7. Let $\{\omega_n\}$ be a log-convex sequence, $n=0,1,\ldots, \omega_n \geqq 1$, and let $A\{\omega_n\}$ be the space of all $f \in C(\mathbf{T})$ such that $\| f \|_{\{\omega_n\}} = \sum |\hat{f}(n)| \, \omega_{|n|} < \infty$. Show that with the norm $\| \ \|_{\{\omega_n\}}$, $A\{\omega_n\}$ is a Banach space. Show that a necessary and sufficient condition for $A\{\omega_n\}$ to contain functions with arbitrarily small support is $\sum_1^\infty \frac{\log \omega_n}{n^2} < \infty$.

8. Let $\{\omega_n\}$ be log convex, $\omega_n \geqq 1$ and assume that $\omega_n \to \infty$ faster than any power of n. The sequences $\sigma_k = \left\{\frac{n^k}{\omega_{|n|}}\right\}_{n=-\infty}^{\infty}$ tend to zero as $|n| \to \infty$. Show that the subspace that σ_k, $k = 0, 1, \ldots$ generate in c_0 (the space of sequences tending to zero at ∞) is uniformly dense in c_0 if, and only if, $\sum \frac{\log \omega_n}{n^2} = \infty$. *Hint*: The dual space of c_0 is ℓ^1. If $\{a_n\} \in \ell^1$ is orthogonal to σ_k, $k = 0, \ldots$, the function $f(t) = \sum \frac{a_n}{\omega_n} e^{int}$, which clearly belongs to $A\{\omega_n\}$, has a zero of infinite order.

9. Let f be as in exercise 1.3. Show that $f = 0$ identically.

Chapter VI

Fourier Transforms on the Line

In the preceding chapters we studied objects (functions, measures, and so on) defined on $\mathbf{T}$. Our aim in this chapter is to extend the study to objects defined on the real line $\mathbf{R}$. Much of the theory, especially the L^1 theory, extends almost verbatim and with only trivial modifications of the proofs; such results, analogous in statement and in proof to theorems that we have proved for $\mathbf{T}$, often are stated without a proof. The difference between the circle and the line becomes more obvious when we try to see what happens for L^p with $p > 1$. The (Lebesgue) measure of $\mathbf{R}$ being infinite entails that, unlike $L^1(\mathbf{T})$ which contains most of the "natural" function spaces on $\mathbf{T}$, $L^1(\mathbf{R})$ is relatively small; in particular $L^p(\mathbf{R}) \not\subseteq L^1(\mathbf{R})$ for $p > 1$. The definition of Fourier transforms in $L^1(\mathbf{R})$ has now a much more special character and a new definition (i.e., an extension of the definition) is needed for $L^p(\mathbf{R})$, $p > 1$. The situation turns out to be quite different for $p \leqq 2$ and for $p > 2$. If $p \leqq 2$, Fourier transforms of functions in $L^p(\mathbf{R})$ can be defined by continuity as functions in $L^q(\mathbf{R})$, $q = p/(p-1)$; however, if $p > 2$, the only reasonable way to define the Fourier transform on $L^p(\mathbf{R})$ is through duality and Fourier transforms are now defined as distributions. The plan of this chapter is as follows: in section 1 we define the Fourier transform in $L^1(\mathbf{R})$ and discuss its elementary properties. We also mention the connection between Fourier transforms and Fourier coefficients and prove Poisson's formula. In section 2 we define Fourier-Stieltjes transforms and obtain various characterizations of Fourier-Stieltjes transforms of arbitrary and positive measures. In section 3 we prove Plancherel's theorem and the Hausdorff-Young inequality, thus defining Fourier transforms in $L^p(\mathbf{R})$, $1 < p \leqq 2$. In section 4 we use Parseval's formula, that is, duality, to define the Fourier transforms of tempered distributions and study some of the properties of Fourier transforms of functions in $L^p(\mathbf{R})$, $p \leqq \infty$. Sections 5 and 6 deal with spectral analysis and synthesis in $L^\infty(\mathbf{R})$. In section 5 we consider the problems relative to the norm topology and show that the class of functions for which we have satisfactory theory is precisely that of Bohr's almost periodic functions. In section 6 we study the analogous problems for the weak-star topology. Sections 7 and 8 are devoted to relations between Fourier transforms and analytic functions. Finally, section 9 contains Kronecker's theorem (which we have already used in chapter II) and some variations on the same theme.

1. FOURIER TRANSFORMS FOR $L^1(\mathbf{R})$

1.1 We denote by $L^1(\mathbf{R})$ the space of Lebesgue integrable functions on the real line. For $f \in L^1(\mathbf{R})$ we write

$$\|f\|_{L^1(\mathbf{R})} = \int_{-\infty}^{\infty} |f(x)|\, dx,$$

and when there is no risk of confusion, we write $\|f\|_{L^1}$ or simply $\|f\|$ instead of $\|f\|_{L^1(\mathbf{R})}$.

The Fourier transform $\hat{f}$ of f is defined by

$$\hat{f}(\xi) = \int f(x) e^{-i\xi x} dx \qquad \text{for all real } \xi.\dagger \tag{1.1}$$

This definition is analogous to I.(1.5), and the disappearance of the factor $1/2\pi$ is due to none other than our (arbitrary) choice to remove it. It was a natural normalizing factor for the Lebesgue measure on $\mathbf{T}$; but, at this point, it seems arbitrary for $\mathbf{R}$. The factor $1/2\pi$ will reappear in the inversion formula and some authors, seeking more symmetry for the inversion formula, write $1/\sqrt{2\pi}$ in front of the integral (1.1) so that the same factor appear in the Fourier transform and its inverse. The added symmetry, however, may increase the possibility of confusion between the domains of definition of a function and its transform. In $L^1(\mathbf{T})$ the functions are defined on $\mathbf{T}$ whereas

† Throughout this chapter, integrals with unspecified limits of integration are always to be taken over the entire real line.

the Fourier transforms are defined on the integers; in $L^1(\mathbf{R})$ the functions are defined on $\mathbf{R}$ and the domain of definition of the Fourier transforms is again the real line. It may be helpful to consider two copies of the real line: one is $\mathbf{R}$ and the other, which will serve as the domain of definition of Fourier transforms of functions in $L^1(\mathbf{R})$, we denote by $\hat{\mathbf{R}}$. This notation is in accordance with that of chapter VII.

Most of the elementary properties of Fourier coefficients are valid for Fourier transforms.

Theorem: *Let $f, g \in L^1(\mathbf{R})$. Then*

(a)
$$\widehat{(f+g)}(\xi) = \hat{f}(\xi) + \hat{g}(\xi)$$

(b) *For any complex number α*
$$\widehat{(\alpha f)}(\xi) = \alpha \hat{f}(\xi)$$

(c) *If $\bar{f}$ is the complex conjugate of f, then*
$$\hat{\bar{f}}(\xi) = \overline{\hat{f}(-\xi)}$$

(d) *Denote $f_y(x) = f(x-y)$, $y \in \mathbf{R}$. Then*
$$\hat{f}_y(\xi) = \hat{f}(\xi)\, e^{-i\xi y}$$

(e)
$$|\hat{f}(\xi)| \leqq \int |f(x)|\, dx = \|f\|$$

(f) *For real-valued λ, $\lambda > 0$, denote*
$$\varphi(x) = \lambda f(\lambda x);$$
then
$$\hat{\varphi}(\xi) = \hat{f}\left(\frac{\xi}{\lambda}\right).$$

PROOF: The theorem follows immediately from (1.1). Parts (a) through (e) are analogous to the corresponding parts of I.1.4. Part (f) is obtained by a change of variable $y = \lambda x$:

$$\hat{\varphi}(\xi) = \int f(\lambda x)\, e^{-i(\xi/\lambda)\lambda x} d\lambda x = \int f(y)\, e^{-i(\xi/\lambda)y} dy = \hat{f}\left(\frac{\xi}{\lambda}\right). \quad ◀$$

1.2 ***Theorem***: *Let $f \in L^1(\mathbf{R})$. Then $\hat{f}$ is uniformly continuous in $-\infty < \xi < \infty$.*

PROOF:

$$\hat{f}(\xi+\eta) - \hat{f}(\xi) = \int f(x)(e^{-i(\xi+\eta)x} - e^{-i\xi x})\, dx,$$

hence

$$|\hat{f}(\xi+\eta)-\hat{f}(\xi)| \leqq \int |f(x)|\,|e^{i\eta x}-1|\,dx. \tag{1.2}$$

The integral on the right of (1.2) is independent of ξ, the integrand is bounded by $2|f(x)|$ and tends to zero everywhere as $\eta \to 0$. ◀

1.3 The following are immediate adaptations of the corresponding theorems in chapter I.

Theorem: *Let $f, g \in L^1(\mathbf{R})$. For almost all x, $f(x-y)\,g(y)$ is integrable (as a function of y) and, if we write*

$$h(x) = \int f(x-y)\,g(y)\,dy,$$

then $h \in L^1(\mathbf{R})$ and

$$\|h\| \leqq \|f\|\,\|g\|;$$

moreover,

$$\hat{h}(\xi) = \hat{f}(\xi)\,\hat{g}(\xi) \qquad \text{for all } \xi.$$

As in chapter I we denote $h = f * g$, call h the *convolution* of f and g, and notice that the convolution operation is commutative, associative, and distributive.

1.4 ***Theorem***: *Let f, $h \in L^1(\mathbf{R})$ and*

$$h(x) = \frac{1}{2\pi}\int H(\xi)\,e^{i\xi x}\,d\xi$$

with integrable $H(\xi)$. Then

$$(h*f)(x) = \frac{1}{2\pi}\int H(\xi)\hat{f}(\xi)\,e^{ix\xi}\,d\xi. \tag{1.3}$$

PROOF The function $H(\xi)f(y)$ is integrable in (ξ, y), hence, by Fubini's theorem,

$$(h*f)(x) = \int h(x-y)f(y)\,dy = \frac{1}{2\pi}\iint H(\xi)\,e^{i\xi x}\,e^{-i\xi y}f(y)\,d\xi\,dy$$

$$= \frac{1}{2\pi}\int H(\xi)\,e^{i\xi x}\int e^{-i\xi y}f(y)\,dy\,d\xi = \frac{1}{2\pi}\int H(\xi)\hat{f}(\xi)e^{i\xi x}d\xi.$$

◀

1.5 *Theorem*: *Let* $f \in L^1(\mathbf{R})$ *and define*

$$F(x) = \int_{-\infty}^{x} f(y)\,dy\,.$$

Then, if $F \in L^1(\mathbf{R})$ *we have*

$$\hat{F}(\xi) = \frac{1}{i\xi}\hat{f}(\xi) \qquad \text{all real } \xi \neq 0. \tag{1.4}$$

An equivalent statement of the theorem is: if $F, F' \in L^1(\mathbf{R})$, then $\widehat{F'}(\xi) = i\xi\hat{F}(\xi)$.

1.6 *Theorem*: *Let* $f \in L^1(\mathbf{R})$ *and* $xf(x) \in L^1(\mathbf{R})$. *Then* $\hat{f}$ *is differentiable and*

$$\frac{d}{d\xi}\hat{f}(\xi) = \widehat{(-ixf)}(\xi)\,. \tag{1.5}$$

PROOF:

$$\frac{\hat{f}(\xi + h) - \hat{f}(\xi)}{h} = \int f(x)\,e^{-i\xi x}\left(\frac{e^{-ihx} - 1}{h}\right)dx\,. \tag{1.6}$$

The integrand in (1.6) is bounded by $|xf(x)|$ (which is in $L^1(\mathbf{R})$ by assumption) and tends to $-ixf(x)\,e^{-i\xi x}$ pointwise, hence (Lebesgue) it converges to $-ixf(x)\,e^{-i\xi x}$ in the $L^1(\mathbf{R})$ norm. This implies that as $h \to 0$ the right-hand side of (1.6) converges to $\widehat{(-ixf)}(\xi)$ and the theorem follows. ◀

1.7 *Theorem* (*Riemann-Lebesgue lemma*): *For* $f \in L^1(\mathbf{R})$

$$\lim_{|\xi| \to \infty} \hat{f}(\xi) = 0\,.$$

PROOF: If g is continuously differentiable and with compact support we have, by 1.5 and 1.1, $|\xi\hat{g}(\xi)| \leq \|g'\|_{L^1(\mathbf{R})}$; hence $\lim_{|\xi|\to\infty} |\hat{g}(\xi)| = 0$. For arbitrary $f \in L^1(\mathbf{R})$, let $\varepsilon > 0$ and g be a continuously differentiable, compactly supported function such that $\|f - g\|_{L^1(\mathbf{R})} < \varepsilon$. We have $|\hat{f}(\xi) - \hat{g}(\xi)| < \varepsilon$ and, as $|\xi| \to \infty$, $\lim |\hat{g}(\xi)| = 0$; consequently, $\limsup_{|\xi|\to\infty} |\hat{f}(\xi)| < \varepsilon$, and, this being true for all $\varepsilon > 0$, we obtain $\lim_{|\xi|\to\infty}\hat{f}(\xi) = 0$. ◀

1.8 We denote by $A(\hat{\mathbf{R}})$ the space of all functions φ on $\hat{\mathbf{R}}$, which are the Fourier transforms of functions in $L^1(\mathbf{R})$. By the results above, $A(\hat{\mathbf{R}})$ is an algebra of continuous functions vanishing at infinity, that is, a subalgebra of $C_0(\hat{\mathbf{R}})$, the algebra of all continuous functions on $\hat{\mathbf{R}}$ which vanish at infinity. We introduce a norm to $A(\hat{\mathbf{R}})$ by transferring to it the norm of $L^1(\mathbf{R})$, that is, we write

$$\| \hat{f} \|_{A(\hat{\mathbf{R}})} = \| f \|_{L^1(\mathbf{R})} .$$

It follows from 1.3 that the norm $\| \ \|_{A(\hat{\mathbf{R}})}$ is multiplicative, that is, satisfies the inequality: $\| \varphi_1 \varphi_2 \|_{A(\hat{\mathbf{R}})} \leqq \| \varphi_1 \|_{A(\hat{\mathbf{R}})} \| \varphi_2 \|_{A(\hat{\mathbf{R}})}$. The norm $\| \ \|_{A(\hat{\mathbf{R}})}$ is not equivalent to the supremum norm; consequently, $A(\hat{\mathbf{R}})$ is a proper subalgebra of $C_0(\hat{\mathbf{R}})$.

1.9 A summability kernel on the real line is a family of continuous functions $\{k_\lambda\}$ on **R**, with either discrete or continuous parameter† λ satisfying the following:

$$\int k_\lambda(x)\,dx = 1$$

$$\text{(1.7)} \qquad \| k_\lambda \|_{L^1(\mathbf{R})} = O(1) \quad \text{as} \quad \lambda \to \infty$$

$$\lim_{\lambda \to \infty} \int_{|x|>\delta} | k_\lambda(x) |\,dx = 0, \quad \text{for all } \delta > 0.$$

A common way to produce summability kernels on **R** is to take a function $f \in L^1(\mathbf{R})$ such that $\int f(x)\,dx = 1$ and to write $k_\lambda(x) = \lambda f(\lambda x)$ for $\lambda > 0$. Condition (1.7) is satisfied since, introducing the change of variable $y = \lambda x$, we obtain

$$\int k_\lambda(x)\,dx = \int f(y)\,dy = 1$$

$$\| k_\lambda \| = \int | k_\lambda(x) |\,dx = \int | f(y) |\,dy = \| f \|$$

and

$$\int_{|x|>\delta} | k_\lambda(x) |\,dx = \int_{|y|>\lambda\delta} | f(y) |\,dy \to 0 \qquad \text{as } \lambda \to \infty .$$

The Fejér kernel on **R** is defined by

$$\mathbf{K}_\lambda(x) = \lambda \mathbf{K}(\lambda x), \qquad \lambda > 0,$$

where

$$\text{(1.8)} \qquad \mathbf{K}(x) = \frac{1}{2\pi} \left(\frac{\sin x/2}{x/2} \right)^2 = \frac{1}{2\pi} \int_{-1}^{1} (1 - | \xi |) e^{i\xi x}\,d\xi .$$

The second equality in (1.8) is obtained directly by integration. By the previous remark it is clear that the only thing that we have to check, in

† The indexing parameter λ is often continuous, that is, real valued; however, it should not be considered as an element of **R** so that no confusion with the notation of 1.1.d should arise.

order to establish that $\{\mathbf{K}_\lambda\}$ is a summability kernel, is that $\int K(x)\,dx = 1$. This can be done directly, for example, by contour integration, or using the information that we have about the Fejér kernel of the circle, that is, that for all $0 < \delta < \pi$

$$\lim_{n\to\infty} \frac{1}{2\pi} \int_{-\delta}^{\delta} \frac{1}{n+1} \left(\frac{\sin (n+1)x/2}{\sin x/2}\right)^2 dx = 1. \tag{1.9}$$

Since $\int \mathbf{K}(x)\,dx = \int \mathbf{K}_\lambda(x)\,dx$, we may take $\lambda = n+1$, in which case $\mathbf{K}_\lambda(x) = \frac{1}{2\pi(n+1)} \left(\frac{\sin (n+1)x/2}{x/2}\right)^2$, and notice that if $\delta > 0$ is small enough, the ratio of $2\pi \mathbf{K}_\lambda(x)$ to the integrand in (1.9) is arbitrarily close to one in $|x| < \delta$. More precisely, we obtain ($\lambda = n+1$):

$$\left(\frac{\sin \delta}{\delta}\right)^2 \frac{1}{2\pi} \int_{-\delta}^{\delta} \frac{1}{n+1} \left(\frac{\sin (n+1)x/2}{\sin x/2}\right)^2 dx < \int_{-\delta}^{\delta} \mathbf{K}_\lambda(x)\,dx$$
$$< \frac{1}{2\pi} \int_{-\pi}^{\pi} \frac{1}{n+1} \left(\frac{\sin (n+1)x/2}{\sin x/2}\right)^2 dx.$$

Letting $n \to \infty$ we see that $\int \mathbf{K}(x)\,dx = \lim_{\lambda\to\infty} \int_{-\delta}^{\delta} \mathbf{K}_\lambda(x)\,dx$ is a number between $\sin^2\delta/\delta^2$ and 1; since $\delta > 0$ is arbitrary $\int \mathbf{K}(x)\,dx = 1$.

1.10 ***Theorem***: *Let $f \in L^1(\mathbf{R})$ and let $\{k_\lambda\}$ be a summability kernel on $\mathbf{R}$, then*

$$\lim_{\lambda\to\infty} \| f - k_\lambda * f \|_{L^1(\mathbf{R})} = 0.$$

PROOF: Repeat the proof of theorem I.2.3 and lemma I.2.4. ◄

1.11 Specifying theorem 1.10 to the Fejér kernel and using theorem 1.4, we obtain

Theorem: *Let $f \in L^1(\mathbf{R})$, then*

$$f = \lim_{\lambda\to\infty} \frac{1}{2\pi} \int_{-\lambda}^{\lambda} \left(1 - \frac{|\xi|}{\lambda}\right) \hat{f}(\xi)\, e^{i\xi x}\, d\xi \tag{1.10}$$

in the $L^1(\mathbf{R})$ norm.

Corollary (*the uniqueness theorem*): *Let $f \in L^1(\mathbf{R})$ and assume $\hat{f}(\xi) = 0$ for all $\xi \in \hat{\mathbf{R}}$; then $f = 0$.*

1.12 If it happens that $\hat{f}$ is Lebesgue integrable, the integral on the right-hand side of (1.10) converges, uniformly in x, to $1/2\pi \int \hat{f}(\xi) e^{i\xi x} d\xi$. We see that f is equivalent to a uniformly continuous function and obtain the so-called "inversion formula":

$$f(x) = \frac{1}{2\pi} \int \hat{f}(\xi) e^{i\xi x} d\xi. \tag{1.11}$$

An immediate consequence of (1.11) is

$$\widehat{\mathbf{K}_\lambda}(\xi) = \max\left(1 - \frac{|\xi|}{\lambda}, 0\right) \tag{1.12}$$

and, by theorem 1.3,

$$\widehat{\mathbf{K}_\lambda * f}(\xi) = \begin{cases} \left(1 - \dfrac{|\xi|}{\lambda}\right)\hat{f}(\xi), & |\xi| \leqq \lambda \\ 0, & |\xi| > \lambda. \end{cases} \tag{1.13}$$

Combining this with theorem 1.10, we obtain

Theorem: *The functions with compactly carried Fourier transforms form a dense subspace of* $L^1(\mathbf{R})$.

This theorem is analogous to the statement that trigonometric polynomials form a dense subspace of $L^1(\mathbf{T})$.

1.13 Besides the Fejér kernel we mention the following:

De la Vallée Poussin's kernel

$$\mathbf{V}_\lambda(x) = 2\mathbf{K}_{2\lambda}(x) - \mathbf{K}_\lambda(x), \tag{1.14}$$

whose Fourier transform is given by

$$\hat{\mathbf{V}}_\lambda(\xi) = \begin{cases} 1, & |\xi| \leqq \lambda \\ 2 - \dfrac{|\xi|}{\lambda}, & \lambda \leqq |\xi| \leqq 2\lambda \\ 0, & 2\lambda \leqq |\xi|. \end{cases} \tag{1.15}$$

Poisson's kernel

$$\mathbf{P}_\lambda(x) = \lambda \mathbf{P}(\lambda x),$$

where

$$\mathbf{P}(x) = \frac{1}{\pi(1 + x^2)} \tag{1.16}$$

and

(1.17) $$\hat{\mathbf{P}}(\xi) = e^{-|\xi|}$$

And finally Gauss' kernel

$$\mathbf{G}_\lambda(x) = \lambda \mathbf{G}(\lambda x),$$

where

(1.18) $$\mathbf{G}(x) = \pi^{-1/2} e^{-x^2}$$

and

(1.19) $$\hat{\mathbf{G}}(\xi) = e^{-\xi^2/4}.$$

To the inversion formula (1.11) and the summability in norm (theorems 1.10 and 1.11), one should add results about pointwise summability. Both the statements and the proofs of section I.3 can be adapted to $L^1(\mathbf{R})$ almost verbatim and we avoid the repetition.

1.14 As in chapter I, we can replace the $L^1(\mathbf{R})$ norm, in the statement of theorems 1.10 and 1.11, by the norm of any homogeneous Banach space $B \subset L^1(\mathbf{R})$. As in chapter I, a homogeneous Banach space is a space of functions which is invariant under translation and such that for every $f \in B$, f_y (defined by $f_y(x) = f(x - y)$) depends continuously on y. The assumption $B \subset L^1(\mathbf{R})$ is more restrictive than was the assumption $B \subset L^1(\mathbf{T})$ in chapter I; it excludes such natural spaces as $L^p(\mathbf{R})$, $p > 1$. We can obtain a reasonably general theory by considering homogeneous Banach space of locally summable functions, that is, functions which are Lebesgue integrable on every finite interval. We denote by $\mathscr{L}$ the space of all measurable functions f on $\mathbf{R}$ such that

$$\|f\|_{\mathscr{L}} = \sup_y \int_y^{y+1} |f(x)|\, dx < \infty$$

and by $\mathscr{L}_c$ the subspace of $\mathscr{L}$ consisting of all the functions f which satisfy

$$\|f_y - f\|_{\mathscr{L}} \to 0 \qquad \text{as } y \to 0.$$

Theorem: *If B is a homogeneous Banach space of locally summable functions on $\mathbf{R}$ and if convergence in B implies convergence in measure, then the $\mathscr{L}$ norm is majorized by the B norm and, in particular, $B \subset \mathscr{L}_c$.*

PROOF: If the $\mathscr{L}$ norm is not majorized by $\| \ \|_B$, we can choose a sequence $f_n \in B$ such that $\|f_n\|_B < 2^{-n}$ and $\|f_n\|_{\mathscr{L}} > 3^n$. Replacing f_n by $f_n(x - y_n)$ (if necessary) we may assume $\int_0^1 |f_n(x)|\,dx > 3^n$. Since $\|f_n\|_B \to 0$, f_n converges to zero in measure and it follows that if $n_j \to \infty$ fast enough $\sum f_{n_j}$, which belongs to B, is not integrable on $(0, 1)$. ◀

We can now extend theorem 1.10 to homogeneous Banach spaces of locally summable functions (see exercises 11–14 at the end of this section); theorem 1.11 can be generalized only after we extend the definition of the Fourier transformation.

1.15 We finish this section with a remark concerning the relation between Fourier coefficients and Fourier transforms.

Let $f \in L^1(\mathbf{R})$ and define φ by

$$\varphi(t) = 2\pi \sum_{j=-\infty}^{\infty} f(t + 2\pi j).$$

t is here a real number, but it is clear that $\varphi(t)$ depends only on $t \bmod 2\pi$ so that we can consider φ as defined on $\mathbf{T}$. φ is clearly summable on $\mathbf{T}$ and we have

$$\|\varphi\|_{L^1(\mathbf{T})} \leqq \|f\|_{L^1(\mathbf{R})}.$$

If n is a rational integer, we have

$$\hat{\varphi}(n) = \frac{1}{2\pi}\int_0^{2\pi} \varphi(t)\, e^{-int}\, dt = \sum_{j=-\infty}^{\infty} \int_0^{2\pi} f(t + 2\pi j)\, e^{-int}\, dt$$
$$= \int f(x)\, e^{-inx}\, dx = \hat{f}(n),$$

so that $\hat{\varphi}$ is simply the restriction to the integers of $\hat{f}$. Similarly, if we write $f_\lambda(x) = \lambda f(\lambda x)$ and:

$$\varphi_\lambda(t) = 2\pi \sum_{j=-\infty}^{\infty} f_\lambda(t + 2\pi j), \tag{1.20}$$

we obtain, using 1.1,

$$\hat{\varphi}_\lambda(n) = \hat{f}\left(\frac{n}{\lambda}\right). \tag{1.21}$$

The preceding remarks, as simple as they sound, link the theory of Fourier integrals to that of Fourier series, and we can obtain a great many facts about Fourier integrals from the corresponding facts about Fourier series. (For examples, see exercises 5 and 6 at the end of this section.)

An application to the above procedure is the very important *formula of Poisson*:

$$2\pi\lambda \sum_{n=-\infty}^{\infty} f(2\pi\lambda n) = \sum_{n=-\infty}^{\infty} \hat{f}\left(\frac{n}{\lambda}\right). \tag{1.22}$$

In order to establish Poisson's formula, to understand its meaning and its domain of validity, all that we need to do is simply rewrite it as

$$\varphi_\lambda(0) = \sum_{n=-\infty}^{\infty} \hat{\varphi}_\lambda(n). \tag{1.23}$$

If $\varphi_\lambda(0)$, as defined by (1.20), is well defined and if the Fourier series of φ_λ converges to $\varphi_\lambda(0)$ for $t = 0$, then (1.23) and (1.22) are valid. One enhances the generality of (1.22) considerably by interpreting the sum on the right as

$$\lim_{N\to\infty} \sum_{-N}^{N} \left(1 - \frac{|n|}{N}\right)\hat{f}\left(\frac{n}{\lambda}\right),$$

that is, using C-1 summability instead of summation. Using Féjer's theorem, for instance, one obtains that, with this interpretation, (1.22) is valid if $t = 0$ is a point of continuity of φ_λ. We remark that the continuity of f and $\hat{f}$ is not sufficient to imply (1.22) even if both sides of (1.22) converge absolutely (see exercise 15).

EXERCISES FOR SECTION 1

1. Perform the integration in (1.8).

2. Prove that $\dfrac{1}{2\pi}\displaystyle\int \left(\frac{\sin x/2}{x/2}\right)^2 dx = 1$ by contour integration.

3. Prove (1.17). *Hint*: Use contour integration.

4. Prove (1.19). *Hint*: Use 1.5 and 1.6 to show that $\hat{\mathbf{G}}(\xi)$ satisfies the equation $d/d\xi\, \hat{\mathbf{G}}(\xi) = -(\xi/2)\hat{\mathbf{G}}(\xi)$.

5. Let $f \in L^1(\mathbf{R})$ and let $\varphi_\lambda(t)$ be defined as in (1.20). Show that $\lim_{\lambda\to\infty} \| \varphi_\lambda \|_{L^1(\mathbf{T})} = \| f \|_{L^1(\mathbf{R})}$; hence deduce the uniqueness theorem from (1.21).

6. Prove theorem 1.7 using (1.21), the uniform continuity of Fourier transforms and the Riemann-Lebesgue lemma for Fourier coefficients.

7. Show that $A(\hat{\mathbf{R}})$ contains every twice continuously differentiable function with compact support on $\hat{\mathbf{R}}$. Deduce that $A(\hat{\mathbf{R}})$ is uniformly dense in $C_0(\hat{\mathbf{R}})$; however, show that $A(\hat{\mathbf{R}}) \neq C_0(\hat{\mathbf{R}})$.

8. Let $f \in L^1(\mathbf{R})$ be continuous at $x = 0$ and assume that $\hat{f}(\xi) \geq 0$, $\xi \in \hat{\mathbf{R}}$. Show that $\hat{f} \in L^1(\hat{\mathbf{R}})$ and $f(0) = 1/2\pi \int \hat{f}(\xi)\, d\xi$. *Hint*: Use the analog to Féjer's theorem and the fact that for positive functions C-1 summability is equivalent to convergence.

9. Show that $C_0 \cap L^1(\mathbf{R})$, with the norm $\| f \| = \sup_x |f(x)| + \| f \|_{L^1(\mathbf{R})}$, is a homogeneous Banach space on $\mathbf{R}$ and conclude that if $f \in C_0 \cap L^1(\mathbf{R})$ then $f(x) = \lim_{\lambda = \infty} 1/2\pi \int_{-\lambda}^{\lambda} (1 - |\xi|/\lambda)\, \hat{f}(\xi)\, e^{i\xi x}\, d\xi$ uniformly.

10. Let f be bounded and continuous on $\mathbf{R}$ and let $\{k_\lambda\}$ be a summability kernel. Show that $k_\lambda * f = \int k_\lambda(x - y) f(y)\, dy$ converges to f uniformly on compact sets on $\mathbf{R}$.

11. Let $f \in \mathscr{L}_c$ and let φ be continuous with compact support; write

$$\varphi * f = \int \varphi(y) f(x - y)\, dy .$$

Interpreting the integral above as an $\mathscr{L}_c$-valued integral, show that $\varphi * f \in \mathscr{L}_c$ and $\| \varphi * f \|_{\mathscr{L}} \leq \| \varphi \|_{L^1(\mathbf{R})} \| f \|_{\mathscr{L}}$. Use this to define $g * f$ for $g \in L^1(\mathbf{R})$ and $f \in \mathscr{L}_c$.

12. Show that if $f \in \mathscr{L}_c$ then $|f| \in \mathscr{L}_c$ (notice, however, that $\exp(ix \log |x|) \notin \mathscr{L}_c$) and, using exercise 11, prove that if $f \in \mathscr{L}_c$ and $g \in L^1(\mathbf{R})$ then for almost all $x \in \mathbf{R}$, $g(y) f(x - y) \in L^1(\mathbf{R})$ and $g * f$, as defined in exercise 11, is equal to $\int g(y) f(x - y)\, dy$.

13. Let $f \in \mathscr{L}$ and let $g \in L^1(\mathbf{R})$. Prove that for almost all $x \in \mathbf{R}$ $g(y) f(x - y) \in L^1(\mathbf{R})$ and that $h(x) = \int g(y) f(x - y)\, dy$ satisfies $\| h \|_{\mathscr{L}} \leq \| g \|_{L^1(\mathbf{R})} \| f \|_{\mathscr{L}}$.

14. Let $\{k_\lambda\}$ be a summability kernel in $L^1(\mathbf{R})$ and let $B \subseteq \mathscr{L}_c$ be a homogeneous Banach space. Show that for every $f \in B$, $\| k_\lambda * f - f \|_B \to 0$, and conclude that if $f \in B \cap L^1(\mathbf{R})$, $f = \lim_{\lambda \to \infty} \int_{-\lambda}^{\lambda} (1 - |\xi|/\lambda) \hat{f}(\xi)\, e^{i\xi x}\, d\xi$ in the B norm.

15. Construct a continuous function $f \in L^1(\mathbf{R})$ such that $\hat{f} \in L^1(\hat{\mathbf{R}})$, $f(2\pi n) = 0$ for all integers n, $\hat{f}(0) = 1$ and $\hat{f}(n) = 0$ for all integers $n \neq 0$. *Hints*:

(a) We denote $\| f \|_{A(\mathbf{R})} = 1/2\pi \int_{-\infty}^{\infty} |\hat{f}(\xi)|\, d\xi$. Let g be continuous with support in $[0, 2\pi]$ and such that $\hat{g} \in L^1(\hat{\mathbf{R}})$. Write

$$g_N(x) = \frac{1}{N + 1} \sum_{j=-N}^{N} \left(1 - \frac{|j|}{N + 1}\right) g(x - 2\pi j) .$$

Show that $\hat{g}_N(\xi) = (N+1)^{-1}\, \mathbf{K}_N(\xi)\, \hat{g}(\xi)$ where $\mathbf{K}_N$ is the 2π-periodic Fejér kernel, and deduce that $\| g_N \|_{A(\mathbf{R})} \to 0$ as $N \to \infty$.

(b) Let $g^{(j)}$ be nonnegative continuous functions such that $\widehat{g^{(j)}} \in L^1(\hat{\mathbf{R}})$ and such that $\sum_1^\infty g^{(j)}(x) = \begin{cases} 1 & 0 < x < 2\pi \\ 0 & \text{otherwise} \end{cases}$. Then if $N_j \to \infty$ fast enough $\| g_{Nj}^{(j)} \|_{A(\mathbf{R})} \leqq 2^{-j}$, and $f = \sum_{j=1}^\infty g_{Nj}^{(j)}$ has the desired properties.

2. FOURIER-STIELTJES TRANSFORMS

2.1 We denote by $M(\mathbf{R})$ the space of all finite Borel measures on $\mathbf{R}$. $M(\mathbf{R})$ is identified with the dual space of $C_0(\mathbf{R})$—the (supremum-normed) space of all continuous functions on $\mathbf{R}$ which vanish at infinity—by means of the coupling

$$\langle f, \mu \rangle = \int f\, \overline{d\mu} \qquad f \in C_0(\mathbf{R}),\ \mu \in M(\mathbf{R}). \tag{2.1}$$

The *total mass* norm on $M(\mathbf{R})$ is defined by $\| \mu \|_{M(\mathbf{R})} = \int |d\mu|$ and is identical to the "dual space" norm defined by means of (2.1). The mapping $f \to f(x)\,dx$ identifies $L^1(\mathbf{R})$ with a closed subspace of $M(\mathbf{R})$.

The convolution of a measure $\mu \in M(\mathbf{R})$ and a function $\varphi \in C_0(\mathbf{R})$ is defined by the integral

$$(\mu * \varphi)(x) = \int \varphi(x-y)\, d\mu(y), \tag{2.2}$$

and it is clear that $\mu * \varphi \in C_0(\mathbf{R})$ and that $\| \mu * \varphi \|_\infty \leqq \| \mu \|_{M(\mathbf{R})} \| \varphi \|_\infty$. The convolution of two measures, $\mu, \nu \in M(\mathbf{R})$, can be defined by means of duality and (2.2), analogously to what we have done in I.7, or directly by defining

$$(\mu * \nu)(E) = \int \mu(E-y)\, d\nu(y)$$

for every Borel set E. Whichever way we do it, we obtain easily that $\| \mu * \nu \|_{M(\mathbf{R})} \leqq \| \mu \|_{M(\mathbf{R})} \| \nu \|_{M(\mathbf{R})}$.

2.2 The Fourier-Stieltjes transform of a measure $\mu \in M(\mathbf{R})$ is defined by:

$$\hat{\mu}(\xi) = \overline{\int e^{i\xi x}\, \overline{d\mu(x)}} = \int e^{-i\xi x}\, d\mu(x) \qquad \xi \in \hat{\mathbf{R}}. \tag{2.3}$$

It is clear that if μ is absolutely continuous with respect to Lebesgue measure, say $d\mu = f(x)\,dx$, then $\hat{\mu}(\xi) = \hat{f}(\xi)$. Many of the properties of L^1 Fourier transforms are shared by Fourier-Stieltjes transforms: if $\mu, \nu \in M(\mathbf{R})$ then $|\hat{\mu}(\xi)| \leqq \|\mu\|_{M(\mathbf{R})}$, $\hat{\mu}$ is uniformly continuous, and $\widehat{\mu * \nu}(\xi) = \hat{\mu}(\xi)\,\hat{\nu}(\xi)$. A departure from the theory of L^1 Fourier transforms is the failing of the Riemann-Lebesgue lemma (the same way it fails for $M(\mathbf{T})$); the Fourier-Stieltjes transform of a measure μ need not vanish at infinity.

Theorem: (*Parseval's formula*):† *Let $\mu \in M(\mathbf{R})$ and let f be a continuous function in $L^1(\mathbf{R})$ such that $\hat{f} \in L^1(\hat{\mathbf{R}})$. Then*

$$\int f(x)\,d\mu(x) = \frac{1}{2\pi}\int \hat{f}(\xi)\,\hat{\mu}(-\xi)\,d\xi. \tag{2.4}$$

PROOF: By (1.11)

$$f(x) = \frac{1}{2\pi}\int \hat{f}(\xi)\,e^{i\xi x}\,d\xi;$$

hence

$$\int f(x)\,d\mu(x) = \frac{1}{2\pi}\iint \hat{f}(\xi)\,e^{i\xi x}\,d\mu(x)\,d\xi = \frac{1}{2\pi}\int \hat{f}(\xi)\,\hat{\mu}(-\xi)\,d\xi. \quad ◀$$

The assumption $\hat{f}(\xi) \in L^1(\hat{\mathbf{R}})$ justifies the change of order of integration (by Fubini's theorem); however, it is not really needed. Formula (2.4) is valid under the weaker assumption $\hat{f}(\xi)\hat{\mu}(-\xi) \in L^1(\hat{\mathbf{R}})$, and is valid for all bounded continuous $f \in L^1(\mathbf{R})$ if we replace the integral on the right by $\lim\limits_{\lambda\to\infty} \frac{1}{2\pi}\int_{-\lambda}^{\lambda}\left(1 - \frac{|\xi|}{\lambda}\right)\hat{f}(\xi)\hat{\mu}(-\xi)\,d\xi$ (cf. exercise 1.10).

2.3 The problem of characterizing Fourier-Stieltjes transforms among bounded and uniformly continuous functions on $\hat{\mathbf{R}}$ is very hard. As far as local behavior is concerned this is equivalent to characterizing $A(\hat{\mathbf{R}})$: every $\hat{f} \in A(\hat{\mathbf{R}})$ is a Fourier-Stieltjes transform, and on the other hand, if $\mu \in M(\mathbf{R})$ and $\mathbf{V}_\lambda$ is de la Vallee Poussin's kernel (1.14), then $\mu * \mathbf{V}_\lambda \in L^1(\mathbf{R})$ and $\widehat{\mu * \mathbf{V}_\lambda}(\xi) = \hat{\mu}(\xi)$ for $|\xi| \leqq \lambda$.

The following theorem is analogous to I.7.3:

Theorem: *Let φ be continuous on $\hat{\mathbf{R}}$, define Φ_λ by*:

$$\Phi_\lambda(x) = \frac{1}{2\pi}\int_{-\lambda}^{\lambda}\left(1 - \frac{|\xi|}{\lambda}\right)\varphi(\xi)\,e^{i\xi x}\,d\xi.$$

† Notice that (2.4) is equivalent to $\int f(x)\,\overline{d\mu(x)} = 1/2\pi \int \hat{f}(\xi)\overline{\hat{\mu}(\xi)}\,d\xi$.

Then φ is a Fourier-Stieltjes transform if, and only if, $\Phi_\lambda \in L^1(\mathbf{R})$ for all $\lambda > 0$, and $\| \Phi_\lambda \|_{L^1(\mathbf{R})}$ is bounded as $\lambda \to \infty$.

PROOF: If $\varphi = \hat{\mu}$ with $\mu \in M(\mathbf{R})$, then $\Phi_\lambda = \mu * \mathbf{K}_\lambda$, $\mathbf{K}_\lambda$ denoting Fejér's kernel. It follows that for all $\lambda > 0$, $\Phi_\lambda \in L^1(\mathbf{R})$ and $\| \Phi_\lambda \|_{L^1} \leqq \| \mu \|_{M(\mathbf{R})}$.

If we assume that $\Phi_\lambda \in L^1(\mathbf{R})$ with uniformly bounded norms, we consider the measures $\Phi_\lambda \, dx$ and denote by μ a weak-star limit point of $\Phi_\lambda \, dx$ as $\lambda \to \infty$. We claim that $\varphi = \hat{\mu}$ and since both functions are continuous, this will follow if we show that

$$\int \varphi(-\xi) g(\xi) \, d\xi = \int \hat{\mu}(-\xi) \, g(\xi) \, d\xi \tag{2.5}$$

for every twice continuously differentiable g with compact support. For such g we have $g = \hat{G}$ with $G \in L^1 \cap C_0(\mathbf{R})$; by Parseval's formula

$$\begin{aligned} \int g(\xi)\varphi(-\xi) \, d\xi &= \lim_{\lambda \to \infty} \int_{-\lambda}^{\lambda} g(\xi)\varphi(-\xi)\left(1 - \frac{|\xi|}{\lambda}\right) d\xi \\ &= \lim_{\lambda \to \infty} 2\pi \int G(x)\Phi_\lambda(x) \, dx = 2\pi \int G(x) \, d\mu(x) \\ &= \int g(\xi)\hat{\mu}(-\xi) \, d\xi \end{aligned}$$

and the proof is complete. ◀

Remark: The application of Parseval's formula above is typical and is the, more or less, standard way to check that weak-star limits in $M(\mathbf{R})$ are what we expect them to be. Nothing like that was needed in the case of $M(\mathbf{T})$ since weak-star convergence in $M(\mathbf{T})$ implies pointwise convergence of the Fourier-Stieltjes coefficients (the exponentials belong to $C(\mathbf{T})$ of which $M(\mathbf{T})$ is the dual). The exponentials on $\mathbf{R}$ do not belong to $C_0(\mathbf{R})$ and it is false that weak-star convergence in $M(\mathbf{R})$ implies pointwise convergence of the Fourier-Stieltjes transforms (cf. exercise 1 at the end of this section.) However, the argument above gives:

Lemma: *Let $\mu_n \in M(\mathbf{R})$ and assume that $\mu_n \to \mu$ in the weak-star topology. Assume also that $\hat{\mu}_n(\xi) \to \varphi(\xi)$ pointwise, φ being continuous on $\hat{\mathbf{R}}$. Then $\hat{\mu} = \varphi$.*

2.4 A similar application of Parseval's formula gives the following useful criterion:

Theorem: *A function φ, defined and continuous on $\hat{\mathbf{R}}$, is a Fourier-Stieltjes transform if, and only if, there exists a constant C such that*

$$\left| \frac{1}{2\pi} \int \hat{f}(\xi)\varphi(-\xi)\, d\xi \right| \leqq C \sup_x |f(x)| \tag{2.6}$$

for every continuous $f \in L^1(\mathbf{R})$ such that $\hat{f}$ has compact support.

PROOF: If $\varphi = \hat{\mu}$, (2.6) follows from (2.4) with $C = \| \mu \|_{M(\mathbf{R})}$. If (2.6) holds, $f \to 1/2\pi \int \hat{f}(\xi)\varphi(-\xi)\, d\xi$ defines a bounded linear functional on a dense subspace of $C_0(\mathbf{R})$, namely on the space of all the functions $f \in C_0 \cap L^1(\mathbf{R})$ such that $\hat{f}$ has a compact support. This functional has a unique bounded extension to $C_0(\mathbf{R})$, which, by the Riesz representation theorem, has the form $f \to \int f(x)\, d\mu(x)$. Moreover, $\| \mu \|_{M(\mathbf{R})} \leqq C$. Using (2.4) again we see that $\hat{\mu} - \varphi$ is orthogonal to all the continuous, compactly supported functions $\hat{f}$ with $f \in L^1(\mathbf{R})$, and consequently $\varphi = \hat{\mu}$. ◀

Remark: The family $\{f\}$ of test functions for which (2.6) should be valid can be taken in many ways. The only properties that have been used are that $\{f\}$ is dense in $C_0(\mathbf{R})$ and $\{\hat{f}\}$ is dense in $C_0(\hat{\mathbf{R}})$. Thus we could require the validity of (2.6) only for (a) functions f such that $\hat{f}$ is infinitely differentiable with compact support; or (b) functions f which are themselves infinitely differentiable with compact support, and so on.

2.5 With measures on $\mathbf{R}$ we can associate measures on $\mathbf{T}$ simply by integrating 2π-periodic functions. Formally: if E is a Borel set on $\mathbf{T}$ ($\mathbf{T}$ being identified with $(-\pi, \pi]$) we denote by E_n the set $E + 2\pi n$ and write $\tilde{E} = \bigcup E_n$; if $\mu \in M(\mathbf{R})$ we define

$$\mu_{\mathbf{T}}(E) = \mu(\tilde{E}).$$

It is clear that $\mu_{\mathbf{T}}$ is a measure on $\mathbf{T}$ and that, identifying continuous functions on $\mathbf{T}$ with 2π-periodic functions on $\mathbf{R}$,

$$\int_{\mathbf{R}} f(x)\, d\mu = \int_{\mathbf{T}} f(t)\, d\mu_{\mathbf{T}}. \tag{2.7}$$

The mapping $\mu \to \mu_{\mathbf{T}}$ is an operator of norm one from $M(\mathbf{R})$ onto $M(\mathbf{T})$, and its restriction to $L^1(\mathbf{R})$ is the mapping that we have discussed in section 1.15. It follows from (2.7) that $\hat{\mu}(n) = \hat{\mu}_{\mathbf{T}}(n)$ for all n;

thus the restriction of a Fourier-Stieltjes transform to the integers gives a sequence of Fourier-Stieltjes coefficients.

Theorem: *A function φ, defined and continuous on $\hat{\mathbf{R}}$, is a Fourier-Stieltjes transform if, and only if, there exists a constant $C > 0$, such that for all $\lambda > 0$, $\{\varphi(\lambda n)\}_{n=-\infty}^{\infty}$ are the Fourier-Stieltjes coefficients of a measure of norm $\leqq C$ on $\mathbf{T}$.*

PROOF: If $\varphi = \hat{\mu}$ with $\mu \in M(\mathbf{R})$ we have $\varphi(n) = \hat{\mu}(n) = \hat{\mu}_{\mathbf{T}}(n)$ with $\| \mu_{\mathbf{T}} \| \leqq \| \mu \|$. Writing $d\mu(x/\lambda)$ for the measure satisfying

$$\int f(x)\, d\mu\left(\frac{x}{\lambda}\right) = \int f(\lambda x)\, d\mu(x)$$

we have $\| \mu(x/\lambda) \|_{M(\mathbf{R})} = \| \mu \|_{M(\mathbf{R})}$ and $\widehat{\mu(x/\lambda)}(\xi) = \hat{\mu}(\xi\lambda)$ so that $\varphi(\lambda n) = \widehat{\mu(x/\lambda)_{\mathbf{T}}}(n)$ and the "only if" part is established.

In order to establish the converse we use 2.4. Let f be continuous and integrable on $\mathbf{R}$ and assume that $\hat{f}$ is infinitely differentiable and compactly supported. We want to estimate the integral $1/2\pi \int \hat{f}(\xi)\hat{\varphi}(-\xi)\, d\xi$ and, since the integrand is continuous and compactly supported, we can approximate the integral by its Riemann sums. Thus, for arbitrary $\varepsilon > 0$, if λ is small enough:

$$\left| \frac{1}{2\pi} \int \hat{f}(\xi)\varphi(-\xi)\, d\xi \right| < \left| \frac{\lambda}{2\pi} \sum \hat{f}(\lambda n)\, \varphi(-\lambda n) \right| + \varepsilon. \tag{2.8}$$

Now, $(\lambda/2\pi)\hat{f}(\lambda n)$ are the Fourier coefficients of the function $\psi_\lambda(t) = \sum_{m=-\infty}^{\infty} f((t + 2\pi m)/\lambda)$ on $\mathbf{T}$, and since the infinite differentiability of $\hat{f}$ implies a very fast decrease of $f(x)$ as $|x| \to \infty$, we see that if λ is sufficiently small

$$\sup |\psi_\lambda(t)| \leqq \sup |f(x)| + \varepsilon. \tag{2.9}$$

Assuming that $\varphi(\lambda n) = \hat{\mu}_\lambda(n)$, $\mu_\lambda \in M(\mathbf{T})$ and $\| \mu_\lambda \|_{M(\mathbf{T})} \leqq C$, we obtain from Parseval's formula

$$\left| \frac{\lambda}{2\pi} \sum \hat{f}(\lambda n)\varphi(-\lambda n) \right| = \left| \sum \hat{\psi}_\lambda(n)\hat{\mu}_\lambda(-n) \right| \leqq C \sup |\psi_\lambda(t)|;$$

by (2.8) and (2.9)

$$\left| \frac{1}{2\pi} \int \hat{f}(\xi)\varphi(-\xi)\, d\xi \right| \leqq C \sup |f(x)| + (C + 1)\, \varepsilon$$

and since $\varepsilon > 0$ is arbitrary, (2.6) is satisfied and the theorem follows from theorem 2.4. ◀

2.6 Parseval's formula also offers an obvious criterion for determining when a function φ is the Fourier-Stieltjes transform of a positive measure. The analog to 2.4 is

Theorem: *A function φ, bounded and continuous on $\hat{\mathbf{R}}$, is the Fourier-Stieltjes transform of a positive measure on* $\mathbf{R}$ *if, and only if,*

$$\int \hat{f}(\xi)\varphi(-\xi)\,d\xi \geqq 0 \tag{2.10}$$

for every nonnegative function f which is infinitely differentiable and compactly supported.

PROOF: Parseval's formula clearly implies the "only if" part and also the fact that if we assume $\varphi = \hat{\mu}$ with $\mu \in M(\mathbf{R})$, then μ is a positive measure. In order to complete the proof we show that (2.10) implies (2.6), with $C = \varphi(0)$, for every real-valued, compactly supported infinitely differentiable f (and consequently with $C = 2\varphi(0)$ for complex-valued f).

As usual, we denote by $\mathbf{K}_\lambda(x)$ the Féjer kernel (1.8) and notice that $\lambda^{-1}\mathbf{K}_\lambda(x) = \mathbf{K}(\lambda x) = \dfrac{1}{2\pi}\left(\dfrac{\sin \lambda x/2}{\lambda x/2}\right)^2$ is nonnegative and tends to $1/2\pi$, as $\lambda \to 0$, uniformly on compact subsets of $\mathbf{R}$. By (1.12) the Fourier transform of $\mathbf{K}(\lambda x)$ is $\lambda^{-1}\max(1 - |\xi|/\lambda, 0)$ and consequently when $\lambda \to 0$ (using the continuity of $\varphi(\xi)$ at $\xi = 0$)

$$\int \frac{1}{\lambda}\hat{\mathbf{K}}_\lambda(\xi)\,\varphi(-\xi)\,d\xi \to \varphi(0). \tag{2.11}$$

If f is real-valued and compactly supported and $\varepsilon > 0$, then, for sufficiently small λ and all x,

$$2\pi(\varepsilon + \sup|f|)\,\mathbf{K}(\lambda x) - f(x) \geqq 0;$$

hence, by (2.10) and (2.11), if $\hat{f} \in L^1(\hat{\mathbf{R}})$

$$\frac{1}{2\pi}\int \hat{f}(\xi)\varphi(-\xi)\,d\xi \leqq \varphi(0)(2\varepsilon + \sup|f|), \tag{2.12}$$

rewriting (2.12) for $-f$ and letting $\varepsilon \to 0$ we obtain:

$$\left|\frac{1}{2\pi}\int \hat{f}(\xi)\,\varphi(-\xi)\,d\xi\right| \leqq \varphi(0)\sup|f|. \qquad ◀$$

Remark: The assumption that φ is bounded is superfluous (cf. exercise 6 at the end of this section).

2.7 The analog to 2.5 is:

Theorem: *A function φ, defined and continuous on $\hat{\mathbf{R}}$, is the Fourier-Stieltjes transform of a positive measure, if and only if, for all $\lambda > 0$, $\{\varphi(\lambda n)\}_{n=-\infty}^{\infty}$ are the Fourier-Stieltjes coefficients of a positive measure on $\mathbf{T}$.*

PROOF: The "only if" part follows as in 2.5. For the "if" part we notice first that if $\varphi(\lambda n) = \hat{\mu}_\lambda(n)$ with $\mu_\lambda \geqq 0$ on $\mathbf{T}$, then $\| \mu_\lambda \| = \varphi(0)$ and consequently, by 2.5, φ is a Fourier-Stieltjes transform. Using the continuity of φ, we can now establish (2.10) by approximating the integral by its Riemann sums as in the proof of 2.5. ◀

2.8 DEFINITION: A function φ defined on $\hat{\mathbf{R}}$ is said to be *positive definite* if, for every choice of $\xi_1, \dots, \xi_N \in \hat{\mathbf{R}}$ and complex numbers $z_1, \dots, z_N$, we have

$$\sum_{j,k=1}^{N} \varphi(\xi_j - \xi_k) z_j \bar{z}_k \geqq 0. \tag{2.13}$$

Immediate consequences of (2.13) are:

$$\varphi(-\xi) = \overline{\varphi(\xi)} \tag{2.14}$$

and

$$|\varphi(\xi)| \leqq \varphi(0). \tag{2.15}$$

In order to prove (2.14) and (2.15), we take $N = 2$, $\xi_1 = 0$, $\xi_2 = \xi$, $z_1 = 1$, $z_2 = z$; then (2.13) reads

$$\varphi(0)(1 + |z|^2) + \varphi(\xi)z + \varphi(-\xi)\bar{z} \geqq 0;$$

taking $z = 1$, we get $\varphi(\xi) + \varphi(-\xi)$ real and taking $z = i$, we get $i(\varphi(\xi) - \varphi(-\xi))$ real, hence (2.14). If we take z such that $z\varphi(\xi) = -|\varphi(\xi)|$ we obtain:

$$2\varphi(0) - 2|\varphi(\xi)| \geqq 0$$

which establishes (2.15).

Theorem (*Bochner*): *A function φ defined on $\hat{\mathbf{R}}$ is a Fourier-Stieltjes transform of a positive measure if, and only if, it is positive definite and continuous.*

PROOF: Assume first $\varphi = \hat{\mu}$ with $\mu \geqq 0$. Let $\xi_1, \ldots, \xi_N \in \hat{\mathbf{R}}$ and $z_1, \ldots, z_N$ be complex numbers; then

$$\sum_{j,k} \varphi(\xi_j - \xi_k) z_j \overline{z_k} = \int \sum e^{-i\xi_j x} z_j e^{i\xi_k x} \overline{z_k}\, d\mu(x)$$

(2.16)

$$= \int \left| \sum_1^N z_j e^{-i\xi_j x} \right|^2 d\mu(x) \geqq 0$$

so that Fourier-Stieltjes transforms of positive measures are positive definite.

If, on the other hand, we assume that φ is positive definite, it follows that for all $\lambda > 0$, $\{\varphi(\lambda n)\}$ is a positive definite sequence (cf. I.7.6)· By Herglotz' theorem I.7.6, $\varphi(\lambda n) = \hat{\mu}_\lambda(n)$ for some positive measure μ_λ on $\mathbf{T}$, and by theorem 2.7, $\varphi = \hat{\mu}$ for some positive $\mu \in M(\mathbf{R})$. ◄

2.9 For the convenience of future reference we state here the analog to Wiener's theorem I.7.11. The theorem can be proved either by essentially repeating the proof of I.7.11 or by reducing it to I.7.11. We leave the proof as an exercise (exercise 7 at the end of this section) to the reader.

Theorem: *Let* $\mu \in M(\mathbf{R})$. *Then*

$$\sum |\mu(\{x\})|^2 = \lim_{\lambda\to\infty} \frac{1}{2\lambda} \int_{-\lambda}^{\lambda} |\hat{\mu}(\xi)|^2 \, d\xi .$$

In particular, a necessary and sufficient condition for the continuity of μ *is*

$$\lim_{\lambda\to\infty} \frac{1}{2\lambda} \int_{-\lambda}^{\lambda} |\hat{\mu}(\xi)|^2 \, d\xi = 0 .$$

EXERCISES FOR SECTION 2

1. Denote by δ_n the measure of mass one on $\mathbf{R}$ concentrated at $x = n$. Show that $\lim_{n\to\infty} \delta_n = 0$ in the weak-star topology of $M(\mathbf{R})$ and conclude that weak-star convergence of a sequence of measures *does not* imply pointwise convergence of the Fourier-Stieltjes transforms.

2. Writing $\mu_n = n^{-1}(\delta_1 + \delta_2 + \cdots + \delta_n)$ show that $\mu_n \to 0$ in the weak-star topology and $\hat{\mu}_n(\xi)$ converges for every $\xi \in \hat{\mathbf{R}}$; however, $\lim \hat{\mu}_n(\xi)$ is not identically zero.

3. Let $\mu_n \in M(\mathbf{R})$ and assume $\lim_{\lambda\to\infty}\sup_n \int_{|x|>\lambda} |d\mu_n| = 0$. Show that if $\mu_n \to \mu$ in the weak-star topology, $\hat{\mu}_n(\xi) \to \hat{\mu}(\xi)$ uniformly on compact subsets of $\hat{\mathbf{R}}$.

4. Let $\mu_n \in M(\mathbf{R})$ such that $\| \mu_n \| \leqq 1$. Assume that $\hat{\mu}_n$ converges pointwise to a continuous function φ. Show that $\varphi = \hat{\mu}$ for some $\mu \in M(\mathbf{R})$ such that $\| \mu \| \leqq 1$.

5. Show that there exists a uniformly continuous bounded function φ which is not a Fourier-Stieltjes transform, and such that $\{\varphi(\lambda n)\}$ is a sequence of Fourier-Stieltjes coefficients for every $\lambda > 0$. *Hint*: Construct a continuous function with compact support which is not a Fourier transform of a summable function.

6. Show that if φ is continuous on $\hat{\mathbf{R}}$ and (2.10) is valid, then φ is positive definite. Conclude from (2.15) that the boundedness assumption of 2.6 is superfluous.

7. Prove theorem 2.9.

8. Express $\mu\{[a, b]\}$ and $\mu\{(a, b)\}$ in terms of $\hat{\mu}$. ($[a, b]$ is the closed interval with endpoints a and b, and (a, b) is the open one.)

3. FOURIER TRANSFORMS IN $L^p(\mathbf{R})$; $1 < p \leqq 2$

The definition of the Fourier transform (i.e., Fourier coefficients) for functions in various function spaces on $\mathbf{T}$, was largely simplified by the fact that all these spaces were contained in $L^1(\mathbf{T})$. The fact that the Lebesgue measure of $\mathbf{R}$ is infinite changes the situation radically. If $p > 1$ we no longer have $L^p \subset L^1$, and, if we want to have Fourier transforms for functions in $L^p(\mathbf{R})$ (or other function spaces on $\mathbf{R}$), we have to find a new way to define them. In this section we consider the case $1 < p \leqq 2$ and obtain a reasonably satisfactory extension of the Fourier transformation for this case.

3.1 We start with $L^2(\mathbf{R})$.

Lemma: *Let f be continuous and with compact support on* $\mathbf{R}$; *then*

$$\frac{1}{2\pi}\int |\hat{f}(\xi)|^2\, d\xi = \int |f(x)|^2\, dx\,.$$

We give two proofs.

PROOF I: Assume first that the support of f is included in $(-\pi, \pi)$. By theorem I.5.5,

$$\frac{1}{2\pi}\int |f(x)|^2\, dx = \sum_{n=-\infty}^{\infty} \left| \frac{1}{2\pi}\hat{f}(n) \right|^2$$

and replacing f by $e^{-i\alpha x}f$ we have

$$\int |f(x)|^2\,dx = \frac{1}{2\pi}\sum_{n=-\infty}^{\infty} |\hat{f}(n+\alpha)|^2; \tag{3.1}$$

integrating both sides of (3.1) with respect to α on $0 \leqq \alpha < 1$, we obtain

$$\int |f(x)|^2\,dx = \frac{1}{2\pi}\int |\hat{f}(\xi)|^2\,d\xi.$$

If the support of f is not included in $(-\pi,\pi)$, we consider $g(x) = \lambda^{1/2}f(\lambda x)$. If λ is sufficiently big, the support of g is included in $(-\pi,\pi)$ and, since $\hat{g}(\xi) = \lambda^{-1/2}\hat{f}(\xi/\lambda)$, we obtain

$$\int |f(x)|^2\,dx = \int |g(x)|^2\,dx = \frac{1}{2\pi}\int |\hat{g}(\xi)|^2\,d\xi = \frac{1}{2\pi}\int |\hat{f}(\xi)|^2\,d\xi.$$

◂

PROOF II: Write $g = f * \overline{f(-x)}$; we have $g(0) = \int f(x)\bar{f}(x)\,dx = \int |f(x)|^2\,dx$ and $\hat{g}(\xi) = |\hat{f}(\xi)|^2$. If we know that $\int |\hat{f}(\xi)|^2\,d\xi < \infty$ (e.g., if we assume that f is differentiable), it follows from the inversion formula (1.11) that

$$\frac{1}{2\pi}\int |\hat{f}(\xi)|^2\,d\xi = g(0) = \int |f(x)|^2\,dx.$$

In the general case we may apply Féjer's theorem and obtain

$$\lim_{\lambda\to\infty}\frac{1}{2\pi}\int_{-\lambda}^{\lambda}\left(1 - \frac{|\xi|}{\lambda}\right)|\hat{f}(\xi)|^2\,d\xi = g(0)$$

and, since the integrand is nonnegative, its C-1 summability is equivalent to its convergence and the proof is complete. ◂

DEFINITION: For $g \in L^2(\hat{\mathbf{R}})$ we write

$$\|g\|_{L^2(\hat{\mathbf{R}})} = \left(\frac{1}{2\pi}\int |g(\xi)|^2\,d\xi\right)^{1/2}.$$

Theorem *(Plancherel): There exists a unique operator $\mathcal{F}$ from $L^2(\mathbf{R})$ onto $L^2(\hat{\mathbf{R}})$ having the properties:*

$$\mathcal{F}f = \hat{f} \qquad \text{for } f \in L^1 \cap L^2(\mathbf{R}), \tag{3.2}$$

$$\|\mathcal{F}f\|_{L^2(\hat{\mathbf{R}})} = \|f\|_{L^2(\mathbf{R})}. \tag{3.3}$$

Remark: In view of (3.2) we shall often write $\hat{f}$ instead of $\mathcal{F}f$.

PROOF: We notice first that $L^1 \cap L^2(\mathbf{R})$ is dense in $L^2(\mathbf{R})$ and consequently any continuous operator defined on $L^2(\mathbf{R})$ is determined by its values on $L^1 \cap L^2(\mathbf{R})$. This shows that there exists at most one operator satisfying (3.2) and (3.3). By the lemma, (3.3) is satisfied if f is continuous with compact support, and since continuous functions with compact support are dense in $L^1 \cap L^2(\mathbf{R})$ (with respect to the norm $\| \ \|_{L^1(\mathbf{R})} + \| \ \|_{L^2(\mathbf{R})}$), (3.3) holds for all $f \in L^1 \cap L^2(\mathbf{R})$. The mapping $f \to \hat{f}$ clearly can be extended by continuity to an isometry from $L^2(\mathbf{R})$ into $L^2(\hat{\mathbf{R}})$. Finally, since every twice differentiable compactly supported function on $\hat{\mathbf{R}}$ is the Fourier transform of a bounded integrable function on $\mathbf{R}$ (1.5 and the inversion formula), it follows that the range of $f \to \hat{f}$ is dense in $L^2(\hat{\mathbf{R}})$ and hence coincides with it. ◀

Remarks: (a) Given a function $f \in L^2(\mathbf{R})$ we define $\hat{f}$ as the limit (in $L^2(\hat{\mathbf{R}})$) of $\hat{f}_n$, where f_n is any sequence in $L^1 \cap L^2(\mathbf{R})$ which converges to f in $L^2(\mathbf{R})$. As such a sequence we can take

$$f_n(x) = \begin{cases} f(x) & |x| < n \\ 0 & |x| \geqq n \end{cases}$$

and obtain the following form of Plancherel's theorem: the sequence

$$\hat{f}_n(\xi) = \int_{-n}^{n} f(x)\, e^{-i\xi x}\, dx \tag{3.4}$$

converges, in $L^2(\hat{\mathbf{R}})$, to a function which we denote by $\hat{f}$, and for which (3.2) and (3.3) are valid.

(b) The mapping $f \to \hat{f}$ being an isometry of $L^2(\mathbf{R})$ onto $L^2(\hat{\mathbf{R}})$, clearly has an inverse. Using theorem 1.11 and the fact that we have an isometry, we obtain the inverse map by $f = \lim f_{(n)}$ in $L^2(\mathbf{R})$ where

$$f_{(n)}(x) = \frac{1}{2\pi} \int_{-n}^{n} \hat{f}(\xi)\, e^{ix\xi}\, d\xi \tag{3.5}$$

(c) Parseval's formula

$$\int f(x)\overline{g(x)}\, dx = \frac{1}{2\pi} \int \hat{f}(\xi)\overline{\hat{g}(\xi)} d\xi \tag{3.6}$$

for $f, g \in L^2(\mathbf{R})$, follows immediately from (3.3) (and in fact is equivalent to it).

3.2 We turn now to define Fourier transforms for functions in $L^p(\mathbf{R})$, $1 < p < 2$. Using the Riesz-Thorin theorem and the fact that $\mathcal{F}: f \to \hat{f}$

has norm 1 as operator from $L^1(\mathbf{R})$ into $L^\infty(\hat{\mathbf{R}})$ and from $L^2(\mathbf{R})$ onto $L^2(\hat{\mathbf{R}})$, we obtain as in IV.2:

Theorem: (*Hausdorff-Young*): *Let* $1 < p < 2$, $q = p/(p-1)$ *and* $f \in L^1 \cap L^2(\mathbf{R})$. *Then*

$$\left(\frac{1}{2\pi}\int |\hat{f}(\xi)|^q d\xi\right)^{1/q} \leqq \left(\int |f(x)|^p dx\right)^{1/p}.$$

For $f \in L^p(\mathbf{R})$, $1 < p < 2$, we now define $\hat{f}$ by continuity; for example, as the limit in $L^q(\hat{\mathbf{R}})$ of $\int_{-n}^{n} e^{-i\xi x} f(x)\,dx$. The mapping $\mathcal{F}: f \to \hat{f}$ so defined is an operator of norm 1 from $L^p(\mathbf{R})$ into $L^q(\hat{\mathbf{R}})$; however, it is no longer an isometry and the range is not the whole of $L^q(\hat{\mathbf{R}})$. (see exercise 10 at the end of this section).

3.3 The fact that for $p < 2$, $\mathcal{F}$ is not an invertible operator from $L^p(\mathbf{R})$ onto $L^q(\mathbf{R})$ makes the inversion problem more delicate than it is for L^2. The situation in the case of $L^p(\mathbf{R})$ is similar to that which we encountered for $L^p(\mathbf{T})$. We have inversion formulas both in terms of summability and in terms of convergence. The summability result can be stated in terms of general summability kernels without reference to the Fourier transform as we did in 1.10 for $L^1(\mathbf{R})$; and in fact the statement of theorem 1.10 remains valid if we replace in it $L^1(\mathbf{R})$ by $L^p(\mathbf{R})$, $1 \leqq p < \infty$. For $p \leqq 2$ we can generalize theorem 1.11. We first check (see exercise 9 at the end of this section) that if $g \in L^p(\mathbf{R})$ and $f \in L^1(\mathbf{R})$ then $f * g$ is a well-defined element in $L^p(\mathbf{R})$ and $\widehat{f * g} = \hat{f}\hat{g}$. This is particularly simple if we take for f the Féjer kernel $\mathbf{K}_\lambda$: we have

$$\widehat{\mathbf{K}_\lambda * g} = \left(1 - \frac{|\xi|}{\lambda}\right)\hat{g}$$

and, since $\mathbf{K}_\lambda * g$ is clearly bounded ($\mathbf{K}_\lambda \in L^q(\mathbf{R})$, $q = p/(p-1)$) and hence belongs to $L^p \cap L^\infty(\mathbf{R}) \subset L^2(\mathbf{R})$, it follows that

$$\mathbf{K}_\lambda * g = \frac{1}{2\pi}\int_{-\lambda}^{\lambda}\left(1 - \frac{|\xi|}{\lambda}\right)\hat{g}(\xi)\,e^{i\xi x}\,d\xi$$

and from the general form of theorem 1.10 we obtain:

Theorem: *Let* $g \in L^p(\mathbf{R})$, $1 \leqq p \leqq 2$; then

$$g = \lim_{\lambda\to\infty}\frac{1}{2\pi}\int_{-\lambda}^{\lambda}\left(1 - \frac{|\xi|}{\lambda}\right)\hat{g}(\xi)\,e^{i\xi x}\,d\xi$$

in the $L^p(\mathbf{R})$ *norm.*

Corollary: *The functions with compactly supported Fourier transforms form a dense subspace of* $L^p(\mathbf{R})$.

***3.4** The analog to the inversion given by 3.1, remark (b) (i.e., convergence rather than summability) is valid for $1 < p < 2$ but not as easy to prove as for $p = 2$; it corresponds to theorem II.1.5 and can be proved either through the study of conjugate harmonic functions in the half-plane, analogous to that done for the disc in chapter III, or directly from II.1.5. The idea needed in order to obtain the norm inversion formula for $L^p(\mathbf{R})$, $1 < p < 2$, from II.1.5 is basically the one we have used in proof I of lemma 3.1.

For $f \in \bigcup_{1 \leqq p \leqq 2} L^p(\mathbf{R})$ we write $S_N(f, x) = \frac{1}{2\pi}\int_{-N}^{N} \hat{f}(\xi)\, e^{i\xi x}\, d\xi$.

Lemma: *For $1 < p < \infty$, there exist constants C_p such that*

$$\| S_N(f) \|_{L^p(\mathbf{R})} \leqq C_p \| f \|_{L^p(\mathbf{R})} \tag{3.7}$$

for every function f with compact support and every $N > 0$.

PROOF: Inequality (3.7) is equivalent to the statement that, for $M \to \infty$

$$\left(\int_{-M}^{M} | S_N(f, x) |^p dx\right)^{1/p} \leqq C_p \| f \|_{L^p(\mathbf{R})}. \tag{3.8}$$

Writing $\varphi_M(x) = M^{1/p} f(Mx)$ we see that $\| \varphi_M \|_{L^p(\mathbf{R})} = \| f \|_{L^p(\mathbf{R})}$ and check that

$$S_{MN}(\varphi_M, x) = M^{1/p} S_N(f, Mx). \tag{3.9}$$

In view of (3.9), (3.8) is equivalent to

$$\left(\int_{-1}^{1} | S_{MN}(\varphi_M, x) |^p dx\right)^{1/p} \leqq C_p \| \varphi_M \|_{L^p(\mathbf{R})}. \tag{3.10}$$

As $M \to \infty$, the support of φ_M shrinks to zero and consequently the lemma will be proved if we show that (3.8) is valid, with an appropriate C_p, for all f with support contained in $(-\pi, \pi)$ for $M = 1$ (or any other fixed positive number) and for all integers N.

We now write

$$S_N(f, x) = \int_0^1 \sum_{-N}^{N-1} \frac{1}{2\pi} \hat{f}(n + \alpha)\, e^{i(n+\alpha)x}\, d\alpha \tag{3.11}$$

and notice that $\sum_{-N}^{N-1}(1/2\pi)\hat{f}(n+\alpha)e^{inx}$ is a partial sum of the Fourier series of $f(x)e^{-i\alpha x}$ (it is carried by $(-\pi,\pi)$ which we now identify with $\mathbf{T}$). As an $L^p(\mathbf{T})$-valued function of α, the integrand in (3.11) is clearly continuous† and, by II.1.2 and II.1.5, it is bounded in $L^p(\mathbf{T})$ by a constant multiple of $\| fe^{i\alpha x}\|_{L^p(\mathbf{T})} = (2\pi)^{-1/p}\| f\|_{L^p(\mathbf{R})}$. We therefore obtain

$$\left(\int_{-1}^{1} |S_N(f,x)|^p dx\right)^{1/p} \leqq \left(\int_{-\pi}^{\pi} |S_N(f,x)|^p dx\right)^{1/p} \leqq C_p \| f\|_{L^p(\mathbf{R})}$$

and the proof is complete. ◀

Corollary: *For* $1 < p \leqq 2$, *inequality* (3.7) *is valid for all* $f \in L^p(\mathbf{R})$.

PROOF: Write $f = \lim_{n\to\infty} f_n$ with $f_n \in L^p(\mathbf{R})$, f_n having compact supports and the limit being taken in the $L^p(\mathbf{R})$ norm. By theorem 3.2, $\hat{f} = \lim \hat{f}_n$ in $L^q(\hat{\mathbf{R}})$ and consequently, for fixed $N > 0$, $S_N(f,x) = \lim_n S_N(f_n, x)$ uniformly in x. It follows that

$$\| S_N(f)\|_{L^p(\mathbf{R})} \leqq \liminf_n \| S_N(f_n)\|_{L^p(\mathbf{R})} \leqq C_p \lim_n \| f_n\|_{L^p(\mathbf{R})} = C_p\| f\|_{L^p(\mathbf{R})}.$$

◀

Theorem: *Let* $f \in L^p(\mathbf{R})$, $1 < p \leqq 2$. *Then*

$$\lim_{N\to\infty} \| S_N(f) - f\|_{L^p(\mathbf{R})} = 0.$$

PROOF: $\{S_N\}$ is a uniformly bounded family of operators which converge to the identity, as $N \to \infty$, on all functions with compactly supported Fourier transform, and hence, by corollary 3.3, it converges strongly to the identity. ◀

EXERCISES FOR SECTION 3

1. Let $B \subseteq \mathscr{L}_c$ be a homogeneous Banach space on $\mathbf{R}$ and let $f \in B$. Show that: (a) for every $\varphi \in L^1(\mathbf{R})$, $\varphi * f$ can be approximated (in the B norm) by linear combinations of translates of f, that is, given $\varepsilon > 0$, there exist numbers $y_1, \cdots, y_n \in \mathbf{R}$ and complex numbers $A_1, \cdots, A_n$ such that $\| \varphi * f - \sum_{j=1}^n A_j f_{y_j}\|_B < \varepsilon$, where $f_y(x) = f(x-y)$.

(b) For every $y \in \mathbf{R}$, f_y can be approximated by functions of the form $\varphi * f$ with $\varphi \in L^1(\mathbf{R})$. Deduce that a closed subspace H of B is *translation invariant* (i.e., $f \in H$ implies $f_y \in H$ for all $y \in \mathbf{R}$) if, and only if, $f \in H$ implies $\varphi * f \in H$ for every $\varphi \in L^1(\mathbf{R})$.

† Note that f, having a compact a support, is in $L^1(\mathbf{R})$ and $\hat{f}$ is therefore continuous.

2. Let $F, G \in L^2(\hat{\mathbf{R}})$ and assume $F(\xi) = 0$ implies $G(\xi) = 0$ for almost all $\xi \in \hat{\mathbf{R}}$. Show that, given $\varepsilon > 0$, there exists a twice-differentiable compactly supported function Φ such that

$$\| \Phi F - G \|_{L^2(\hat{\mathbf{R}})} < \varepsilon .$$

3. Let $f, g \in L^2(\mathbf{R})$ and assume that $\hat{f}(\xi) = 0$ implies $\hat{g}(\xi) = 0$ for almost all $\xi \in \hat{\mathbf{R}}$. Show that g can be approximated on $L^2(\mathbf{R})$ by linear combinations of translates of f. *Hint*: Use exercises 1 and 2 and Plancherel's theorem.

4. A measurable set $E \subset \hat{\mathbf{R}}$ is a D *set* if it coincides with its set of points of density. Show that finite intersections of D sets are again D sets. For a subspace $H \subseteq L^2(\mathbf{R})$, define $E(H)$ as the union of all the D sets which are supports of some $\hat{f}$ with $f \in H$. Show that this establishes a one-to-one correspondence between the closed translation invariant subspaces of $L^2(\mathbf{R})$ and the D sets on $\hat{\mathbf{R}}$, and that

$$E(H_1 \cap H_2) = E(H_1) \cap E(H_2),$$

$$E(H_1 + H_2) = E(H_1) \cup E(H_2) \cup \text{ a set of measure zero},$$

$$H_1 \perp H_2 \Leftrightarrow E(H_1) \cap E(H_2) = \varnothing .$$

5. Show that every closed translation invariant subspace of $L^2(\mathbf{R})$ is singly generated.

6. Let $f \in L^2(\mathbf{R})$. Show that the translates of f generate $L^2(\mathbf{R})$ if, and only if, $\hat{f}(\xi) \neq 0$ almost everywhere.

7. The information obtained through exercises 2 through 6 can be obtained very easily by duality arguments (i.e., using the Hahn-Banach theorem). For instance, by Plancherel's theorem, exercise 6 is equivalent to the statement that $\{\hat{f}(\xi) e^{i\xi x}\}_{x \in \mathbf{R}}$ spans $L^2(\hat{\mathbf{R}})$ if, and only if, $\hat{f}(\xi) \neq 0$ almost everywhere. By the Hahn-Banach theorem $\{\hat{f}(\xi) e^{i\xi x}\}_{x \in \mathbf{R}}$ does not span $L^2(\hat{\mathbf{R}})$ if, and only if, there exists a function $\psi \in L^2(\hat{\mathbf{R}})$ which does not vanish identically, such that $\int \hat{f}(\xi) \overline{\psi(\xi)}\, e^{i\xi x}\, d\xi = 0$ for all $x \in \mathbf{R}$, which, by the uniqueness theorem, is equivalent to: $\hat{f}\bar{\psi} = 0$ identically, that is, $\hat{f}$ vanishes on the support of ψ. Use the same method to prove exercises 3 through 5.

8. Both the "if" and the "only if" parts of exercise 6 are based on Plancherel's theorem and are both false for $L^p(\mathbf{R})$, $p < 2$. Assuming the existence of a measure μ carried by a closed set of measure zero and such that $\hat{\mu} \in L^q$ for all $q > 2$, construct a function $f \in L^1 \cap L^\infty(\mathbf{R})$ such that $\hat{f}(\xi) \neq 0$ almost everywhere and such that the translates of f do not span $L^p(\mathbf{R})$ for any $p < 2$. *Hint*: Put μ on $\hat{\mathbf{R}}$.

9. Show that if $f \in L^1(\mathbf{R})$ and $g \in L^p(\mathbf{R})$, then $\widehat{f * g} = \hat{f}\hat{g}$.

We denote by $\mathscr{F}L^p$ the space of all functions $\hat{f}$ such that $f \in L^p(\mathbf{R})$, (thus $\mathscr{F}L^1 = A(\hat{\mathbf{R}})$). By definition:

$$\|\hat{f}\|_{\mathscr{F}L^p} = \|f\|_{L^p(\mathbf{R})}.$$

10. If $\mu \in M(\mathbf{R})$ and $\varphi \in \mathscr{F}L^p$, $1 \leqq p \leqq 2$, then $\hat{\mu}\varphi \in \mathscr{F}L^p$.

11. Show that if $\varphi \in \mathscr{F}L^p$, $1 < p \leqq 2$, and if we write

$$\psi(\xi) = \begin{cases} \varphi(\xi) & \xi > 0 \\ 0 & \xi \leqq 0, \end{cases}$$

then $\psi \in \mathscr{F}L^p$, and $\|\psi\|_{\mathscr{F}L^p} \leqq C_p \|\varphi\|_{\mathscr{F}L^p}$.

12. Let α and β be real numbers, $\alpha\beta \neq 0$. Show that if $\varphi \in \mathscr{F}L^p$, $1 < p \leqq 2$, and if we write

$$\psi_{\alpha,\beta}(\xi) = \begin{cases} \varphi(\alpha\xi) & \xi > 0 \\ \varphi(\beta\xi) & \xi \leqq 0, \end{cases}$$

then $\psi_{\alpha,\beta} \in \mathscr{F}L^p$.

13. Let $f \in L^p(\mathbf{R})$, $1 < p \leqq 2$. Show that $h(x) = \pi^{-1} \int f(x-y) \sin y/y \, dy$ is well defined and continuous on $\mathbf{R}$, $h \in L^p(\mathbf{R})$ and $\|h\|_{L^p(\mathbf{R})} \leqq C_p \|f\|_{L^p(\mathbf{R})}$.

14. Show that, for $1 \leqq p < 2$, the norms $\|\varphi\|_{\mathscr{F}L^p}$ and $\|\varphi\|_{L^q(\hat{\mathbf{R}})}$ are not equivalent. Deduce that $\mathscr{F}L^p \neq L^q(\hat{\mathbf{R}})$ $(q = p/(p-1))$. *Hint*: See IV.2.

4. TEMPERED DISTRIBUTIONS AND PSEUDO-MEASURES

In the previous section we defined the Fourier transforms for functions in $L^p(\mathbf{R})$, $1 < p \leqq 2$, by showing that on dense subspace on which $\mathcal{F}: f \to \hat{f}$ is already well defined (e.g., on $L^1 \cap L^2(\mathbf{R})$), we have the norm inequality

$$\|\hat{f}\|_{L^q(\hat{\mathbf{R}})} \leqq \|f\|_{L^p(\mathbf{R})}$$

and consequently there exists a unique continuous extension of $\mathcal{F}$, as an operator from $L^p(\mathbf{R})$ into $L^q(\hat{\mathbf{R}})$. If $p > 2$ this procedure fails. It is not hard to see that not only is it impossible to extend the validity of the Hausdorff-Young theorem for $p > 2$, but also there is no homogeneous Banach space B on $\mathbf{R}$ such that for some $p > 2$, some constant C and all $f \in L^1 \cap L^\infty(\mathbf{R})$,

$$\|\hat{f}\|_B \leqq C \|f\|_{L^p(\mathbf{R})}.$$

So, a different procedure is needed if we want to extend the notion of Fourier transforms to $L^p(\mathbf{R})$, $p > 2$. Clearly, we try to extend the notion, keeping as many of its properties as possible; in particular, we would like to keep some form of the inversion formula and the very useful Parseval's formula. We realize immediately that, since the Fourier transforms of measures are bounded functions, if any reasonable form of inversion is to be valid, the Fourier transforms of some bounded functions will have to be measures; and once we accept the idea that Fourier transforms need not be functions but could be other objects, such as measures, the procedure that we look for is given to us by Parseval's formula.

So far we have established Parseval's formula for various function spaces as a theorem following the definition of the Fourier transforms of functions in the corresponding spaces. In this section we consider Parseval's formula as a definition of Fourier transform for a much larger class of objects. Having proved Parseval's formula for $L^p(\mathbf{R})$, $1 \leqq p \leqq 2$, we are assured that our new definition is consistent with the previous ones.

4.1 We denote by $S(\mathbf{R})$ the space of all infinitely differentiable functions on $\mathbf{R}$ which satisfy:

$$\lim_{|x| \to \infty} x^n f^{(j)}(x) = 0 \qquad \text{for all } n \geqq 0,\ j \geqq 0. \tag{4.1}$$

$S(\mathbf{R})$ is a topological vector space, the topology being given† by the family of seminorms

$$\| f \|_{j,n} = \sup_x \left| x^n f^{(j)}(x) \right|. \tag{4.2}$$

This topology on $S(\mathbf{R})$ is clearly metrizable and $S(\mathbf{R})$ is complete, in other words, $S(\mathbf{R})$ is a Frechet space.

DEFINITION: A *tempered distribution* on $\mathbf{R}$ is a continuous linear functional on $S(\mathbf{R})$.

We denote the space of tempered distributions on $\mathbf{R}$, that is, the dual of $S(\mathbf{R})$, by $S^*(\mathbf{R})$.

† A sequence of functions $f_m \in S(\mathbf{R})$ converges to f if $\lim_{m \to \infty} \| f_m - f \|_{j,n} = 0$ for all $j \geqq 0$ and $n \geqq 0$. The metric in $S(\mathbf{R})$ can be defined by:

$$\text{dist(f, g)} = \sum_{j,n \geqq 0} \frac{1}{2^{j+n}} \frac{\| f - g \|_{j,n}}{1 + \| f - g \|_{j,n}}$$

$S(\mathbf{R})$ is a natural space to study within the theory of Fourier transforms. By theorems 1.5 and 1.6 we see that if $f \in S(\mathbf{R})$ then $\hat{f} \in S(\hat{\mathbf{R}})$ (the analogous space on $\hat{\mathbf{R}}$) and, $\xi^n \hat{f}^{(j)}(\xi)$ being the Fourier transform $(-i)^{n+j} \dfrac{d^n(x^j f(x))}{dx^n}$, we see that the mapping $f \to \hat{f}$ is continuous from $S(\mathbf{R})$ into $S(\hat{\mathbf{R}})$. By the inversion formula this mapping is onto $S(\hat{\mathbf{R}})$ and is bicontinuous.

We now define $\hat{\mu}$, for $\mu \in S^*(\mathbf{R})$, as the tempered distribution on $\hat{\mathbf{R}}$ satisfying

$$(4.3) \qquad \langle \hat{f}, \hat{\mu} \rangle = \langle f, \mu \rangle$$

for all $f \in S(\mathbf{R})$.

The space of tempered distributions on $\mathbf{R}$ is quite large. Every function g which is measurable and locally summable, and which is bounded at infinity by a power of x can be identified with a tempered distribution by means of:

$$\langle f, g \rangle = \int f(x)\overline{g(x)}\,dx \qquad f \in S(\mathbf{R})$$

and so can every $g \in L^p(\mathbf{R})$, for any $p \geqq 1$, and every measure $\mu \in M(\mathbf{R})$; thus our definition has a very satisfactory domain. However, the range of the definition is as large and this is clearly a disadvantage; it gives relatively little information about the Fourier transform. We thus have to supplement this definition with studies of the following general problem: knowing that a distribution $\mu \in S^*(\mathbf{R})$ has some special properties, what can we say about $\hat{\mu}$?

Much of what we have done in the first three sections of this chapter falls into this category: if μ is (identified with) a summable function, then $\hat{\mu}$ is (identified with) a function in $C_0(\hat{\mathbf{R}})$; if μ is a measure, $\hat{\mu}$ is a uniformly continuous bounded function; if $\mu \in L^p(\mathbf{R})$ with $1 < p \leqq 2$, then $\hat{\mu} \in L^q(\hat{\mathbf{R}})$, $q = p/(p-1)$.

We shall presently obtain some information about Fourier transforms of functions in $L^p(\mathbf{R})$ with $2 < p \leqq \infty$, but we should not leave the general setup without mentioning the notion of the support of a tempered distribution.

4.2 DEFINITION: A distribution $\nu \in S^*(\mathbf{R})$ *vanishes on an open set* $O \subset \mathbf{R}$, if $\langle \varphi, \nu \rangle = 0$ for all $\varphi \in S(\mathbf{R})$ with compact support contained in O.

Lemma: *Let O_1, O_2 be open on* **R** *and let K be a compact set, $K \subset O_1 \cup O_2$. Then there exist two compactly supported C^∞ functions φ_1 and φ_2 satisfying: support of $\varphi_j \subset O_j$ and $\varphi_1 + \varphi_2 \equiv 1$ on K.*

PROOF: Let $U_j \subset O_j$ have the following properties: U_j is open, $\bar{U}_j$ is compact and included in O_j, and $K \subset U_1 \cup U_2$. Denote the characteristic function of U_1 by ψ_1 and that of $U_2 \setminus U_1$ by ψ_2. Let $\varepsilon > 0$ be smaller than the distance of K to the boundary of $U_1 \cup U_2$ and also smaller than the distance of U_j to the complement of O_j, $j = 1, 2$. Let $\delta(x)$ be an infinitely differentiable function carried by $(-\varepsilon, \varepsilon)$ and whose integral is 1. Then we can take $\varphi_j = \psi_j * \delta$. ◀

Corollary: *If $v \in S^*(\mathbf{R})$ vanishes on O_1 and on O_2, it vanishes on $O_1 \cup O_2$.*

PROOF: Let $f \in S(\mathbf{R})$ have a compact support included in $O_1 \cup O_2$. Denote the support of f by K and let φ_1, φ_2 be the functions described in the lemma. Then $\varphi_j \in S(\mathbf{R})$, $f = f(\varphi_1 + \varphi_2) = f\varphi_1 + f\varphi_2$. Now $\langle f\varphi_1, v\rangle = 0$, $\langle f\varphi_2, v\rangle = 0$, and consequently $\langle f, v\rangle = 0$. ◀

Our corollary clearly implies that the union of any finite number of open sets on which v vanishes has the same property, and since our test functions all have compact support, the same is valid for arbitrary unions. The union of all the open sets on which v vanishes is clearly the largest such set.

4.3 DEFINITION: The *support* $\Sigma(v)$ of $v \in S^*(\mathbf{R})$ is the complement of the largest open set $O \subset \mathbf{R}$ on which v vanishes.

Remarks: (a) If v is (identified with) a continuous function g then $\Sigma(v)$ is the closure of $\{x; g(x) \neq 0\}$. If v is a measurable function g then $\Sigma(v)$ is the *closure* of the set of points of density of $\{x; g(x) \neq 0\}$. The set of points of density of $\{x; g(x) \neq 0\}$ is a finer notion of support which may be useful (cf. exercise 3.4).

(b) The definition of $\Sigma(v)$ implies that if $\varphi \in S(\mathbf{R})$ and if the support of φ is compact and disjoint from $\Sigma(v)$ then $\langle \varphi, v\rangle = 0$. It may be useful to notice that if $\psi \in S(\mathbf{R})$ and if δ is infinitely differentiable with compact support and $\delta(0) = 1$, then $\psi = \lim_{\lambda \to 0} \delta(\lambda x)\psi$ in $S(\mathbf{R})$, and consequently if the support of ψ is disjoint from $\Sigma(v)$ (but not necessarily compact) we have $\langle \psi, v\rangle = \lim_{\lambda \to 0} \langle \delta(\lambda x)\psi, v\rangle = 0$. In particular, if $\Sigma(v) = \varnothing$ then $v = 0$.

(c) Let $B \supseteq S(\mathbf{R})$ be a function space and assume that every $f \in B$ with compact support can be approximated in the topology of B by functions $\varphi_n \in S(\mathbf{R})$ such that the supports of φ_n tend to that of f. Let $v \in S^*(\mathbf{R})$ and assume that v can be extended to a continuous linear functional on B. If $f \in B$ has a compact support disjoint from $\Sigma(v)$, then $\langle f, v \rangle = 0$.

4.4 $S(\mathbf{R})$ is an algebra under pointwise multiplication. The product fv of a function $f \in S(\mathbf{R})$ and a distribution $v \in S^*(\mathbf{R})$ is defined by

$$\langle g, fv \rangle = \langle g\bar{f}, v \rangle, \qquad g \in S(\mathbf{R}),$$

that is, the multiplication by f in $S^*(\mathbf{R})$ is the adjoint of the multiplication by f in $S(\mathbf{R})$. From the definitions above, it is clear that $\Sigma(fv) \subseteq \Sigma(f) \cap \Sigma(v)$.

4.5 We denote by $\mathscr{F}L^p = \mathscr{F}L^p(\hat{\mathbf{R}})$ the space of distributions on $\hat{\mathbf{R}}$ which are Fourier transforms of functions in $L^p(\mathbf{R})$, $1 < p \leqq \infty$ (we keep the notation $A(\hat{\mathbf{R}})$ for $\mathscr{F}L^1(\hat{\mathbf{R}})$). $\mathscr{F}L^p$ inherits from $L^p(\mathbf{R})$ its Banach space structure; we simply put $\|\hat{f}\|_{\mathscr{F}L^p} = \|f\|_{L^p(\mathbf{R})}$; and we can identify $\mathscr{F}L^p$ with the dual of $\mathscr{F}L^q$ if $q = p/(p-1) < \infty$. In particular, $\mathscr{F}L^\infty$ is the dual of $A(\hat{\mathbf{R}})$. This identification may be considered as purely formal: writing $\langle \hat{f}, \hat{g} \rangle = \langle f, g \rangle$ carries the duality from $(L^p(\mathbf{R}), L^q(\mathbf{R}))$ to $(\mathscr{F}L^p, \mathscr{F}L^q)$; however, we have already made enough formal identifications to allow a somewhat clearer meaning to the one above. Having identified functions with the corresponding distributions, we clearly have $S(\hat{\mathbf{R}}) \subset \mathscr{F}L^p$ and, if $p < \infty$, $S(\hat{\mathbf{R}})$ is dense in $\mathscr{F}L^p$; consequently, every continuous linear functional on $\mathscr{F}L^p$ is *canonically* identified with a tempered distribution. The identification of $\mathscr{F}L^q$ as the dual of $\mathscr{F}L^p$ now becomes a theorem stating that a distribution $v \in S^*(\hat{\mathbf{R}})$ is continuous on $S(\hat{\mathbf{R}})$ with respect to the norm induced by $\mathscr{F}L^p$ if, and only if, $v \in \mathscr{F}L^q$. We leave the proof as an exercise to the reader.

We now confine our attention to $\mathscr{F}L^\infty$. If μ is a measure in $M(\hat{\mathbf{R}})$ then μ is the Fourier transform of the bounded function $h(x) = \int e^{i\xi x} d\mu(\xi)$; thus $M(\hat{\mathbf{R}}) \subset \mathscr{F}L^\infty$. The elements of $\mathscr{F}L^\infty$ are commonly referred to as *pseudo-measures*. It is clear that $M(\hat{\mathbf{R}})$ is a relatively small part of $\mathscr{F}L^\infty$; for instance, if $\varphi \in L^\infty$ is not uniformly continuous on $\mathbf{R}$, $\hat{\varphi}$ cannot be a measure.

DEFINITION: *The convolution* $\hat{h}_1 * \hat{h}_2$ of the pseudo-measures $\hat{h}_1$ and $\hat{h}_2$, $(h_j \in L^\infty(\mathbf{R}))$, is the Fourier transform of $h_1 h_2$.

Again we reverse the roles; we take something which we have proved for measures, as a definition for the larger class of pseudo-measures. Thus, if $\hat{h}_1$ and $\hat{h}_2$ happen to be measures, $\hat{h}_1 * \hat{h}_2$ is their (measure theoretic) convolution.

4.6 Another case in which we can identify the convolution is given by

Lemma: *Let* $h_1 \in L^\infty(\mathbf{R})$ *and* $h_2 \in L^1 \cap L^\infty(\mathbf{R})$; *then*

$$(\hat{h}_1 * \hat{h}_2)(\xi) = \langle \hat{h}_2(\xi - \eta), \overline{\hat{h}_1(\eta)} \rangle . \tag{4.4}$$

PROOF: We remark first that $h_1 h_2 \in L^1 \cap L^\infty(\mathbf{R})$ and consequently $\hat{h}_1 * \hat{h}_2 = \widehat{h_1 h_2} \in A(\mathbf{R})$ so that we can talk about its value at $\xi \in \hat{\mathbf{R}}$. If $h_1 \in S(\mathbf{R})$ we have

$$(\hat{h}_1 * \hat{h}_2)(\xi) = \int h_1(x) h_2(x) e^{-i\xi x} dx = \frac{1}{2\pi} \iint \hat{h}_1(\eta) h_2(x) e^{i(\eta - \xi)x} dx\, d\eta$$

$$= \frac{1}{2\pi} \int \hat{h}_2(\xi - \eta) \hat{h}_1(\eta)\, d\eta = \langle \hat{h}_2(\xi - \eta), \overline{\hat{h}_1(\eta)} \rangle .$$

Since $S(\mathbf{R})$ is dense in $L^\infty(\mathbf{R})$ in the weak-star topology (as dual of $L^1(\mathbf{R})$), and since both sides of (4.4) depend on h_1 continuously with respect to the weak-star topology, (4.4) is valid for arbitrary $h_1 \in L^\infty(\mathbf{R})$. ◀

Corollary: *If* $h_1 \in L^\infty(\mathbf{R})$ *and* $h_2 \in L^1 \cap L^\infty(\mathbf{R})$, *then*

$$\Sigma(\hat{h}_1 * \hat{h}_2) \subseteq \Sigma(\hat{h}_1) + \Sigma(\hat{h}_2).$$

4.7 This corollary can be improved:

Lemma: *Assume* $h_1, h_2 \in L^\infty(\mathbf{R})$. *Then*

$$\Sigma(\hat{h}_1 * \hat{h}_2) \subseteq \Sigma(\hat{h}_1) + \Sigma(\hat{h}_2).$$

PROOF: Consider a smooth function $\hat{f} \in A(\hat{\mathbf{R}})$ with compact support disjoint from $\Sigma(\hat{h}_1) + \Sigma(\hat{h}_2)$. We have to show that $\langle \hat{f}, \widehat{h_1 h_2} \rangle = 0$ which is the same as $\int f h_1 h_2\, dx = \langle \widehat{f h_1}, \hat{h}_2 \rangle = 0$. Now $\Sigma(\hat{\bar{h}}_1)$ is the set $\{\xi;\ -\xi \in \Sigma(\hat{h}_1)\}$ and, by 4.6, $\Sigma(\widehat{f\bar{h}_1}) \subseteq \Sigma(\hat{f}) + \Sigma(\hat{\bar{h}}_1)$. If $\xi_0 \in \Sigma(\widehat{f\bar{h}_1}) \cap \Sigma(\hat{h}_2)$, then there exist $\eta_0 \in \Sigma(\hat{f})$ and $\eta_1 \in \Sigma(\hat{h}_1)$ such that $\xi_0 = \eta_0 - \eta_1$, that is, $\eta_0 = \xi_0 + \eta_1$ which contradicts $\Sigma(\hat{f}) \cap (\Sigma(\hat{h}_1) + \Sigma(\hat{h}_2)) = \varnothing$. It follows that $\Sigma(\widehat{f\bar{h}_1})$ is disjoint from $\Sigma(\hat{h}_2)$, hence $\langle \hat{f}, \widehat{h_1 h_2} \rangle = \langle \widehat{f h_1}, h_2 \rangle = 0$ and the lemma is proved. ◀

4.8 The reader might have noticed that we were using not only the duality between $L^1(\mathbf{R})$ and $L^\infty(\mathbf{R})$ but also the fact that a multiplication by a bounded function is a bounded operator on $L^1(\mathbf{R})$. Another operation between $L^1(\mathbf{R})$ and $L^\infty(\mathbf{R})$ which we have used is the convolution that takes $L^1 \times L^\infty$ into $L^\infty(\mathbf{R})$. Passing to Fourier transforms we see that $\mathscr{F}L^\infty$ is a module over $A(\hat{\mathbf{R}})$ the multiplication of a pseudo-measure by a function in $A(\hat{\mathbf{R}})$ being the adjoint of the multiplication in $A(\hat{\mathbf{R}})$. This extends the notion of multiplication introduced in 4.4.

4.9 Let k be an infinitely differentiable function on $\hat{\mathbf{R}}$, carried by $[-1,1]$ and such that $\int k(\xi)\,d\xi = 1$. For $\hat{f} \in A(\hat{\mathbf{R}})$ we set $\hat{f}_\lambda = \lambda k(\lambda\xi) * \hat{f} = \lambda \int k(\lambda\eta)\hat{f}(\xi - \eta)\,d\eta$. $\hat{f}_\lambda$ is infinitely differentiable, $\Sigma(\hat{f}_\lambda) \subseteq \Sigma(f) + [-1/\lambda, 1/\lambda]$, and as $\lambda \to \infty$, $\hat{f}_\lambda \to \hat{f}$ in $A(\hat{\mathbf{R}})$.

By 4.3, remark (c) it follows that if $v \in \mathscr{F}L^\infty$ and $\hat{f} \in A(\hat{\mathbf{R}})$ has a compact support disjoint from $\Sigma(v)$, we have $\langle \hat{f}, v\rangle = 0$. Further, if $\hat{f} \in A(\hat{\mathbf{R}})$ and $\Sigma(\hat{f}) \cap \Sigma(v) = \varnothing$, it follows that $\langle (1 - |\xi|/\lambda)\hat{f}, v\rangle = 0$ for all $\lambda > 0$, and letting $\lambda \to \infty$, we obtain $\langle \hat{f}, v\rangle = 0$.

For convenient reference we state this as:

***Lemma*:** *Let* $v \in \mathscr{F}L^\infty$ *and* $\hat{f} \in A(\hat{\mathbf{R}})$. *If* $\Sigma(\hat{f}) \cap \Sigma(v) = \varnothing$ *then* $\langle \hat{f}, v\rangle = 0$.

4.10 We leave the proof of the following lemma as an exercise to the reader.

***Lemma*:** *Let* $v \in \mathscr{F}L^\infty$ *and* $\hat{f} \in A(\hat{\mathbf{R}})$; *then*

$$\Sigma(\hat{f}v) \subseteq \Sigma(\hat{f}) \cap \Sigma(v).$$

4.11 We show now that a pseudo-measure with finite support is a measure. Using the multiplication by elements of $A(\hat{\mathbf{R}})$ we see that a pseudo-measure with finite support is a linear combination of pseudo-measures carried by one point each; thus it would be sufficient to prove:

***Theorem*:** *A pseudo-measure carried by one point is a measure.*

PROOF: Let $h \in L^\infty(\mathbf{R})$ and assume $\Sigma(\hat{h}) = \{0\}$. If $\varphi_1, \varphi_2 \in A(\hat{\mathbf{R}})$ and $\varphi_1(\xi) = \varphi_2(\xi)$ in a neighborhood of $\xi = 0$, then $\langle \varphi_1, \hat{h}\rangle = \langle \varphi_2, \hat{h}\rangle$. Put $c = \langle \varphi, \hat{h}\rangle$ where φ is any function in $A(\hat{\mathbf{R}})$ such that $\varphi(\xi) = 1$ near $\xi = 0$. As usual we denote by $\mathbf{K}$ the Féjer kernel and recall that $\hat{\mathbf{K}}(\xi) = \sup(1 - |\xi|, 0)$. By lemma 4.6 we have

(4.5) $$\widehat{h\mathbf{K}}(\xi) = \langle \hat{\mathbf{K}}(\xi - \eta), \overline{\hat{h}(\eta)} \rangle .$$

For $|\xi| \geqq 1$ we clearly have $\widehat{h\mathbf{K}}(\xi) = 0$. If $-1 < \xi_1 < \xi_2 < 0$ we have $\hat{\mathbf{K}}(\xi_2 - \eta) - \hat{\mathbf{K}}(\xi_1 - \eta) = \xi_2 - \xi_1$ for η near zero. By (4.5) and the definition of c we conclude that $\widehat{h\mathbf{K}}(\xi_2) - \widehat{h\mathbf{K}}(\xi_1) = c(\xi_2 - \xi_1)$, and since $\widehat{h\mathbf{K}}(\xi)$ is continuous, upon letting $\xi_1 \to -1$ we obtain $\widehat{h\mathbf{K}}(\xi) = c(1 + \xi)$ for $-1 \leqq \xi \leqq 0$. Repeating the argument for $0 \leqq \xi \leqq 1$ we obtain $\widehat{h\mathbf{K}}(\xi) = c\hat{\mathbf{K}}(\xi)$ and by the uniqueness theorem $h(x) = c$ a.e. It follows that $\hat{h}$ is the measure of mass c concentrated at the origin. ◄

4.12 We add just a few remarks concerning distributions in $\mathscr{F}L^p$, $2 < p < \infty$. There is clearly no inclusion relation between $\mathscr{F}L^p$ and $\mathscr{F}L^\infty$ but it might be useful to notice that locally $\mathscr{F}L^p \subseteq \mathscr{F}L^{p'}$ if $p \leqq p'$ and in particular all distributions in $\mathscr{F}L^p$ are locally pseudo-measures. (We recall that a tempered distribution ν *belongs locally to a set* $G \subset S^*(\hat{\mathbf{R}})$ if for every $\xi \in \hat{\mathbf{R}}$ there exists $\mu \in G$ such that $\Sigma(\mu - \nu)$ does not contain ξ.) If $\nu \in \mathscr{F}L^p$ and $\xi \in \hat{\mathbf{R}}$ we may take $\lambda > |\xi|$ and consider $\mu = \hat{\mathbf{V}}_\lambda \nu$ where $\mathbf{V}_\lambda$ is de la Vallée Poussin's kernel ($\hat{\mathbf{V}}_\lambda(\xi) = 1$ for $|\xi| \leqq \lambda$, $= 2 - |\xi|\lambda^{-1}$ for $\lambda < |\xi| < 2\lambda$, and $= 0$ for $|\xi| > 2\lambda$). It is clear that $\nu = \mu$ on $(-\lambda, \lambda)$, that is, $\Sigma(\mu - \nu) \cap (-\lambda, \lambda) = \varnothing$ and if $\nu = \hat{f}$ with $f \in L^p(\mathbf{R})$, then $\mu = \widehat{\mathbf{V}_\lambda * f}$ and $\mathbf{V}_\lambda * f \in L^p \cap L^\infty(\mathbf{R})$ since $\mathbf{V}_\lambda \in L^1 \cap L^q(\mathbf{R})$, $q = p/(p-1)$. In particular, if $\Sigma(\nu)$ is compact, say $\Sigma(\nu) \subset (-\lambda, \lambda)$. then $\mu = \nu$; we have thus proved:

Theorem: *If* $\nu \in \mathscr{F}L^p$ *and* $\Sigma(\nu)$ *is compact, then* $\nu \in \mathscr{F}L^\infty$.

4.13 If $\nu \in \mathscr{F}L^p \cap \mathscr{F}L^\infty$ we can consider the repeated convolution of ν with itself; writing $\nu = \hat{f}$ with $f \in L^p \cap L^\infty(\mathbf{R})$, the convolution of ν with itself m times is the Fourier transform of f^m, and if $m > p$, $f^m \in L^1(\mathbf{R})$ so that $\overbrace{\nu * \cdots * \nu}^{m \text{ times}} \in A(\hat{\mathbf{R}})$. In particular, assuming $\nu \neq 0$, $\Sigma(\nu * \cdots * \nu) \subseteq \Sigma(\nu) + \cdots + \Sigma(\nu)$ contains an interval. As an immediate consequence we obtain:

Theorem: *Let* $\nu \in \mathscr{F}L^p$, $p < \infty$, *and let* J *be an open interval on* $\hat{\mathbf{R}}$ *such that* $J \cap \Sigma(\nu) \neq \varnothing$. *Then* $J \cap \Sigma(\nu)$ *is a basis for* $\hat{\mathbf{R}}$.

Theorems 4.11 and 4.13 are equivalent to the following approximation theorems:

4.11′ Theorem: *Let $\xi \in \hat{\mathbf{R}}$ and denote*

$$I(\xi) = \{f; f \in A(\hat{\mathbf{R}}), f(\xi) = 0\}$$
$$I_0(\xi) = \{f; f \in S(\hat{\mathbf{R}}), \xi \notin \Sigma(f)\}.$$

Then $I_0(\xi)$ is dense in $I(\xi)$ in the $A(\hat{\mathbf{R}})$ topology.

4.13′ Theorem: *Let $E \subset \hat{\mathbf{R}}$ be closed, and denote*

$$I_0(E) = \{f; f \in S(\hat{\mathbf{R}}), \Sigma(f) \cap E = \varnothing\}.$$

Assume that $E + E + \cdots + E$ (m times) has no interior. Let $1 < p \leqq m$ and $q = p/(p-1)$. Then $I_0(E)$ is (norm) dense in $\mathscr{F}L^q$.

The proofs of 4.11′ and 4.13′ are essentially the same and follow immediately from the Hahn-Banach theorem (and 4.11, 4.13, respectively). A linear functional on $A(\hat{\mathbf{R}})$ which annihilates $I_0(\xi)$ is a pseudo-measure supported by $\{\xi\}$, hence is constant multiple of the Dirac measure at ξ, and hence annihilates $I(\xi)$. A linear functional on $\mathscr{F}L^q$ which annihilates $I_0(E)$ is an element of $\mathscr{F}L^p$ supported by E; hence it must be zero. ◀

EXERCISES FOR SECTION 4

1. Deduce 4.11 from 4.11′.
2. Deduce 4.13 from 4.13′.
3. What is a function $h \in L^\infty$ such that $\Sigma\ (\hat{h})$ is finite?
4. If $f \in A(\hat{\mathbf{R}})$ and $\nu \in \mathscr{F}L^\infty$, then

$$\|f\nu\|_{\mathscr{F}L^\infty} \leqq \|f\|_{A(\hat{\mathbf{R}})} \|\nu\|_{\mathscr{F}L^\infty}.$$

5. Let $f \in L^\infty(\mathbf{R})$. Show that $\Sigma(\widehat{\mathrm{Re}(f)}) \subseteq \Sigma(\hat{f}) \cup (-\Sigma(\hat{f}))$.
6. (Bernstein) Let $h \in L^\infty(\mathbf{R})$ and assume that $\Sigma(\hat{h}) \subseteq [-k, k]$. Show that h is infinitely differentiable and that $\|h^{(m)}\|_\infty \leqq k^m \|h\|_\infty$.
7. Let $h \in L^\infty(\mathbf{R})$, $h \neq 0$. Assume that $h(x+y) = h(x)h(y)$ a.e. in (x, y). Show that $h(x) = e^{i\xi_0 x}$ for some $\xi_0 \in \hat{\mathbf{R}}$.
8. Assume $\nu_j \in \mathscr{F}L^\infty$, $j = 0, 1, \cdots$, and $\nu_j \to \nu_0$ in the weak-star topology. Let U be an open set such that $U \cap \Sigma(\nu_j) = \varnothing$ for infinitely many j's. Show that $U \cap \Sigma(\nu_0) = \varnothing$.
9. Fourier transforms of functions in $C_0(\mathbf{R})$ are called *pseudo-functions* (on $\hat{\mathbf{R}}$).

(a) Show that if $f \in L^1(\hat{\mathbf{R}})$, then f is a pseudo-function.

(b) Show that if f_j are pseudo-functions on $\hat{\mathbf{R}}$, $\|f_j\|_{\mathscr{F}L^\infty} < c$ and $\lambda_j \to \infty$ fast enough, then $\sum e^{i\lambda_j \xi} f_j$ converges (weak-star) in $\mathscr{F}L^\infty$.

10. Let $h(x) = \sin x^2$. Show that $e^{i\lambda\xi}\hat{h} \to 0$ (weak-star) as $|\lambda| \to \infty$.

5. ALMOST-PERIODIC FUNCTIONS ON THE LINE

The usefulness of Fourier series of functions on **T** is largely due to the information they offer about approximation of the functions by trigonometric polynomials. On the line, trigonometric polynomials do not belong to many of the function spaces in which we are interested, for example, to $L^p(\mathbf{R})$ for $p < \infty$; and the positive results, which we had for $L^p(\mathbf{R})$, $1 \leqq p \leqq 2$, were in terms of trigonometric integrals rather than polynomials. Trigonometric polynomials do belong to $L^\infty(\mathbf{R})$, and in this section we characterize the functions that are uniform limits of trigonometric polynomials.

5.1 DEFINITION: Let f be a complex-valued function on **R** and let $\varepsilon > 0$. An *ε-almost-period* of f is a number τ such that

$$\sup_x |f(x-\tau) - f(x)| < \varepsilon .$$

Examples: $\tau = 0$ is a trivial ε-almost-period for all $\varepsilon > 0$; if f is periodic then its period, or any integral multiple thereof, is an ε-almost-period for all $\varepsilon > 0$; if f is uniformly continuous, every sufficiently small τ is an ε-almost-period.

5.2 DEFINITION: A function f is *(uniformly) almost-periodic on* **R** if it is continuous and if for every $\varepsilon > 0$ there exists a number $\Lambda = \Lambda(\varepsilon, f)$ such that every interval of length Λ on **R** contains an ε-almost-period of f.

We denote by $AP(\mathbf{R})$ the set of all almost-periodic functions on **R**.

Examples: (a) Continuous periodic functions are almost-periodic.
(b) We shall show (see 5.7) that the sum of two almost-periodic functions is almost-periodic; hence $f = \cos x + \cos \pi x$ is almost-periodic (see also exercise 1 at the end of this section); noticing, however, that $f(x) = 2$ only for $x = 0$, we see that f is not periodic.
(c) If f is almost-periodic, so are $|f|$, $\bar{f}$, af for any complex number a, and $f(\lambda x)$ for any real λ.

5.3 ***Lemma***: *Almost-periodic functions are bounded.*

PROOF: Let f be almost-periodic. Take $\varepsilon = 1$ and let $\Lambda = \Lambda(1, f)$. For arbitrary $x \in \mathbf{R}$ let τ be a 1-almost-period in the interval $[x - \Lambda, x]$. We have $0 \leqq x - \tau \leqq \Lambda$ and $|f(x) - f(x - \tau)| < 1$, consequently $|f(x)| \leqq \sup_{0 \leqq y \leqq \Lambda} |f(y)| + 1$. ◀

Corollary: *If f is almost-periodic, so is f^2.*

PROOF: Without loss of generality we may assume $|f(x)| \leqq \frac{1}{2}$ for all $x \in \mathbf{R}$. We have $f^2(x-\tau) - f^2(x) = (f(x-\tau) + f(x))(f(x-\tau) - f(x))$ which implies that, for every $\varepsilon > 0$, ε-almost-periods of f are also ε-almost-periods of f^2. ◀

5.4 ***Lemma***: *Almost-periodic functions are uniformly continuous.*

PROOF: Let f be almost-periodic, $\varepsilon > 0$, $\Lambda = \Lambda(\varepsilon/3, f)$. Since f is uniformly continuous on $[0, \Lambda]$, there exists $\eta_0 > 0$ such that for all $|\eta| < \eta_0$

$$\sup_{0 \leqq x \leqq \Lambda} |f(x+\eta) - f(x)| < \frac{\varepsilon}{3}.$$

Let $y \in \mathbf{R}$; we can find an $\varepsilon/3$-almost-period of f say τ, within the interval $[y-\Lambda, y]$, and writing

$$f(y+\eta) - f(y) = (f(y+\eta) - f(y-\tau+\eta)) + \\ +(f(y-\tau+\eta) - f(y-\tau)) + (f(y-\tau) - f(y)),$$

we see that each of the three summands is bounded by $\varepsilon/3$; the first and the third since τ is an $\varepsilon/3$-almost-period, and the second since $0 \leqq y - \tau \leqq \Lambda$ and $|\eta| < \eta_0$. Thus, if $|\eta| < \eta_0$, $|f(x+\eta) - f(x)| < \varepsilon$ for all x, and the proof is complete. ◀

5.5 For a function $f \in L^\infty(\mathbf{R})$ we denote by $W_0(f)$ the set of all translates of f; $W_0(f) = \{f_y\}_{y \in \mathbf{R}}$.†

Theorem: *A function $f \in L^\infty(\mathbf{R})$ is almost-periodic if, and only if, $W_0(f)$ is precompact (in the norm topology of $L^\infty(\mathbf{R})$).*

PROOF: We recall that a set in a complete metric space is precompact (i.e., has a compact closure) if, and only if, it is totally bounded, that is, if for every $\varepsilon > 0$, it can be covered by a union of a finite number of balls of radius ε. Assume first that f is almost-periodic and let us show that $W_0(f)$ is totally bounded in $L^\infty(\mathbf{R})$. Let $\varepsilon > 0$ be given and let $\Lambda = \Lambda(\varepsilon/2, f)$; by the uniform continuity of f we can find numbers $\eta_1, \cdots, \eta_M$ in $[0, \Lambda]$ such that if $0 \leqq y_0 \leqq \Lambda$, $\inf_{1 \leqq j \leqq M} \|f_{y_0} - f_{\eta_j}\|_\infty < \varepsilon/2$. For arbitrary $y \in \mathbf{R}$ let τ be an $\varepsilon/2$-almost-period of f in $[y-\Lambda, y]$; writing $y_0 = y - \tau$ we obtain $0 \leqq y_0 \leqq \Lambda$ and $\|f_y - f_{y_0}\|_\infty < \varepsilon/2$; con-

† Remember the notation $f_y(x) = f(x-y)$.

sequently, $\inf_{1 \leqq j \leqq M} \|f_y - f_{\eta_j}\|_\infty < \varepsilon$ and $W_0(f)$ is covered by the union of balls of radius ε, centered at f_{η_j}, $j = 1, \ldots, M$.

Assume now that $W_0(f)$ is precompact. Let $\varepsilon > 0$ and let $O_1, \ldots, O_M$ be balls of radius $\varepsilon/2$ such that $W_0(f) \subset \bigcup_1^M O_j$. We may clearly assume that $O_j \cap W_0(f) \neq \emptyset$ and hence pick $f_{y_j} \in O_j$, $j = 1, \ldots, M$; the balls of radius ε centered at f_{y_j} clearly cover $W_0(f)$. We claim that every interval J of length $\Lambda = 2 \max_{1 \leqq j \leqq M} |y_j|$ contains an ε-almost-period of f. In fact, denoting by y the middle of J, there exists a j_0 such that $\|f_y - f_{y_{j_0}}\|_\infty < \varepsilon$; writing $\tau = y - y_{j_0}$ it is obvious that $\tau \in J$ and, on the other hand,

$$\|f_\tau - f\|_\infty = \|f_{\tau + y_{j_0}} - f_{y_{j_0}}\|_\infty < \varepsilon .$$

Hence, all that we have to do in order to complete the proof is show that, under the assumption that $W_0(f)$ is precompact, f is continuous.† We show directly that it is uniformly continuous, that is, that $\lim_{\eta \to 0} \|f_\eta - f\|_\infty = 0$. If this were false, there would be an $\varepsilon > 0$ and a sequence $\eta_n \to 0$ such that $\|f_{\eta_n} - f\|_\infty > \varepsilon$; by the precompactness, there would exist a subsequence $\{\eta_{n_j}\}$ such that $f_{\eta_{n_j}}$ converges in $L^\infty(\mathbf{R})$ to some function g which would clearly satisfy $\|f - g\|_\infty \geqq \varepsilon$. But, since for arbitrary $f \in L^\infty(\mathbf{R})$ (actually for arbitrary measurable f), $\eta \to 0$ implies $f(x - \eta) \to f(x)$ in measure,‡ we see that $g(x) = f(x)$ a.e. and this contradiction completes the proof. ◀

5.6 DEFINITION: *The translation convex hull, $W(f)$, of a function $f \in L^\infty(\mathbf{R})$ is the closed convex hull of $\bigcup_{|a| \leqq 1} W_0(af)$. Equivalently, it is the set of uniform limits of functions of the form*

$$\sum a_k f_{x_k}, \qquad x_k \in \mathbf{R}, \sum |a_k| \leqq 1 . \tag{5.1}$$

Remark: If f is uniformly continuous we can define $W(f)$ as the closure of the set of all functions of the form

$$\varphi * f \quad \text{with} \quad \varphi \in L^1(\mathbf{R}), \ \|\varphi\|_{L^1(\mathbf{R})} \leqq 1 . \tag{5.1'}$$

Another observation that will be useful later is:

$$W(e^{i\xi x} f) = \{e^{i\xi x} g;\ g \in W(f)\} . \tag{5.2}$$

† That is: f is equal a.e. to a continuous function.

‡ That is: for every finite interval I and every $\varepsilon > 0$, the measure of

$$I \cap \{x; |f(x - \eta) - f(x)| > \varepsilon\}$$

tends to zero as $\eta \to 0$.

By its very definition $W(f)$ is convex and closed in $L^\infty(\mathbf{R})$. Since $W(f) \supset W_0(f)$, it is clear that if $W(f)$ is compact then $W_0(f)$ is precompact; the converse is also true: if $W_0(f)$ is precompact, there exist for every $\varepsilon > 0$, a finite number of translates $\{f_{y_j}\}_{j=1}^{M}$ such that every translate of f lies within less than ε from f_{y_j} for some $1 \leqq j \leqq M$. Thus, every function of the form (5.1) lies within ε of a function having the form $\sum_{j=1}^{M} b_j f_{y_j}$ with $\sum_1^M |b_j| \leqq 1$. In the unit disc $|b| \leqq 1$ we can pick a finite number of points $\{c_k\}_{k=1}^{N}$ such that every b in the unit disc lies within $\varepsilon M^{-1} \|f\|_{L^\infty(\mathbf{R})}^{-1}$ from one of the c_k's; thus every combination $\sum b_j f_{y_j}$, $\sum |b_j| \leqq 1$ lies within ε of some

$$(5.3) \qquad \sum_1^M b_j' f_{y_j}, \qquad b_j' \in \{c_k\}_{k=1}^{N}.$$

It follows that $W(f)$ is covered by the union of MN balls of radius 3ε centered at the functions of the form (5.3); hence $W(f)$ is precompact and being closed it is compact. We have proved:

Lemma: *$W(f)$ is compact if, and only if, $W_0(f)$ is precompact, that is, if, and only if, $f \in AP(\mathbf{R})$.*

5.7 ***Theorem***: *$AP(\mathbf{R})$ is a closed subalgebra of $L^\infty(\mathbf{R})$.*

PROOF: In order to show that $AP(\mathbf{R})$ is a subspace, we have to show that if $f, g \in AP(\mathbf{R})$ so does $f + g$. We clearly have $W(f + g) \subseteq W(f) + W(g)$ and since, by 5.6, $W(f)$ and $W(g)$ are both compact, $W(f) + W(g)$ is compact and hence $W(f + g)$ is precompact. Since $W(f + g)$ is closed, it is compact, and by 5.6, $f + g \in AP(\mathbf{R})$.

It follows from the corollary 5.3 that $f^2, g^2, (f + g)^2 \in AP(\mathbf{R})$ and consequently $fg = \frac{1}{2}((f + g)^2 - f^2 - g^2)$ is almost-periodic and we have proved that $AP(\mathbf{R})$ is a subalgebra of $L^\infty(\mathbf{R})$. In order to show that it is closed, we consider a function f in its closure. Since f is the uniform limit of continuous functions, it is continuous. Given $\varepsilon > 0$ we can find a $g \in AP(\mathbf{R})$ such that $\|f - g\|_\infty < \varepsilon/3$, and if τ is an $\varepsilon/3$ almost-period of g we have

$$f_\tau - f = (f_\tau - g_\tau) + (g_\tau - g) + (g - f),$$

hence $\|f_\tau - f\|_\infty < \varepsilon/3 + \varepsilon/3 + \varepsilon/3 = \varepsilon$ and τ is an ε-almost-period of f. Thus every interval of length $\Lambda(\varepsilon/3, g)$ contains an ε-almost-period of f, and f is almost-periodic. ◀

5.8 DEFINITION: *A trigonometric polynomial* on $\mathbf{R}$ is a function of the form

$$f(x) = \sum_1^n a_j e^{i\xi_j x}, \qquad \xi_j \in \hat{\mathbf{R}}.$$

The numbers ξ_j are called *the frequencies* of f.

By theorem 5.7, all trigonometric polynomials and all uniform limits of trigonometric polynomials are almost-periodic. The main theorem in the theory of almost-periodic functions states that every almost-periodic function is the uniform limit of trigonometric polynomials, and actually gives a recipe, analogous to Féjer's theorem for periodic functions, for finding the approximating polynomials (see 5.18).

5.9 DEFINITION: The *norm spectrum* of a function $h \in L^\infty(\mathbf{R})$ is the set

$$\sigma(h) = \{\xi; \xi \in \hat{\mathbf{R}},\ a\, e^{i\xi x} \in W(h) \text{ for sufficiently small } a \neq 0\}.$$

$\sigma(h)$ may well be empty even if $h \neq 0$; for instance, if $h \in C_0(\mathbf{R})$ we have $W(h) \subset C_0(\mathbf{R})$ and consequently $\sigma(h) = \varnothing$. We notice that from (5.2) and our definition above it follows immediately that

$$\text{(5.4)} \qquad \sigma(e^{i\xi x} h) = \xi + \sigma(h) = \{\xi + \eta; \eta \in \sigma(h)\}.$$

***Lemma*:** *If* $h \in L^\infty(\mathbf{R})$ *then* $\sigma(h) \subseteq \Sigma(\hat{h})$.

PROOF: Since $\hat{h}_y = e^{i\xi y}\hat{h}$ it is clear that $\Sigma(\hat{h}_y) = \Sigma(\hat{h})$ and consequently $\Sigma(\hat{f}) \subseteq \Sigma(\hat{h})$ for any $f \in W(h)$. If $f = a\,e^{i\xi x}$, then $\hat{f} = a\delta_\xi$ (δ_ξ being the measure of mass one concentrated at ξ) and $\Sigma(\hat{f}) = \{\xi\}$; thus if $\xi \in \sigma(h)$ then $\xi \in \Sigma(\hat{h})$. ◀

5.10 ***Lemma*:** *Let h be bounded and uniformly continuous. Assume that $\eta\mathbf{K}(\eta x) * h$ converges uniformly as $\eta \to 0$ to a limit which is not identically zero. Then $0 \in \sigma(h)$.*

PROOF: Writing $g_\eta = \eta\mathbf{K}(\eta x) * h$ we have $\hat{g}_\eta = \hat{\mathbf{K}}(\xi/\eta)\hat{h}$, hence $\Sigma(\hat{g}_\eta) \subset [-\eta, \eta]$ and it follows that $\Sigma(\widehat{\lim_{\eta\to 0} g_\eta}) = \{0\}$. By 4.11, $\lim_{\eta\to 0} g_\eta$ is a constant, and by the remark following definition 5.6, $g_\eta \in W(h)$ and hence $\lim_{\eta\to 0} g_\eta \in W(h)$; now, $\lim g_\eta$ being a constant different from zero, we obtain $0 \in \sigma(h)$. ◀

***Corollary*:** *Let μ be a measure on $\hat{\mathbf{R}}$ and assume $\mu(\{0\}) \neq 0$. Let $h(x) = \int e^{i\xi x}\, d\mu(\xi)$ (so that $\mu = \hat{h}$); then $0 \in \sigma(h)$.*

PROOF: Keeping the notations above, we have $\hat{g}_\eta = \hat{\mathbf{K}}(\xi/\eta)\mu$ and consequently $\hat{g}_\eta$ tends to $\mu(\{0\})\delta_0$ in $M(\hat{\mathbf{R}})$ which implies $g_\eta \to \mu(\{0\})$ uniformly. ◀

5.11 *Remarks*: It is clear that $0 \in \hat{\mathbf{R}}$ plays no specific role in 5.10; if $\mu(\{\xi\}) \neq 0$ we have $\xi \in \sigma(h)$ (h as above). Also, it is not essential to use Féjer's kernel: if $F \in L^1(\mathbf{R})$, and if we assume that $\eta F(\eta x) * h$ converges uniformly to a nonvanishing limit, it follows that $0 \in \sigma(h)$. This can be seen as follows: given a sequence $\varepsilon_n \to 0$, we can write $F = G_n + H_n$ such that $G_n, H_n \in L^1(\mathbf{R}), \hat{G}_n$ has compact support, say included in $(-c_n, c_n)$, and $\| H_n \|_{L^1(\mathbf{R})} < \varepsilon_n$. Writing $G_{n,\eta}(x) = \eta G_n(\eta x)$, $H_{n,\eta} = \eta H_n(\eta x)$ and noticing that $\| H_{n,\eta} * h \|_{L^\infty(\mathbf{R})} < \varepsilon_n \| h \|$, we obtain $\lim_{n\to\infty} G_{n,\eta} * h = \lim_{\eta\to 0} F_\eta * h$. Remembering that $\Sigma(\widehat{G_{n,\eta} * h}) \subset (-\eta c_n, \eta c_n)$ we obtain, letting $\eta \to 0$ faster than $c_n \to \infty$, $\Sigma(\widehat{\lim F_\eta * h}) = \{0\}$ as before.

The condition of existence of a uniform limit of $F_\eta * h$ as $\eta \to 0$ can clearly be replaced by the less stringent condition of the existence of a nonvanishing limit point, that is, a limit of some sequence $F_{\eta_n} * h$ with $\eta_n \to 0$. We restate these remarks as:

Lemma: *Let $f \in AP(\mathbf{R})$ and assume $0 \notin \sigma(f)$; then for all $F \in L^1(\mathbf{R})$* $\lim_{\eta\to 0} \| \eta F(\eta x) * f \|_{L^\infty(\mathbf{R})} = 0$.

PROOF: Let $F \in L^1(\mathbf{R})$; with no loss of generality we assume $\| F \|_{L^1(\mathbf{R})} \leqq 1$. It follows that $\eta F(\eta x) * f \in W(f)$ and, if it did not tend to zero as $\eta \to 0$, it would have, $W(f)$ being compact, other limit points. By the preceding remarks this would imply $0 \in \sigma(f)$. ◀

5.12 Lemma 5.11 has the following converse:

Lemma: *Let $f \in AP(\mathbf{R})$, $F \in L^1(\mathbf{R})$ and $\int F(x)\,dx \neq 0$. If for some sequence $\eta_n \to 0$, $\lim_{n\to\infty} \| \eta_n F(\eta_n x) * f \|_{L^\infty(\mathbf{R})} = 0$, then $0 \notin \sigma(f)$.*

PROOF: We notice first that for any translate of f, hence for any linear combination of translates, and hence for any $g \in W(f)$, we have $\lim_{n\to\infty} \| \eta_n F(\eta_n x) * g \|_{L^\infty(\mathbf{R})} = 0$. If $g = \text{const}$, $\eta_n F(\eta_n x) * g = \hat{F}(0) g$ and consequently the only constant in $W(f)$ is zero, that is, $0 \notin \sigma(f)$. ◀

5.13 ***Theorem***: *To every $f \in AP(\mathbf{R})$ there corresponds a unique number $M(f)$, called the mean value of f, having the property that $0 \notin \sigma(f - M(f))$.*

PROOF: We have seen before that uniform limit points of $\eta\mathbf{K}(\eta x)*f$ as $\eta\to 0$ are necessarily constants. Since $\eta\mathbf{K}(\eta x)*f\in W(f)$ and since $W(f)$ is compact, there exists a number α such that for an appropriate sequence $\eta_n\to 0$, $\eta_n\mathbf{K}(\eta_n x)*f$ converges uniformly to α. Since $\hat{\mathbf{K}}(0)=1$, $\eta_n\mathbf{K}(\eta_n x)*(f-\alpha)\to 0$ uniformly; hence, by 5.12, $0\notin\sigma(f-\alpha)$. If β is another number such that $0\notin\sigma(f-\beta)$ we obtain, using 5.11, that

$$\eta\mathbf{K}(\eta x)*[(f-\alpha)-(f-\beta)] = \eta\mathbf{K}(\eta x)*(f-\alpha)-\eta\mathbf{K}(\eta x)*(f-\beta)$$

converges to zero uniformly as $\eta\to 0$. However, $\eta\mathbf{K}(\eta x)*[(f-\alpha)-(f-\beta)]$ $=\beta-\alpha$ identically and consequently $\beta=\alpha$. Thus the property $0\notin\sigma(f-\alpha)$ determines α uniquely and we set $M(f)=\alpha$. ◀

Corollary: *If $f\in AP(\mathbf{R})$ and $F\in L^1(\mathbf{R})$, then $\eta F(\eta x)*f$ converges uniformly as $\eta\to 0$ to $\hat{F}(0)M(f)$.*

In particular, taking $F(x)=\begin{cases}\frac{1}{2} & |x|<1\\ 0 & |x|\geqq 1\end{cases}$ writing $T=\eta^{-1}$, and evaluating the convolution at the origin, we obtain:

Corollary: *For $f\in AP(\mathbf{R})$,*

$$M(f) = \lim_{T\to\infty}\frac{1}{2T}\int_{-T}^{T} f(x)\,dx. \tag{5.5}$$

Using the mean value we can determine the norm spectrum of f completely. By (5.4) it is clear that $\xi\in\sigma(f)$ if, and only if, $0\in\sigma(fe^{-i\xi x})$ and consequently

$$\xi\in\sigma(f) \Leftrightarrow M(fe^{-i\xi x})\neq 0. \tag{5.6}$$

By our definition of $M(f)$ and by corollary 5.10 it is clear that if $\hat{f}$ is a measure then $\hat{f}(\{0\})=M(f)$ and similarly

$$\hat{f}(\{\xi\}) = M(fe^{-i\xi x}); \tag{5.6'}$$

thus we can recover the discrete part of $\hat{f}$. We shall soon see that $\hat{f}$ has no continuous part when $f\in AP(\mathbf{R})$.

5.14 The mean value clearly has the basic properties of a translation invariant integral, namely:

$$M(f+g) = M(f)+M(g), \tag{5.7}$$

$$M(af) = aM(f), \tag{5.8}$$

(5.9) $$M(f_y) = M(f) \qquad (\text{where } f_y(x) = f(x-y)).$$

It is also positive:

Lemma: *Assume $f \in AP(\mathbf{R})$, $f(x) \geqq 0$ on $\mathbf{R}$, and f not identically zero. Then $M(f) > 0$.*

PROOF: By (5.9) we may assume $f(0) > 0$ and consequently, if $\alpha > 0$ is small enough, $f(x) > \alpha$ on $-\alpha < x < \alpha$. Let $\Lambda = \Lambda(\alpha/2, f)$; every interval of length Λ contains an $\alpha/2$-almost-period of f, say τ, and $f(x) > \alpha/2$ in $(\tau - \alpha, \tau + \alpha)$. It follows that the integral of f over any interval of length Λ is at least α^2; hence $M(f) \geqq \alpha^2/\Lambda$. ◀

5.15 We define the inner product of almost-periodic functions by:

(5.10) $$\langle f, g \rangle_M = M(f\bar{g})$$

and claim that with the inner product so defined, $AP(\mathbf{R})$ is a pre-Hilbert space, that is, satisfies all the axioms of a Hilbert space except for completeness. The bilinearity of $\langle f, g \rangle_M$ is obvious and the fact that $\langle f, f \rangle_M > 0$ unless $f = 0$ has been established in 5.14. In this pre-Hilbert space, the exponentials $\{e^{i\xi x}\}_{\xi \in \hat{\mathbf{R}}}$ form an orthonormal family, since

$$\langle e^{i\xi x}, e^{i\eta x} \rangle_M = \lim_{T \to \infty} \frac{1}{2T} \int_{-T}^{T} e^{i(\xi - \eta)x} dx = \begin{cases} 1 & \text{if } \xi = \eta \\ 0 & \text{if } \xi \neq \eta . \end{cases}$$

We now introduce the notation†

(5.11) $$\hat{f}(\{\xi\}) = \langle f, e^{i\xi x} \rangle_M = M(f e^{-i\xi x}),$$

that is, $\hat{f}(\{\xi\})$ are the Fourier coefficients of f relative to the orthonormal family $\{e^{i\xi x}\}_{\xi \in \hat{\mathbf{R}}}$. Bessel's inequality now reads

(5.12) $$\sum_{\xi \in \hat{\mathbf{R}}} |\hat{f}(\{\xi\})|^2 \leqq \langle f, f \rangle_M = M(|f|^2)$$

and it follows that $\hat{f}(\{\xi\}) = 0$ except possibly for a countable set of ξ's. Combining this with (5.6) we obtain that for all $f \in AP(\mathbf{R})$, $\sigma(f)$ is countable.

† If $\hat{f}$ is a measure on $\hat{\mathbf{R}}$, (5.11) agrees with (5.6′). By abuse of language we shall sometimes refer to $\hat{f}(\{\xi\})$ for arbitrary $f \in AP(\hat{\mathbf{R}})$, as the mass of the pseudomeasure $\hat{f}$ at ξ.

We have not discussed yet the question whether the orthonormal family $\{e^{i\xi x}\}_{\xi \in \hat{\mathbf{R}}}$ is complete or not. It is well known that the completeness of an orthonormal family is equivalent to the uniqueness theorem, that is, to the statement that zero is the only element all of whose Fourier coefficients (relative to the orthonormal family under consideration) vanish. Thus in our case the completeness of $\{e^{i\xi x}\}_{\xi \in \hat{\mathbf{R}}}$ follows from

Theorem (*uniqueness theorem*): *Let $f \in AP(\mathbf{R})$, $f \neq 0$. Then $\sigma(f) \neq \varnothing$.*

PROOF: Assume first that $\hat{f}$ is a measure on $\hat{\mathbf{R}}$. Saying that $\sigma(f) \neq \varnothing$ is equivalent to saying that $\hat{f}$ has a nonvanishing discrete part. By Wiener's theorem 2.9, if $\hat{f}$ is continuous then $\lim (2T)^{-1}\int_{-T}^{T} |f(x)|^2\, dx = 0$ and, by 5.14, $f = 0$.

We now show that if $f \in AP(\mathbf{R})$ there is always a (nonzero) function $h \in W(f)$ such that $\hat{h}$ is a positive measure on $\hat{\mathbf{R}}$. Since by the previous argument $\sigma(h)$ would then be nonempty, and since $\sigma(h) \subseteq \sigma(f)$, the theorem will follow.

We construct h as follows: assuming $\|f\|_\infty \leqq 1$ we put

$$h_T(x) = \frac{1}{2T} \int_{-T}^{T} \bar{f}(y) f(x+y)\, dy$$

and, noticing that $h_T \in W(f)$ for all $T > 0$ and remembering that $W(f)$ is compact, we take h to be a limit point of h_T as $T \to \infty$, say $h = \lim h_{T_n}$. We have $h(0) = \lim h_{T_n}(0) = M(|f|^2) \neq 0$, hence $h \neq 0$. h is clearly continuous and we claim that it is positive definite: if $x_j \in \mathbf{R}$ and z_j are complex numbers, $j = 1, \dots, N$, then

$$\sum h(x_j - x_k) z_j \bar{z}_k = \lim_{n \to \infty} \frac{1}{2T_n} \int_{-T_n}^{T_n} \sum f(x_j + y) \overline{f(x_k + y)} z_j \bar{z}_k\, dy$$

$$= \lim_{n \to \infty} \frac{1}{2T_n} \int_{-T_n}^{T_n} \left|\sum z_j f(x_j + y)\right|^2 dy \geqq 0.$$

Thus, by Bochner's theorem 2.8, h is the Fourier transform of a positive measure or, equivalently, $\hat{h}$ is a positive measure and the proof is complete. ◀

Corollary: *For $f \in AP(\mathbf{R})$*

$$\sum |\hat{f}(\{\xi\})|^2 = M(|f|^2). \tag{5.13}$$

5.16 ***Theorem***: *If* $f \in AP(\mathbf{R})$ *and* $\hat{f} \in M(\hat{\mathbf{R}})$, *then* $\hat{f} = \sum \hat{f}(\{\xi\})\delta_\xi$, $\|\hat{f}\|_{M(\hat{\mathbf{R}})} = \sum |\hat{f}(\{\xi\})|$, *and* $f(x) = \sum \hat{f}(\{\xi\})\, e^{i\xi x}$.

PROOF: By (5.6)′, $\sum \hat{f}(\{\xi\})\delta_\xi$ is the discrete part of $\hat{f}$, hence $\sum |\hat{f}(\{\xi\})| \leqq \|\hat{f}\|_{M(\hat{\mathbf{R}})}$. If we denote the continuous part of $\hat{f}$ by μ, then μ is the Fourier transform of the almost-periodic function $f - \sum \hat{f}(\{\xi\})e^{i\xi x}$ and, as in the first part of the proof of 5.15, $\mu = 0$; hence $f = \sum \hat{f}(\{\xi\})\, e^{i\xi x}$ and $\|\hat{f}\|_{M(\hat{\mathbf{R}})} = \sum |\hat{f}(\{\xi\})|$. ◀

5.17 For arbitrary $f \in AP(\mathbf{R})$, the series $\sum \hat{f}(\{\xi\})\, e^{i\xi x}$, to which we refer as the Fourier series of f, converges to f in the norm induced by the bilinear form $\langle \cdot, \cdot \rangle_M$. Our next goal is to show that, as in the case of periodic functions, the Fourier series of any $f \in AP(\mathbf{R})$ is summable to f in the uniform norm.

Before describing the summablity process we introduce the mean convolution $f \underset{M}{*} g$ of two almost-periodic functions.

Lemma: *Let* $f, g \in AP(\mathbf{R})$. *Define*:

(5.14) $$(f \underset{M}{*} g)(x) = M_y(f(x-y)g(y)) = \lim_{T\to\infty} \frac{1}{2T}\int_{-T}^{T} f(x-y)g(y)\,dy.$$

Then $f \underset{M}{*} g$ *is almost-periodic and, if* $M(|g|) \leqq 1$, *then* $f * g \in W(f)$. *Furthermore, if g is a trigonometric polynomial, then*

$$(f \underset{M}{*} g)(x) = \sum \hat{f}(\{\xi\})\hat{g}(\{\xi\})\, e^{i\xi x}.$$

PROOF: We may clearly assume from the beginning that $M(|g|) < 1$. It follows that for all T sufficiently large

$$\frac{1}{2T}\int_{-T}^{T} f(x-y)\, g(y)\, dy \in W(f)$$

and, combining the compactness of $W(f)$ with the fact that pointwise (5.14) is well defined, we obtain $f \underset{M}{*} g$ as the uniform limit of $(2T)^{-1}\int_{-T}^{T} f(x-y)\, g(y)\, dy$. If $g(x) = \sum \hat{g}(\{\xi\})\, e^{i\xi x}$ (the sum being finite) then

$$f \underset{M}{*} g = \sum \hat{g}(\{\xi\}) f \underset{M}{*} e^{i\xi x}$$

and since

$$f \underset{M}{*} e^{i\xi x} = M_y(f(x-y)\, e^{i\xi y}) = M_t(f(t)\, e^{i\xi(x-t)}) = \hat{f}(\{\xi\})\, e^{i\xi x},$$

the proof is complete. ◀

Remark: Notice that the function h introduced in the proof of 5.15 is nothing but $f \underset{M}{*} \bar{f}(-x)$.

5.18 Lemma: *Given a finite number of points $\xi_1, \ldots, \xi_N \in \hat{\mathbf{R}}$ and an $\varepsilon > 0$, there exists a trigonometric polynomial* $\mathbf{B}$ *having the following properties:*

(a) $\mathbf{B}(x) \geqq 0$
(b) $M(\mathbf{B}) = 1$
(c) $\hat{\mathbf{B}}(\{\xi_j\}) > 1 - \varepsilon \qquad$ for $j = 1, \ldots, N$.

PROOF: We notice first that if $\xi_1, \ldots, \xi_N$ happen to be integers and if m is an integer larger than $\varepsilon^{-1} \max|\xi_j|$, then the Féjer kernel of order m, namely $\mathbf{K}_m = \sum_{-m}^{m} \left(1 - \frac{|k|}{m+1}\right) e^{ikx}$ has all the properties mentioned. In the general case let $\lambda_1, \ldots, \lambda_q$ be a basis for $\xi_1, \ldots, \xi_N$; that is, $\lambda_1, \ldots, \lambda_q$ are linearly independent over the rationals and every ξ_j can be written in the form $\xi_j = \sum_1^q A_{j,k}\lambda_k$ with integral $A_{j,k}$. Let $\varepsilon_1 > 0$ be such that $(1 - \varepsilon_1)^q > 1 - \varepsilon$, and let $m > \varepsilon_1^{-1} \max_{j,k}|A_{j,k}|$; we contend that $\mathbf{B} = \prod_{k=1}^q \mathbf{K}_m(\lambda_k x)$ has all the required properties. Property (a) is obvious since $\mathbf{B}$ is a product of nonnegative functions. In order to check (b) and (c) we rewrite $\mathbf{B}$ as

$$(5.15) \quad \mathbf{B}(x) = \sum \left(1 - \frac{|k_1|}{m+1}\right) \cdots \left(1 - \frac{|k_q|}{m+1}\right) e^{i(k_1\lambda_1 + \cdots + k_q\lambda_q)x},$$

the summation extending over $|k_1| \leqq m, \ldots, |k_q| \leqq m$. Because of the independence of the λ_j's there is no regrouping of terms having the same frequency and we conclude from (5.15) that $\hat{\mathbf{B}}(\{0\}) =$ the constant term in (5.15) $= M(\mathbf{B}) = 1$, which establishes (b), and

$$\hat{\mathbf{B}}(\{\xi_j\}) = \hat{\mathbf{B}}(\{\textstyle\sum A_{j,k}\lambda_k\}) = \prod_{k=1}^{q} \left(1 - \frac{|A_{j,k}|}{m+1}\right) > (1 - \varepsilon_1)^q > 1 - \varepsilon,$$

which establishes (c). ◀

Theorem: *Let $f \in AP(\mathbf{R})$. Then f can be approximated uniformly by trigonometric polynomials $P_n \in W(f)$.*

PROOF: Since $\sigma(f)$ is countable we can write it as $\{\xi_j\}_{j=1}^{\infty}$. For each n let $\mathbf{B}_n$ be the polynomial described in the lemma for $\xi_1, \ldots, \xi_n$ and $\varepsilon = 1/n$. Write $P_n = f \underset{M}{*} \mathbf{B}_n$. By 5.17, $P_n \in W(f)$ and taking account of (c) above, $\lim \hat{P}_n(\{\xi_j\}) = \hat{f}(\{\xi_j\})$ for every $\xi_j \in \sigma(f)$. If $\xi \notin \sigma(f)$ we have $\hat{P}_n(\{\xi\}) = \hat{f}(\{\xi\}) = 0$ for all n. It follows that if g is a limit point of P_n in $W(f)$, then $\hat{g}(\{\xi\}) = \hat{f}(\{\xi\})$ for all ξ and by the uniqueness theorem $g = f$. Thus, f is the only limit point of the sequence P_n in the compact space $W(f)$ and it follows that P_n converge to f (in norm, i.e., uniformly.) ◀

Corollary: *Every closed translation invariant subspace of* $AP(\mathbf{R})$ *is spanned by exponentials.*

5.19 We finish this section with two theorems providing sufficient conditions for functions to be almost-periodic. Though apparently different they are essentially equivalent and both are derived from the same principle. We start with some preliminary definitions and lemmas.

For $h \in AP(\mathbf{R})$, we say, by abuse of language, that $\hat{h}$ is an *almost-periodic pseudo-measure.*

DEFINITION: A pseudo-measure ν is *almost-periodic at a point* $\xi_0 \in \hat{\mathbf{R}}$, if there exists a function $\varphi \in A(\hat{\mathbf{R}})$, $\varphi(\xi) = 1$ in some neighborhood of ξ_0, such that $\varphi\nu$ is almost-periodic.

It is clear that ν is almost-periodic at ξ_0 if, and only if, $\psi\nu$ is almost-periodic for every $\psi \in A(\hat{\mathbf{R}})$ whose support is sufficiently close to ξ_0 (e.g., within the neighborhood of ξ_0 on which the function φ above is equal to one). In particular, ν is almost-periodic at every $\xi_0 \notin \Sigma(\nu)$.

Lemma: *Let* $\nu \in \mathscr{F}L^\infty$ *and assume that* $\Sigma(\nu)$ *is compact and that* ν *is almost-periodic at every point of* $\Sigma(\nu)$. *Then* ν *is almost-periodic.*

PROOF: By a standard compactness argument we see that there exists an $\eta > 0$ such that $\psi\nu$ is almost-periodic for every $\psi \in A(\hat{\mathbf{R}})$ which is supported by an interval of length η. Let $\psi_j \in A(\hat{\mathbf{R}})$ have their supports contained in intervals of length η, $j = 1, 2, \dots, N$, and such that $\sum_1^k \psi_j = 1$ on a neighborhood of $\Sigma(\nu)$. By the assumption concerning the supports of ψ_j, $\psi_j\nu$ is almost-periodic for all j, and consequently

$$\sum_1^N (\psi_j\nu) = \left(\sum_1^N \psi_j\right)\nu = \nu$$

is almost-periodic. ◀

5.20 Theorem: *Let $h \in L^\infty(\mathbf{R})$ and assume that $\Sigma(\hat{h})$ is compact and that $\hat{h}$ is almost-periodic at every $\xi \in \hat{\mathbf{R}}$ except, possibly, at $\xi = 0$. Then $h \in AP(\mathbf{R})$.*

5.21 Theorem (*Bohr*): *Let $h \in L^\infty(\mathbf{R})$ and assume that it is differentiable and that $h' \in AP(\mathbf{R})$. Then $h \in AP(\mathbf{R})$*

These two theorems are very closely related. We shall first show how theorem 5.20 follows from 5.21, and then prove 5.21.

PROOF OF 5.20: We begin by showing that if $\Sigma(\hat{h})$ is compact, then h is differentiable and $\widehat{h'} = i\xi\hat{h}$ (see exercise 4.6). Let $f \in S(\mathbf{R})$ be such that $\hat{f}(\xi) = 1$ in a neighborhood of $\Sigma(\hat{h})$. We have $\hat{h} = \hat{f}\hat{h}$ and consequently $h = f * h$ or $h(x) = \int f(x-y)h(y)\,dy$. Since h is bounded and $f \in S(\mathbf{R})$ we can differentiate under the integral sign and obtain that h is (infinitely) differentiable and that $h' = f' * h$. Remembering that $\widehat{f'}(\xi) = i\xi$ in a neighborhood of $\Sigma(\hat{h})$, we obtain $\widehat{h'} = \widehat{f'}\hat{h} = i\xi\hat{h}$.

By theorem 4.11′ there exists a sequence $\{\varphi_n\}$ in $L^1(\mathbf{R})$ such that $\hat{\varphi}_n(\xi) = 0$ in a neighborhood of $\xi = 0$, and such that $\|\hat{\varphi}_n - \widehat{f'}\|_{A(\mathbf{R})} \to 0$ This implies (exercise 4.4) that $\|\hat{\varphi}_n\hat{h} - \widehat{h'}\|_{\widehat{\mathscr{F}L^\infty}} \to 0$, that is, h' is the uniform limit of $\varphi_n * h$. Now, since $\hat{\varphi}_n$ vanishes in a neighborhood of $\xi = 0$, it follows from 5.19 that $\varphi_n * h \in AP(\mathbf{R})$; by 5.7, $h' \in AP(\mathbf{R})$, and by 5.21 $h \in AP(\mathbf{R})$. ◀

PROOF OF 5.21: Since h is clearly continuous we only have to show that for every $\varepsilon > 0$ there exists a constant $\Lambda(\varepsilon, h)$ such that every interval of length $\Lambda(\varepsilon, h)$ contains an ε-almost period of h. In view of 5.7 we may consider the real and the imaginary parts of h separately, so that we may assume that h is real valued. Denote

$$(5.16) \qquad M = \sup_x h(x), \qquad m = \inf_x h(x).$$

Let $\varepsilon > 0$. Let x_0 and x_1 be real numbers such that

$$(5.17) \qquad h(x_0) < m + \frac{\varepsilon}{8}, \quad h(x_1) > M - \frac{\varepsilon}{8};$$

we put $\varepsilon_1 = \dfrac{\varepsilon}{4|x_1 - x_0|}$ and claim that if τ is an ε_1-almost period of h' then $h(x_0 - \tau) < m + \varepsilon/2$. In order to see this we write

$$
\begin{aligned}
h(x_1-\tau)-h(x_0-\tau) &= \int_{x_0}^{x_1} h'(x-\tau)\,dx \\
&= \int_{x_0}^{x_1} h'(x)\,dx + \int_{x_0}^{x_1} (h'(x-\tau)-h'(x))\,dx \qquad (5.18)\\
&= h(x_1)-h(x_0)+\int_{x_0}^{x_1} (h'(x-\tau)-h'(x))\,dx,
\end{aligned}
$$

and, since the last integral is bounded by $|x_1-x_0|\varepsilon_1=\varepsilon/4$ it follows from (5.17) and (5.18) that

$$h(x_1-\tau)-h(x_0-\tau) > M-m-\frac{\varepsilon}{2}$$

and, remembering that $h(x_1-\tau)\leqq M$ we obtain that $h(x_0-\tau)<m+\varepsilon/2$.

We now use the points $\{x_0-\tau\}$, where τ is an $\varepsilon_1/2$-almost-period of h' as reference points. Let $\Lambda_1=\Lambda(\varepsilon_1/2,h')$ and define ε_2 by $\varepsilon_2=\min(\varepsilon_1/2,\varepsilon/2\Lambda_1)$. We claim that every ε_2-almost-period of h' is an ε-almost-period of h. In order to prove it let $x\in\mathbf{R}$ and let τ_1 be an ε_2-almost-period of h'; we take τ_0 to be an $\varepsilon_1/2$-almost-period of h' such that $x\leqq x_0-\tau_0\leqq x+\Lambda_1$, and write

$$
\begin{aligned}
h(x-\tau_1)-h(x) &= h(x-\tau_1)-h(x_0-\tau_0-\tau_1)\\
&\quad + h(x_0-\tau_0-\tau_1)-h(x_0-\tau_0)+h(x_0-\tau_0)-h(x) \qquad (5.19)\\
&= h(x_0-\tau_0-\tau_1)-h(x_0-\tau_0)+\int_x^{x_0-\tau_0}(h'(y)-h'(y-\tau_1))\,dy.
\end{aligned}
$$

Since τ_0 and $\tau_0+\tau_1$ are both ε_1-almost-periods we have $m\leqq h(x_0-\tau_0-\tau_1)\leqq m+\varepsilon/2$ and $m\leqq h(x_0-\tau_0)\leqq m+\varepsilon/2$ and consequently $|h(x_0-\tau_0-\tau_1)-h(x_0-\tau_0)|\leqq\varepsilon/2$. The integral in (5.19) is bounded by $\varepsilon_2\Lambda_1\leqq\varepsilon/2$ and it follows that $|h(x-\tau_1)-h(x)|<\varepsilon$. Thus, every interval of length $\Lambda(\varepsilon_2,h')$ contains an ε-almost-period of h and the proof is complete. ◀

5.22 ***Theorem:*** *Let $h\in L^\infty(\mathbf{R})$ and assume that $\Sigma(\hat{h})$ is compact and countable. Then $h\in AP(\mathbf{R})$.*

PROOF: This is a corollary of 5.20. The set of points ξ such that $\hat{h}$ is not almost-periodic at ξ is a subset of $\Sigma(\hat{h})$ and, by 5.20, has no isolated points. Since a countable set contains no nonempty perfect sets, $\hat{h}$ is almost-periodic at every $\xi\in\hat{\mathbf{R}}$ and, by 5.19, $h\in AP(\mathbf{R})$. ◀

EXERCISES FOR SECTION 5

1. Show that $f = \cos 2\pi x + \cos x$ is almost-periodic by showing directly that given $\varepsilon > 0$, there exists an integer M such that at least one of any M consecutive integers lies within ε from an integeral multiple of 2π.

2. Let $h \in L^{\infty}(\mathbf{R})$. Show that $\sigma(h)$ contains every isolated point of $\Sigma(\hat{h})$.

3. Let $f, g \in AP(\mathbf{R})$. Show that $f \underset{M}{*} g = \sum \hat{f}(\{\xi\})\hat{g}(\{\xi\})\, e^{i\xi x}$.

4. Let $f \in AP(\mathbf{R})$ and assume that $W(f)$ is minimal in the sense that if $h \in W(f)$ and $h \neq 0$ then $af \in \mathrm{W}(h)$ for sufficiently small a. Show that f is a constant multiple of an exponential.

5. Let $f \in AP(\mathbf{R})$ and assume that f' is uniformly continuous. Show that $f' \in AP(\mathbf{R})$.

6. Show that the assumption that $\Sigma(\hat{h})$ is compact is essential in the statement of theorem 5.22. *Hint*: Consider discontinuous periodic functions.

7. Show that in the statement of theorem 5.22, the assumption that $\Sigma(\hat{h})$ is compact can be replaced by the weaker condition that h be uniformly continuous.

8. Deduce 5.21 from 5.20.

9. Let P be a trigonometric polynomial on $\mathbf{R}$, and let $\varepsilon > 0$. Show that there exists a positive $\eta = \eta(P, \varepsilon)$ such that if $Q \in L^{\infty}(\mathbf{R})$, $\| Q \| \leqq 1$ and $\Sigma(\hat{Q}) \subset (-\eta, \eta)$, then

$$\mathrm{range}(P + Q) + (-\varepsilon, \varepsilon) \supseteq \mathrm{range}(P) + \mathrm{range}(Q).$$

Hint: The conditions on Q imply that $\| Q' \|_{\infty} \leqq \eta$; see exercise 4.6.

10. Let $\hat{h} \in \mathscr{F}L^{\infty}$, $\xi_0 \in \hat{\mathbf{R}}$, and $\{\eta_n\}$ a sequence tending to zero. Show that if $\hat{\mathbf{K}}(\eta_n^{-1}(\xi - \xi_0))\hat{h}$ tends to a limit (weak-star), then the limit has the form $a\delta_{\xi_0}$. Introducing the notation $a = \hat{h}(\{\xi_0\}, \mathbf{K}, \{\eta_n\})$ show that $\sum | \hat{h}(\{\xi_0\}, \mathbf{K}, \{\eta_n\}) |^2 < \infty$ where the summation extends over all $\xi_0 \in \hat{\mathbf{R}}$ such that weak-star-$\lim_{n \to \infty} \mathbf{K}(\eta_n^{-1}(\xi - \xi_0))\hat{h}$ exists.

11. Let $\hat{h} \in \mathscr{F}L^{\infty}$. Show that for all $\xi_0 \in \hat{\mathbf{R}}$, except possibly countably many, weak-star-$\lim_{\eta \to 0} \mathbf{K}(\eta^{-1}(\xi - \xi_0))\hat{h}$ exists and is equal to zero.

12. Show that if $h \in L^{\infty}(\mathbf{R})$, $\sigma(h)$ is countable.

13. Let B be a homogeneous Banach space on $\mathbf{R}$ such that $AP(\mathbf{R}) \subseteq B \subseteq \mathscr{L}_c$ (see 1.14). Describe the closure in B of $AP(\mathbf{R})$.

6. THE WEAK-STAR SPECTRUM OF BOUNDED FUNCTIONS

6.1 Given a function $h \in L^\infty(\mathbf{R})$, we denote by $[h]$ the smallest translation invariant subspace of $L^\infty(\mathbf{R})$ that contains h; that is, the span of $\{h_y\}_{y \in \mathbf{R}}$. We denote by $\overline{[h]}$ the norm closure of $[h]$ in $L^\infty(\mathbf{R})$, and by $\overline{[h]}_{w*}$ the weak-star closure of $[h]$ in $L^\infty(\mathbf{R})$. Our definition 5.9 of the norm spectrum of h is clearly equivalent to

$$\sigma(h) = \{\xi; e^{i\xi x} \in \overline{[h]}\}$$

and we define the weak-star spectrum by

$$\sigma_{w*}(h) = \{\xi; e^{i\xi x} \in \overline{[h]}_{w*}\}.$$

The problem of weak-star spectral analysis is: given $h \in L^\infty(\mathbf{R})$, find $\sigma_{w*}(h)$. The problem of weak-star spectral synthesis is: for $h \in L^\infty(\mathbf{R})$, is it true that h belongs to the weak-star closure of span $\{e^{i\xi x}\}_{\xi \in \sigma_{w*}(h)}$? The corresponding problems for the uniform topology were studied in section 5. We have obtained some information about $\sigma(h)$ for arbitrary h and complete information in the case that h was almost-periodic (see (5.6)); we proved that the norm spectral synthesis is valid for h if, and only if, $h \in AP(\mathbf{R})$. The problem of weak-star spectral analysis admits the following answer:

Theorem: *For* $h \in L^\infty(\mathbf{R})$, $\sigma_{w*}(h) = \Sigma(\hat{h})$.

PROOF: The subspace of $L^1(\mathbf{R})$ orthogonal to $[h]$ is composed of all the functions $f \in L^1(\mathbf{R})$ satisfying

$$\int f(x)\overline{h(x-y)}\,dx = 0 \qquad \text{for all } y \in \mathbf{R}$$

which is equivalent to

$$f * \bar{h}(-x) = 0. \tag{6.1}$$

We denote this subspace of $L^1(\mathbf{R})$ by $[h]^\perp$.

By the Hahn-Banach theorem, $e^{i\xi x} \in \overline{[h]}_{w*}$ if, and only if,

$$\int f(x)\overline{e^{i\xi x}}\,dx = \hat{f}(\xi) = 0$$

for all $f \in [h]^\perp$.

We thus have an equivalent definition of $\sigma_{w*}(h)$ as the set of all common zeros of $\{\hat{f}; f \in [h]^\perp\}$.

Assume $\xi_0 \notin \Sigma(\hat{h})$; if $\varepsilon > 0$ is small enough $(\xi_0 - \varepsilon, \xi_0 + \varepsilon) \cap \Sigma(\hat{h}) = \varnothing$ and hence if $f \in L^1(\mathbf{R})$ and the support of $\hat{f}$ is contained in $(\xi_0 - \varepsilon, \xi_0 + \varepsilon)$, we have

$$\langle \hat{f}, \hat{h} \rangle = \int f(x)\overline{h(x)}\,dx = 0.$$

We claim that f is orthogonal not only to h, but also to all the translates of h, hence to $[h]$. This follows from

$$\int f(x)\overline{h(x-y)}\,dx = \int f(x+y)\overline{h(x)}\,dx, \tag{6.2}$$

and since the Fourier transform of $f(x+y)$ is $e^{i\xi y}\hat{f}$, hence supported by $(\xi_0 - \varepsilon, \xi_0 + \varepsilon)$, both sides of (6.2) must vanish. There are many functions $\hat{f}$ in $A(\hat{\mathbf{R}})$ supported by $(\xi_0 - \varepsilon, \xi_0 + \varepsilon)$ such that $\hat{f}(\xi_0) \neq 0$; it follows that ξ_0 is not a common zero of $\{\hat{f}; f \in [h]^\perp\}$, hence $\xi_0 \notin \sigma_{w^*}(h)$; this proves $\sigma_{w^*}(h) \subseteq \Sigma(\hat{h})$.

In the course of the proof of the converse inclusion we shall need the following lemma, due to Wiener. The proof of the lemma will come in chapter VIII (see VIII.6.3).

Lemma: *Assume $\hat{f}, \hat{f}_1 \in A(\hat{\mathbf{R}})$ and assume that the support of $\hat{f}_1$ is contained in a bounded interval U on which $\hat{f}$ is bounded away from zero. Then*

$$\hat{f}_1 = \hat{g}\hat{f} \qquad \textit{for some } g \in L^1(\mathbf{R}).$$

To prove $\Sigma(\hat{h}) \subseteq \sigma_{w^*}(h)$, we have to show that if $\xi_0 \notin \sigma_{w^*}(h)$, then $\hat{h}$ vanishes in some neighborhood of ξ_0. Now, since $\xi_0 \notin \sigma_{w^*}(h)$, there exists a function $f \in L^1(\mathbf{R})$ satisfying (6.1) and such that $\hat{f}(\xi_0) \neq 0$ and consequently $\hat{f}$ is bounded away from zero on some neighborhood U of ξ_0. We contend that $\hat{h}$ vanishes in U, a contention that will be proved if we show that if $f_1 \in L^1(\mathbf{R})$ and the support of $\hat{f}_1$ is contained in U then $f_1 * \overline{h(-x)} = 0$. By Wiener's lemma there exists a function $g \in L^1(\mathbf{R})$ such that $\hat{f}_1 = \hat{g}\hat{f}$ or equivalently $f_1 = g * f$. Now

$$f_1 * \overline{h(-x)} = (g * f) * \overline{h(-x)} = g * (f * \overline{h(-x)}) = 0$$

and the proof is complete. ◀

6.2 The Hahn-Banach theorem, used as in the foregoing proof, gives a convenient restatement of the problem of spectral synthesis. We introduce first the following notations: if E is a closed set on $\hat{\mathbf{R}}$ we denote

(6.3) $$\check{I}(E) = \{f; f \in L^1(\mathbf{R}), \hat{f}(\xi) = 0 \text{ on } E\}$$

and

(6.4) $$\Omega(E) = \{g; g \in L^\infty(\mathbf{R}) \text{ and } \langle f, g\rangle = 0 \quad \text{for all } f \in \check{I}(E)\}.$$

$\check{I}(E)$ is clearly the orthogonal complement in $L^1(\mathbf{R})$ to the span of $\{e^{i\xi x}\}_{\xi \in E}$ and $\Omega(E)$ is the orthogonal complement in $L^\infty(\mathbf{R})$ of $\check{I}(E)$. By the Hahn-Banach theorem $\Omega(E)$ is precisely the weak-star closure of span $\{e^{i\xi x}\}_{\xi \in E}$ and the problem of (weak-star) spectral synthesis for $h \in L^\infty(\mathbf{R})$ can be formulated as: is it true that $h \in \Omega(\sigma_{w*}(h))$? Equivalently, is it true that for $\hat{f} \in A(\hat{\mathbf{R}})$,

(6.5) $$\hat{f}(\xi) = 0 \text{ on } \sigma_{w*}(h) \;\Rightarrow\; \langle f, h\rangle = 0?$$

or: is it true that, ($\hat{f} \in A(\hat{\mathbf{R}})$)

(6.6) $$\hat{f}(\xi) = 0 \text{ on } \Sigma(\hat{h}) \;\Rightarrow\; \hat{f}\hat{h} = 0?$$

(The equivalence of (6.5) and (6.6) follows from (6.2).)

Theorem: *Let $\hat{f} \in A(\hat{\mathbf{R}})$ and $\hat{h} \in \mathscr{F}L^\infty$ and assume that $\hat{f}(\xi) = 0$ on $\Sigma(\hat{h})$. Then $\Sigma(\hat{f}\hat{h})$ is a perfect subset of $\Sigma(\hat{f}) \cap \text{bdry}(\Sigma(\hat{h}))$.*

PROOF: By 4.10, $\Sigma(\hat{f}\hat{h}) \subseteq \Sigma(\hat{f}) \cap \Sigma(\hat{h})$ and since $\hat{f}$ vanishes on $\Sigma(\hat{h})$, no interior point of $\Sigma(\hat{h})$ is in $\Sigma(\hat{f})$. Let ξ_0 be an isolated point of $\Sigma(\hat{f}\hat{h})$; with no loss of generality we may assume $\xi_0 = 0$ and that $[-\eta, \eta]$ contains no other point of $\Sigma(\hat{f}\hat{h})$. Writing $\widehat{\mathbf{K}_\eta}(\xi) = \hat{\mathbf{K}}(\eta^{-1}\xi) = \sup(0, 1 - |\eta^{-1}\xi|)$ we see that $\Sigma(\widehat{\mathbf{K}_\eta}\hat{f}\hat{h}) = \{0\}$ and consequently (see 4.11) $\widehat{\mathbf{K}_\eta}\hat{f}\hat{h} = a\delta$, a being a constant $\neq 0$ and δ being the unit mass concentrated at $\xi = 0$. By 4.11′ there exists a function $g \in L^1(\mathbf{R})$ such that $\hat{g}$ vanishes in a neighborhood of $\xi = 0$, say in $(-\eta_1, \eta_1)$, and such that $\|g - f\|_{L^1(\mathbf{R})} < (|a|/2)\,\|h\|^{-1}_{L^\infty(\mathbf{R})}$ (remember that $\hat{f}(0) = 0$). Since $\|\widehat{\mathbf{K}_\eta}\| = 1$, we have $\|\widehat{\mathbf{K}_\eta}(\hat{f} - \hat{g})\hat{h}\|_{\mathscr{F}L^\infty} < |a|/2$ and, multiplying everything by $\widehat{\mathbf{K}_{\eta_1}}$, we obtain, (remembering that $\widehat{\mathbf{K}_{\eta_1}}\hat{g} = 0$) $|a| = \|a\delta\|_{\mathscr{F}L^\infty} < |a|/2$ which is a contradiction. Thus $\Sigma(\hat{f}\hat{h})$ has no isolated points and the proof is complete. ◀

Corollary: *If $\Sigma(\hat{h})$ has countable boundary then h admits weak-star spectral synthesis; that is, $h \in \Omega(\sigma_{w*}(h))$.*

We recall that if $\Sigma(\hat{h})$ itself, and not just its boundary, is countable, and if h is uniformly continuous, then $h \in AP(\mathbf{R})$ (theorem 5.22), that is, admits norm spectral synthesis.

Weak-star spectral synthesis is closely related to the structure of closed ideals in $A(\hat{\mathbf{R}})$, and we shall discuss it further in chapter VIII. In particular, we shall show that weak-star spectral synthesis in $\mathscr{F}L^\infty$ is not always possible.

7. THE PALEY-WIENER THEOREMS

7.1 Our purpose in this section is to study the relationship between properties of analyticity and growth of a function on $\mathbf{R}$, and the growth of its Fourier transform on $\hat{\mathbf{R}}$. The situation is similar to, though not as simple as, the case of functions on the circle. We have seen in chapter I (see exercise I.4.4) that a function f, defined on $\mathbf{T}$, is analytic if, and only if, $\hat{f}(n)$ tends to zero exponentially as $|n| \to \infty$. The simplicity of this characterization of analytic functions on $\mathbf{T}$ is due to the compactness of $\mathbf{T}$. If we consider the canonical identification of $\mathbf{T}$ with the unit circle in the complex plane (i.e. $t \leftrightarrow e^{it}$), then a function f is analytic on $\mathbf{T}$ (i.e., is locally the sum of a convergent power series) if, and only if, f is the restriction to $\mathbf{T}$ of a function F, holomorphic in some annulus, concentric and containing the unit circle. This function F is automatically bounded in an annulus containing the unit circle, and the Fourier series of f is simply the restriction to $\mathbf{T}$ of the Laurent expansion of F.

Considering $\mathbf{R}$ as the real axis in the complex plane, it is clear that a function f is analytic on $\mathbf{R}$ if, and only if, it is the restriction to $\mathbf{R}$ of a function F, holomorphic in some domain containing $\mathbf{R}$; however, this domain need not contain a whole strip $\{z; z = x + iy, |y| < a\}$, nor need F be bounded in strips around $\mathbf{R}$ or on $\mathbf{R}$ itself (cf. exercises 1 through 3 at the end of this section). If we assume exponential decrease of $\hat{f}$ at infinity we can deduce more than just the analyticity of f on $\mathbf{R}$; in fact, writing

$$F(z) = \frac{1}{2\pi} \int \hat{f}(\xi)\, e^{i\xi z}\, d\xi,$$

we see that if $\hat{f}(\xi) = O(e^{-a|\xi|})$ for some $a > 0$, F is well defined and holomorphic in the strip $\{z; |y| < a\}$, is bounded in every strip $\{z; |y| < a_1\}, a_1 < a$, and, by the inversion formula†, $F|_{\mathbf{R}} = f$. Under the same assumption we obtain also that, since $\hat{f} \in L^2(\hat{\mathbf{R}})$, $f \in L^2(\mathbf{R})$;

† $F|_{\mathbf{R}}$ denotes the restriction of F to $\mathbf{R}$.

and since for $|y|<a$, $F(x+iy)$ is the inverse Fourier transforms of $e^{-\xi y}\hat{f}$, we see that, as a function of x, $F(x+iy)\in L^2(\mathbf{R})$ for all $|y|<a$. Even with all this added information about the analytic function extending f to a strip, we cannot obtain exponential decrease of $\hat{f}$; we can only obtain that $e^{\xi y}\hat{f}\in L^2(\hat{\mathbf{R}})$ for all $|y|<a$.

Theorem *(Paley-Wiener): For $f\in L^2(\mathbf{R})$, the following two conditions are equivalent:*

(1) *f is the restriction to* **R** *of a function F holomorphic in the strip $\{z; |y|<a\}$ and satisfying*

(7.1) $$\int |F(x+iy)|^2\,dx \leqq \text{const} \qquad |y|<a\,.$$

(2) $$e^{a|\xi|}\hat{f}\in L^2(\hat{\mathbf{R}})\,.$$

PROOF: (2) ⇒ (1); write

(7.2) $$F(z) = \frac{1}{2\pi}\int \hat{f}(\xi)\,e^{i\xi z}\,d\xi;$$

then by the inversion formula $F\big|_{\mathbf{R}} = f$; F is well defined and holomorphic in $\{z; |y|<a\}$, and, by Plancherel's theorem:

$$\int |F(x+iy)|^2\,dx = \frac{1}{2\pi}\int |\hat{f}(\xi)|^2\,e^{2\xi y}d\xi \leqq \|\hat{f}e^{a|\xi|}\|^2_{L^2(\hat{\mathbf{R}})}.$$

(1)⇒(2); write $f_y(x)=F(x+iy)$ (thus $f=f_0$), and consider the Fourier transforms $\hat{f}_y$. We want to show that $\hat{f}_y(\xi)=\hat{f}(\xi)\,e^{-\xi y}$ since then, by Plancherel's theorem and (7.1), we would have $\int|\hat{f}(\xi)|^2e^{2\xi y}d\xi<\text{const}$ for $|y|<a$, which clearly implies (2). Notice that if we assume (2) then, by the first part of the proof, we do have $\hat{f}_y(\xi)=\hat{f}(\xi)\,e^{-\xi y}$.

For $\lambda>0$ and z in the strip $\{z; |y|<a\}$ we put:

(7.3) $$G_\lambda(z) = \mathbf{K}_\lambda * F = \int_{-\infty}^{\infty} F(z-u)\,\mathbf{K}_\lambda(u)\,du\,,$$

$\mathbf{K}$ denoting Féjer's kernel. G_λ is clearly holomorphic in the strip $\{z; |y|<a\}$ and we notice that $g_{\lambda,y}(x) = G_\lambda(x+iy) = \mathbf{K}_\lambda * f_y$ and hence $\hat{g}_{\lambda,y}(\xi) = \widehat{\mathbf{K}_\lambda}(\xi)\hat{f}_y(\xi)$. Now since $\hat{g}_{\lambda,y}(\xi)$ has a compact support (contained in $[-\lambda,\lambda]$) we have $\hat{g}_{\lambda,y}(\xi) = \hat{g}_{\lambda,0}(\xi)\,e^{-\xi y}$ and consequently if $|\xi|<\lambda$, $\hat{f}_y(\xi)=\hat{f}(\xi)\,e^{-\xi y}$. Since $\lambda>0$ is arbitrary, the above holds for all ξ and the proof is complete. ◀

We may clearly replace the "symmetric" conditions of 7.1 by unsymmetric ones. Thus the assumption (7.1) for $-a_1 < y < a$, with $a, a_1 > 0$, is equivalent to: $(e^{a_1 \xi} + e^{-a\xi})\hat{f} \in L^2(\hat{\mathbf{R}})$

7.2 ***Theorem*** *(Paley-Wiener): For $f \in L^2(\mathbf{R})$ the following two conditions are equivalent:*

(1) There exists a function F, holomorphic in the upper half-plane $\{z; y > 0\}$, and satisfying:

(7.4) $$\int |F(x+iy)|^2\,dx < \text{const}, \qquad y > 0$$

and

(7.5) $$\lim_{y \downarrow 0} \int |F(x+iy) - f(x)|^2\,dx = 0.$$

(2) $$\hat{f}(\xi) = 0 \qquad \text{for } \xi < 0.$$

PROOF: (2) $\Rightarrow$ (1); define $F(z)$, for $y > 0$, by (7.2). F is clearly holomorphic, $F(x+iy)$ is the inverse Fourier transform of $e^{-\xi y}\hat{f}$ and, by Plancherel's theorem,

$$\|F(x+iy)\|_{L^2(\mathbf{R})} = \|\hat{f}e^{-\xi y}\|_{L^2(\hat{\mathbf{R}})} \leqq \|\hat{f}\|_{L^2(\hat{\mathbf{R}})},$$

which establishes (7.4), and also

$$\|F(x+iy) - f\|_{L^2(\mathbf{R})} = \|\hat{f}(e^{-\xi y} - 1)\|_{L^2(\hat{\mathbf{R}})} \to 0$$

as $y \downarrow 0$.

(1) $\Rightarrow$ (2); write $f_1(x) = F(x+i)$. By 7.1:

$$\|\hat{f}_1\, e^{-\xi y}\|_{L^2(\hat{\mathbf{R}})} = \|F(x+i+iy)\|_{L^2(\mathbf{R})} \qquad \text{for } -1 < y < \infty$$

and, in particular, by (7.4):

(7.6) $$\int |\hat{f}_1(\xi)|^2 e^{-2\xi y} d\xi \leqq \text{const}.$$

Letting $y \to \infty$, (7.6) clearly implies $\hat{f}_1(\xi) = 0$ for $\xi < 0$. By 7.1, the Fourier transform of $F(x+iy)$ is $\hat{f}_1(\xi)e^{\xi(1-y)}$; hence, by (7.5), $\hat{f}(\xi) = \hat{f}_1(\xi)e^{\xi}$, and $\hat{f}(\xi) = 0$ for $\xi < 0$. ◀

***7.3** The foregoing proofs yield more information than that stated explicitly. The proof of the implication (2) $\Rightarrow$ (1) also shows that F is bounded for $y \geqq \varepsilon > 0$ since $\int |\hat{f}(\xi)e^{-\xi y}|\,d\xi$ is then bounded. In the proof (1) $\Rightarrow$ (2) no mention of f is needed nor is the assumption

(7.5); if we simply assume that F is holomorphic in the upper half-plane and satisfies (7.4), we obtain, keeping the notations of the proof above, that $\hat{f}_1 e^{\xi} \in L^2(\hat{\mathbf{R}})$ and, denoting by f the function in $L^2(\mathbf{R})$ of which $\hat{f}_1 e^{\xi}$ is the Fourier transform, we obtain (7.5) as a consequence (rather than assumption). The Phragmén-Lindelöf theorem allows a further improvement:

Lemma: *Let F be holomorphic in a neighborhood of the closed upper half-plane $\{z; y \geqq 0\}$ and assume that*

$$\int |F(x)|^2\,dx < \infty \tag{7.7}$$

and

$$\lim_{r\to\infty} r^{-1}\log^{+}|F(re^{i\theta})| = 0 \tag{7.8}$$

for all $0 < \theta < \pi$. *Then (7.4) is valid.*

PROOF: Let φ be continuous with compact support on $\mathbf{R}$ and $\|\varphi\|_{L^2} \leqq 1$. Write $G(z) = \varphi * F = \int_{-\infty}^{\infty} F(z-u)\varphi(u)\,du$; then G is holomorphic in $\{z; y \geqq 0\}$, satisfies the condition (7.8), and, on $\mathbf{R}$,†

$$|G(x)| \leqq \|F|_{\mathbf{R}}\|_{L^2}\|\varphi\|_{L^2} \leqq \|F|_{\mathbf{R}}\|_{L^2}$$

By the Phragmén-Lindelöf theorem we have $|G(z)| \leqq \|F|_{\mathbf{R}}\|_{L^2}$ throughout the upper half-plane, which means that, if $y > 0$, $|\int F(x+iy)\varphi(-x)\,dx| \leqq \|F|_{\mathbf{R}}\|_{L^2}$. This being true for every φ (continuous and with compact support) such that $\|\varphi\|_{L^2} \leqq 1$, it follows that

$$\int |F(x+iy)|^2\,dx \leqq \int |F(x)|^2\,dx. \quad ◄$$

7.4 Theorem: *Let F be an entire function and $a > 0$. The following two conditions on F are equivalent:*

(1) $F|_{\mathbf{R}} \in L^2(\mathbf{R})$ *and*

$$|F(z)| = o(e^{a|z|}) \tag{7.9}$$

(2) *There exists a function $\hat{f} \in L^2(\hat{\mathbf{R}})$, $\hat{f}(\xi) = 0$ for $|\xi| > a$, such that*

$$F(z) = \frac{1}{2\pi}\int_{-a}^{a} \hat{f}(\xi)e^{i\xi z}\,d\xi. \tag{7.10}$$

† $F|_{\mathbf{R}}$ denotes the restriction of F to $\mathbf{R}$.

PROOF: (2) ⇒ (1); if (7.10) is valid we have

$$|F(z)| \leqq \|\hat{f}(\xi)e^{-\xi y}\|_{L^1(\hat{\mathbf{R}})} \leqq \|\hat{f}\|_{L^2(\hat{\mathbf{R}})}\left(\frac{1}{2\pi}\int_{-a}^{a} e^{2\xi y}\,d\xi\right)^{1/2}.$$

Now

$$\frac{1}{2\pi}\int_{-a}^{a} e^{2\xi y}d\xi = \frac{1}{4\pi y}(e^{2ay} - e^{-2ay}) \leqq \frac{e^{2a|y|}}{2\pi|y|}$$

and consequently

$$|F(z)| \leqq \frac{e^{a|y|}}{\sqrt{2\pi\,|y|}}\,\|\hat{f}\|_{L^2(\hat{\mathbf{R}})}$$

which is clearly stricter than (7.9). The square summability of $F|_{\mathbf{R}}$ follows from Plancherel's theorem.

(1)⇒(2); assume first that $F|_{\mathbf{R}}$ is bounded. The function $G(z)=e^{iaz}F(z)$ is entire, satisfies (7.9) in the upper half-plane, and $G(iy)\to 0$ as $y\to\infty$. By the Phragmén-Lindelöf theorem, G is bounded in the upper half-plane and, writing $g = G|_{\mathbf{R}}$, it follows from lemma 7.3 and theorem 7.2 that $\hat{g}$ is carried by $(0,\infty)$. Writing $f = F|_{\mathbf{R}}$ we clearly have $\hat{f}(\xi) = \hat{g}(\xi + a)$ which implies $\hat{f}(\xi) = 0$ for $\xi < -a$. Similarly, considering $G_1 = e^{-iaz}F$, we obtain $\hat{f}(\xi) = 0$ for $\xi > a$ and, writing $H(z) = 1/2\pi \int \hat{f}(\xi)\,e^{i\xi z}\,d\xi$, we obtain, by the inversion theorem, $H|_{\mathbf{R}} = F|_{\mathbf{R}}$ so that $H = F$ and (7.10) is established.

In the general case, that is without assuming that F is bounded on $\mathbf{R}$, we consider $F_\varphi(z) = \varphi * F = \int F(z - u)\varphi(u)\,du$ where φ is an arbitrary continuous function with compact support. F_φ satisfies the conditions in (1) and is bounded on $\mathbf{R}$. Writing $f_\varphi = F_\varphi|_{\mathbf{R}}$, we have $\hat{f}_\varphi(\xi) = \hat{f}(\xi)\hat{\varphi}(\xi)$ and $\hat{f}_\varphi(\xi) = 0$ if $|\xi| > a$. Since φ is arbitrary this implies $\hat{f}(\xi) = 0$ for $|\xi| > a$ and the proof is completed as before. ◄

EXERCISES FOR SECTION 7

1. Show that $F(z) = \sum_{n=1}^{\infty} 2^{-n}[(z + n)^2 + n^{-1}]^{-1}$ is analytic on $\mathbf{R}$ and $F|_{\mathbf{R}} \in L^1 \cap L^\infty(\mathbf{R})$; however, F is not holomorphic in any strip $\{z; |y| < a\}$, $a > 0$.

2. Show that for a proper choice of the constants $\{a_n\}$ and $\{b_n\}$ the function

$$G(z) = \sum a_n e^{-b_n(z-n)^2}$$

is entire, $G|_{\mathbf{R}} \in L^1(\mathbf{R})$, but G is unbounded on $\mathbf{R}$.

3. Show that $H(z) = e^{-e^{z^2}}$ is entire, $H|_{\mathbf{R}} \in L^1 \cap L^\infty(\mathbf{R})$; however, H is unbounded on any line $y = \text{const} \neq 0$.

4. Let F be holomorphic in a neighborhood of the strip $\{z; |y| \leqq a\}$ and assume $\int |F(x+iy)|^2 \, dx < \text{const}$ for $|y| \leqq a$. Show that for z in the interior of the strip:

$$F(z) = \frac{1}{2\pi i} \int_{-\infty}^{\infty} \left(\frac{F(u - ia)}{u - ia - z} - \frac{F(u + ia)}{u + ia - z} \right) du.$$

5. Let μ be a measure on $\hat{\mathbf{R}}$, supported by $[-a, a]$. Define:

$$F(z) = \int e^{-i\xi z} \, d\mu(\xi).$$

Show that F is entire and satisfies $F(z) = O(e^{a|y|})$. Give an example to show that F need not satisfy (7.9). *Hint*: $F(z) = \cos az$.

6. Let v be a distribution on $\hat{\mathbf{R}}$, supported by $[-a, a]$. Define $F(z) = \int e^{-i\xi z} dv$. Show that F is entire and that there exists an integer N such that

$$F(z) = O(z^N e^{a|y|}) \qquad \text{as } |z| \to \infty.$$

7. *Titchmarsh's convolution theorem*:

(a) Let F be an entire function of exponential type (i.e., $F(z) = O(e^{a|z|})$ for some $a > 0$) and assume that $|F(x)| \leqq 1$ for all real x and that $F(iy)$ is real valued. Assuming that F is unbounded in the upper half-plane, show that the domain $D = \{z; y > 0, |F(z)| > 2\}$ is symmetric with respect to the imaginary axis, is connected, and its intersection with the imaginary axis is unbounded. *Hint*: Phragmén-Lindelöf.

(b) Let F_1 and F_2 both have the properties of F in part (a) and denote the corresponding domains by D_1, D_2, respectively. Show that $D_1 \cap D_2 \neq \emptyset$ and deduce that $F_1 F_2$ is unbounded in the upper half-plane.

(c) Let $f_j \in L^2(\hat{\mathbf{R}})$, $j = 1, 2$, and assume that f_j are both real valued and carried by $[-a, 0]$. Show that if $f_1 * f_2$ vanishes in a neighborhood of $\xi = 0$, so does at least one of the functions f_j.

Remark: Titchmarsh's theorem is essentially statement (c) above. The assumption that f_j are real valued is introduced to ensure that the corresponding F_j, defined by an integral analogous to (7.10), is real valued on the imaginary axis. This assumption is not essential; in fact, part (c) is an immediate consequence of the Paley-Wiener theorems in the case $f_1 = f_2$ (in which case part (b) is trivial), and the full part (c) can be obtained from it quite simply (see [18]).

*8. THE FOURIER-CARLEMAN TRANSFORM

We sketch briefly another way to extend the domain of the Fourier transformation. There is no aim here at maximum generality and we describe the main ideas using $L^\infty(\mathbf{R})$ as an example, although only minor modifications are needed in order to extend the theory to functions of polynomial growth at infinity or, more generally, to functions whose growth at infinity is slower than exponential. For more details we refer the reader to [3].

8.1 For $h \in L^\infty(\mathbf{R})$ we write

$$(8.1)\qquad \begin{aligned} F_1(h,\zeta) &= \int_{-\infty}^{0} e^{-i\zeta x} h(x)\,dx \qquad &\zeta = \xi + i\eta,\ \eta > 0 \\ F_2(h,\zeta) &= -\int_{0}^{\infty} e^{-i\zeta x} h(x)\,dx \qquad &\zeta = \xi + i\eta,\ \eta < 0. \end{aligned}$$

$F_1(h,\zeta)$ and $F_2(h,\zeta)$ are clearly holomorphic in their respective domains of definition, and it is apparent from (8.1) that if $\eta > 0$, then $F_1(h,\xi+i\eta) - F_2(h,\xi-i\eta)$ is the Fourier transform of $e^{-\eta|x|}h$. Hence if $h \in L^1(\mathbf{R})$ we obtain

$$(8.2)\qquad \lim_{\eta\to 0+} \left(F_1(h,\xi+i\eta) - F_2(h,\xi-i\eta)\right) = \hat{h}(\xi)$$

uniformly. Since $e^{-\eta|x|}h$ tends to h in the weak-star topology for any $h \in L^\infty(\mathbf{R})$, (8.2) is valid for every $h \in L^\infty(\mathbf{R})$ provided $\hat{h}$ is allowed to be a pseudo-measure and the limit is in the weak-star topology of $\mathscr{F}L^\infty$ as dual of $A(\hat{\mathbf{R}})$.

Let us consider the case $h \in L^1(\mathbf{R})$. If I is an interval on $\hat{\mathbf{R}}$ disjoint from the support of $\hat{h}$, and D is the disc of which I is a diameter, and if we define the function F in D by

$$(8.3)\qquad F(h,\zeta) = \begin{cases} F_1(h,\zeta) & \eta \geqq 0 \\ F_2(h,\zeta) & \eta \leqq 0, \end{cases}$$

then it follows from (8.2) that $F(h,\zeta)$ is well defined and continuous in D and it is holomorphic in $D\setminus I$. It is well known that this implies (e.g., by Morera's theorem) that $F(h,\zeta)$ is holomorphic in D. We see that in the case $h \in L^1 \cap L^\infty(\mathbf{R})$, $F_1(h,\zeta)$ and $F_2(h,\zeta)$ are analytic continuations of each other through the complement of $\Sigma(\hat{h})$ on $\hat{\mathbf{R}}$ On the other hand, if $F_1(h,\zeta)$, and $F_2(h,\zeta)$ are the analytic continuation of each other through an open interval I, $\hat{h}(\xi) = F_1(h, \xi) - F_2(h, \xi) = 0$ on I, and $I \cap \Sigma(\hat{h}) = \varnothing$. Denoting by $c(h)$ the set of concordance of $(F_1(h, \zeta), F_2(h, \zeta))$, that is, the set of points on $\hat{\mathbf{R}}$ in the neighborhood of which $F_1(h, \zeta)$ and $F_2(h, \zeta)$ are analytic continuations of each other, we can state our result as

***Lemma*:** *Assume $h \in L^1 \cap L^\infty(\mathbf{R})$; then $\Sigma(\hat{h})$ is the complement of $c(h)$.*

8.2 We now show that the same is true without assuming $h \in L^1(\mathbf{R})$.

***Theorem*:** *For every bounded function h, $\Sigma(\hat{h})$ is the complement of $c(h)$.*

PROOF: Let E be a compact subset of $c(h)$; then, as $\eta \to 0+$ $F_1(h, \xi + i\eta) - F_2(h, \xi - i\eta) \to 0$ uniformly for $\xi \in E$. If $\hat{f} \in A(\hat{\mathbf{R}})$ and the support of $\hat{f}$ is contained in E, then

$$(8.4) \quad \langle \hat{f}, \hat{h} \rangle = \lim_{\eta \to 0+} \frac{1}{2\pi} \int \hat{f}(\xi) \overline{(F_1(h, \xi + i\eta) - F_2(h, \xi - i\eta))}\, d\xi = 0$$

which proves $\Sigma(\hat{h}) \cap c(h) = \varnothing$.

The fact that $\xi_0 \notin \Sigma(\hat{h})$ implies $\xi_0 \in c(h)$ is obtained from lemma 8.1 and the following simple lemma about removable singularities:

8.3 ***Lemma*:** *Let I be an interval on the real line, D the disc in the ζ plane of which I is a diameter, F a holomorphic function defined in $D \setminus I$, satisfying the growth condition*

$$(8.5) \qquad |F(\xi + i\eta)| < \text{const}\, |\eta|^{-n}.$$

Assume that there exists a sequence of functions $\{\Phi_j\}$, holomorphic in D, satisfying (8.5) *(with a constant independent of j) and such that $\Phi_j(\zeta) \to F(\zeta)$ in $D \setminus I$. Then F can be extended to a function holomorphic in D.*

PROOF: Let D_1 be a concentric disc properly included in D and D_2 a concentric disc properly included in D_1. Denote by ξ_1, ξ_2 the points of intersection of the boundary of D_1 with I. The functions $(\zeta - \xi_1)^n(\zeta - \xi_2)^n\Phi_j(\zeta)$ are uniformly bounded on the boundary of D_1, hence in D_1, and consequently Φ_j are uniformly bounded in D_2. The Cauchy integral formula now shows that Φ_j converge uniformly in D_2 to a holomorphic function which agrees with F on $D_2 \setminus I$. Since D_2 is an arbitrary concentric disc in D, the lemma follows. ◀

8.4 ***Lemma***: *Let* $h \in L^\infty(\mathbf{R})$; *then*

$$|F_1(h,\zeta)| \leqq \|h\|_\infty \eta^{-1} \qquad \zeta = \xi + i\eta,\ \eta > 0$$

$$|F_2(h,\zeta)| \leqq \|h\|_\infty |\eta|^{-1} \qquad \zeta = \xi + i\eta,\ \eta < 0.$$

PROOF:

$$|F_1(h,\zeta)| \leqq \int_{-\infty}^0 e^{\eta x}|h(x)|\,dx \leqq \|h\|_\infty \int_{-\infty}^0 e^{\eta x}dx = \|h\|_\infty\ \eta^{-1}$$

and similarly for F_2. ◀

We can now finish the proof of theorem 8.2. We have to show that if $\xi_0 \notin \Sigma(\hat{h})$, then $\xi_0 \in c(h)$. Assume $\xi_0 \notin \Sigma(\hat{h})$; by lemma 4.7, ξ_0 has an interval I about it which does not intersect $\Sigma(\widehat{h\mathbf{K}_\lambda})$ provided λ is large enough ($\mathbf{K}_\lambda =$ the Féjer kernel and $\Sigma(\hat{\mathbf{K}}_\lambda) = [-1/\lambda, 1/\lambda]$). If D is the disc for which I is a diameter, it follows from lemma 8.1 that the pair $(F_1(h\mathbf{K}_\lambda,\zeta),\ F_2(h\mathbf{K}_\lambda,\zeta))$ defines holomorphic functions Φ_λ in D, which clearly converge, as $\lambda \to \infty$, to $(F_1(h,\zeta),\ F_2(h,\zeta))$ on $D \setminus I$. By lemma 8.4 we can apply lemma 8.3 and the theorem follows. ◀

The Fourier-Carleman transform thus gives an alternative definition of the weak-star spectrum of a bounded function. As an illustration we indicate briefly how theorem 4.11 can be obtained by Carleman's method. We assume again $h \in L^\infty(\mathbf{R})$ and $\Sigma(\hat{h}) = \{0\}$. The pair $(F_1(h,\zeta), F_2(h,\zeta))$ defines an analytic function Φ whose only singularity in the finite ζ plane is at the point $\zeta = 0$. By lemma 8.4 and the Phragmén-Lindelöf theorem, Φ tends to zero at infinity and has a simple pole at $\zeta = 0$. Hence, for some constant c, $\Phi(\zeta) = c/i\zeta$, which is the Fourier-Carleman transform of the constant c.

9. KRONECKER'S THEOREM

9.1 ***Theorem*** (*Kronecker*); *Let* $\lambda_1, \lambda_2, \dots, \lambda_n$ *be real numbers, independent over the rationals. Let* $\alpha_1, \dots, \alpha_n$ *be real numbers and* $\varepsilon > 0$. *Then there exists a real number* x *such that*

$$|e^{i\lambda_j x} - e^{i\alpha_j}| < \varepsilon, \qquad j = 1, 2, \dots, n. \tag{9.1}$$

Kronecker's theorem is equivalent to

9.2 ***Theorem***: *Let* $\lambda_1, \dots, \lambda_n$ *be real numbers independent over the rationals,* $\lambda_0 = 0$, *and let* $a_0, a_1, \dots, a_n$ *be any complex numbers. Then*

(9.2) $$\sup_x \left| \sum_{j=0}^{n} a_j e^{i\lambda_j x} \right| = \sum_{j=0}^{n} |a_j|.$$

We first establish the equivalence of 9.1 and 9.2 and then obtain 9.2 as a limit theorem.

PROOF THAT 9.1 $\Rightarrow$ 9.2: Write $a_j = r_j e^{i\alpha_j}$, $r_j \geqq 0$. By 9.1, there exist values of x for which $\left| e^{i\lambda_j x} - e^{i(\alpha_0 - \alpha_j)} \right|$ is small, $j = 1, \ldots, n$. For these values of x, $\sum a_j e^{i\lambda_j x}$ is close to $e^{i\alpha_0} \sum_0^n r_j$. ◀

PROOF THAT 9.2 $\Rightarrow$ 9.1: Consider the polynomial $1 + \sum_1^n e^{-i\alpha_j} e^{i\lambda_j x}$ and notice that its absolute value can be close to $n + 1$ only if all the summands are close to 1, that is, only if (9.1) is satisfied. ◀

Remark: If $\lambda_1, \ldots, \lambda_n, \pi$ are linearly independent over the rationals, we can add the condition $\left| e^{2\pi i x} - 1 \right| < \varepsilon$ which essentially means that we can pick x in (9.1) to be an integer.

Theorem 9.2 is a limiting case of theorem 9.3_N below. The idea in the proof is that used in the proof of lemma V.1.3, that is, the application of Riesz products and of the inequality

(9.3) $$M(fg) \leqq \|f\|_\infty M(|g|)$$

which is clearly valid for $f, g \in AP(\mathbf{R})$ (see (5.5)). Actually, we use (9.3) for polynomials only, in which case the existence of the limit (5.5) and the fact that it equals the constant term are obvious, and this section is essentially independent of section 5.

For the sake of clarity we state 9.3_N first for $N = 1$, as

9.3 ***Theorem***: *Let $\lambda_1, \ldots, \lambda_n$ be real numbers having the following properties*:

(a) $$\sum_1^n \varepsilon_j \lambda_j = 0, \quad \varepsilon_j = -1, 0, 1, \quad \Rightarrow \quad \varepsilon_j = 0 \text{ for all } j.$$

(b) $$\sum_1^n \varepsilon_j \lambda_j = \lambda_k, \quad \varepsilon_j = -1, 0, 1, \quad \Rightarrow \quad \varepsilon_j = 0 \text{ for } j \neq k.$$

Then, for any complex numbers $a_1, \ldots, a_n$,

(9.4) $$\sup_x \left| \sum a_j e^{i\lambda_j x} \right| \geqq \tfrac{1}{2} \sum |a_j|.$$

PROOF: Write $a_j = r_j e^{i\alpha_j}$, $r_j \geqq 0$ and

$$g(x) = \prod_1^n (1 + \cos(\lambda_j x + \alpha_j))$$

$$f(x) = \sum a_j e^{i\lambda_j x}.$$

g is a nonnegative trigonometric polynomial whose frequencies all have the form $\sum \varepsilon_j \lambda_j$, $\varepsilon_j = -1, 0, 1$. By (a), the constant term in g is 1, hence $M(g) = M(|g|) = 1$. By (b), the constant term (which is the same as the mean value) of fg is $\frac{1}{2} \sum_{j=1}^n r_j$ and, by (9.3),

$$\frac{1}{2} \sum_{j=1}^{n} r_j \leqq \sup |f|.$$

9.3$_N$ Theorem: *Let $\lambda_1, \ldots, \lambda_n$ be real numbers having the following properties:*

(a) $\sum_1^n \varepsilon_j \lambda_j = 0, \quad \varepsilon_j$ integers, $|\varepsilon_j| \leqq N \Rightarrow \varepsilon_j = 0$ for all j.

(b) $\sum_1^n \varepsilon_j \lambda_j = \lambda_k, \quad \varepsilon_j$ integers, $|\varepsilon_j| \leqq N \Rightarrow \varepsilon_j = 0$ for $j \neq k$.

Then, for any complex numbers $a_1, \ldots, a_n$,

$$\sup_x \left| \sum a_j e^{i\lambda_j x} \right| \geqq \left(1 - \frac{1}{N+1}\right) \sum |a_j|. \tag{9.4$_N$}$$

PROOF: Virtually identical to that of 9.3; we only have to replace g as defined there by

$$g(x) = \prod_1^n \mathbf{K}_N(\lambda_j x + \alpha_j)$$

where $\mathbf{K}_N(x) = \sum_{-N}^{N} \left(1 - \frac{|j|}{N+1}\right) e^{ijx}$. We leave the details to the reader. ◀

It is clear that if $\lambda_1, \ldots, \lambda_n$ are linearly independent, the conditions of 9.3$_N$ are satisfied for all N and consequently we obtain (9.4$_N$) for all N, hence (9.2). This completes the proof of theorem 9.2 and hence of Kronecker's theorem. ◀

***9.4** The extension of theorem 9.1 to infinite, linearly independent sets presents a certain number of problems, not all of which are solved.

We restrict our attention to compact linearly independent sets E and ask under what conditions is it possible to approximate uniformly on E every function of modulus 1, by an exponential. The obvious answer is that this is possible if, and only if, E is finite; this follows from Kronecker's theorem ("if") and the fact that uniform limits of exponentials must be continuous on E, and if E is infinite (and compact) not all functions of modulus 1 are continuous ("only if"). We therefore modify our questions and ask under what condition is it possible to approximate uniformly on E every *continuous* function of modulus 1 by an exponential. We do not have a satisfactory answer to this question; for some sets E the approximation is possible, for others it is not, and we introduce the following:

DEFINITION: A compact set $E \subset \mathbf{R}$ is a *Kronecker set* if every continuous function of modulus 1 on E can be approximated on E uniformly be exponentials.

The existence of an infinite perfect Kronecker set is not hard to establish by a direct construction. We choose, however, to prove it by a less direct method which also may be used to obtain finer results (see [14]).

Theorem: *Let E be a perfect totally disconnected set on $\mathbf{R}$. Denote by $C_R(E)$ the space of continuous, real-valued functions on E. Then there exists a set G of the first category*† *in $C_R(E)$ such that every $\varphi \in C_R(E) \setminus G$ maps E homeomorphically onto a Kronecker set.*

PROOF: A function $\varphi \in C_R(E)$ maps E homeomorphically onto a Kronecker set if, and only if, for every continuous function h of modulus 1 on E and for every $\varepsilon > 0$, there exists a real number λ such that

$$\sup_{x \in E} \left| e^{i\lambda\varphi(x)} - h(x) \right| < \varepsilon. \tag{9.5}$$

We show first that if we fix h and ε, the set of functions φ for which (9.5) holds for an appropriate λ is everywhere dense in $C_R(E)$. For this, let $\psi \in C_R(E)$ and let $\eta > 0$. We take $\lambda = 10\eta^{-1}$ and write E as a union of disjoint closed subsets E_j, $j = 1, \cdots, N$, the E_j's being small enough so that the variation of either h or $e^{i\lambda\psi}$ on E_j does not exceed

† $C_R(E)$ is a complete metric space, the metric being determined by the norm $\| \varphi \|_\infty = \sup_{x \in E} | \varphi(x) |$.

$\varepsilon/3$. Let $e^{i\alpha_j}$ be a value assumed by h on E_j and $e^{i\beta_j}$ a value assumed by $e^{i\lambda\psi}$ on E_j; we may clearly assume $|\alpha_j| \leqq \pi$ and $|\beta_j| \leqq \pi$ for all j. We now define

$$\text{(9.6)} \qquad \varphi(x) = \psi(x) + \frac{\alpha_j - \beta_j}{\lambda} \qquad \text{for } x \in E_j.$$

We have $\varphi \in C_R(E)$ and $\| \varphi - \psi \|_\infty \leqq 2\pi/\lambda < \eta$; also, checking on each E_j, it is clear that (9.5) holds.

It follows that the set $G(h, \varepsilon)$ of all $\varphi \in C_R(E)$ for which (9.5) holds for *no* $\lambda \in \mathbf{R}$, a set which is clearly closed, is nondense. Taking a sequence of continuous functions of modulus 1, say $\{h_n\}$, which is dense in the set of all such functions, and taking a sequence of positive numbers $\{\varepsilon_m\}$ such that $\varepsilon_m \to 0$, it is clear that $G = \bigcup_{n,m} G(h_n, \varepsilon_m)$ is of the first category. Also, if $\varphi \notin G$, then every h_n can be approximated uniformly on E by $e^{i\lambda\varphi}$ with appropriate λ's hence so can every continuous function of modulus 1 and the theorem follows. ◀

EXERCISES FOR SECTION 9

1. Let $\lambda_1, \ldots, \lambda_n$ be linearly independent over the rationals and let $f_1, \ldots, f_n$ be continuous and periodic on $\mathbf{R}$, having periods $\lambda_1^{-1}, \ldots, \lambda_n^{-1}$, respectively. Show that the closure of the range of $f_1 + f_2 + \cdots + f_n$ is precisely range $(f_1) + \cdots +$ range (f_n). Deduce that if $0 \in \text{range}(f_j)$ for all j, then

$$\left\| \sum f_j \right\|_\infty \geqq \tfrac{1}{6} \sum \| f_j \|_\infty .$$

Hint: Show that if $c_1, \ldots, c_n$ are complex numbers, one can find $\varepsilon_1, \cdots, \varepsilon_n$, the ε_j's being zero or one, such that $\left| \sum \varepsilon_j c_j \right| > \frac{1}{6} \sum |c_j|$.

2. Let $f \in AP(\mathbf{R})$ and assume that $\sigma(f)$ is independent over the rationals. Show that $\hat{f}$ is a measure and that $\| f \|_\infty = \| \hat{f} \|_{M(\hat{\mathbf{R}})}$.

3. Let $f \in AP(\mathbf{R})$ and assume $\sigma(f) \subseteq \{3^{-j}\}_{j=1}^\infty$. Show that $\hat{f}$ is a measure and that $\| \hat{f} \|_{M(\hat{\mathbf{R}})} \leqq 2 \| f \|_\infty$.

4. Let $\lambda_1, \ldots, \lambda_n$ be real numbers. Set $\lambda_0 = 0$ and assume that for any choice of complex numbers $a_0, \ldots, a_n$, (9.2) is valid. Show that $\lambda_1, \ldots, \lambda_n$ are linearly independent over the rationals.

5. Construct a sequence $\{\lambda_j\}$ of linearly independent numbers such that $\lambda_j \to 0$, and such that $\{\lambda_j\} \cup \{0\}$ is not a Kronecker set.

6. Show that every convergent sequence of linearly independent numbers contains an (infinite) subsequence which is a Kronecker set.

Chapter VII

Fourier Analysis on Locally Compact Abelian Groups

We have been dealing so far with spaces of functions defined on the circle group **T**, the group of integers **Z**, or the real line **R** (or $\hat{\mathbf{R}}$). Most of the theory can be carried, without too much effort, to spaces of functions defined on any locally compact abelian group. The interest in such a generalization lies not only in the fact that we have a more general theory, but also in the light it sheds on the "classical" situations. We give only a brief sketch of the theory: proofs, many more facts, and other references can be found in [5], [9], [15] and [24].

1. LOCALLY COMPACT ABELIAN GROUPS

A *locally compact abelian* (LCA) *group* is an abelian group, say G, which is at the same time a locally compact Hausdorff space and such that the group operations are continuous. To be precise: if we write the group operation as addition, the continuity requirement is that both mappings $x \to -x$ of G onto G and $(x, y) \to x + y$ of $G \times G$ onto G are continuous. For a fixed $x \in G$, the mappings $y \to x + y$ is a homeomorphism of G onto itself which takes 0 into x. Thus the topological nature of G at any $x \in G$ is the same as it is at 0.

Examples: (a) Any abelian group G is trivially an LCA group with the discrete topology.

(b) The circle group **T** and the real line **R** with the usual topology.

(c) Let G be an LCA group and H a closed subgroup, then H

with the induced structure is an LCA group. The same is true for the quotient group G/H if we put on it the canonical quotient topology that is, if we agree that a set U in G/H is open if, and only if, its pre-image in G is open.

(d) The direct sum of a finite number of LCA groups is defined as the algebraic direct sum endowed with the product topology; it is again an LCA group.

(e) The *complete direct sum* of a family $\{G_\alpha\}$, $\alpha \in I$, of abelian groups is the group of all "vectors" $\{x_\alpha\}_{\alpha\in I}$, $x_\alpha \in G_\alpha$, where the addition is performed coordinatewise, that is, $\{x_\alpha\} + \{y_\alpha\} = \{x_\alpha + y_\alpha\}$. If for all $\alpha \in I$, G_α is a compact abelian group, the product topology on the complete direct sums make it a compact abelian group. This follows easily from Tychonoff's theorem.

If for every positive integer n, G_n is the group of order two, then the complete direct sum of $\{G_n\}$ is the group of all sequences $\{\varepsilon_n\}$, $\varepsilon_n = 0, 1$ with coordinatewise addition modulo 2, and with the topology that makes the mapping $\{\varepsilon_n\} \to 2 \sum \varepsilon_n 3^{-n}$ a homeomorphism of the group onto the classical cantor set on the line. We denote this particular group by $\mathbf{D}$.

2. THE HAAR MEASURE

Let G be a locally compact abelian group. A Haar measure on G is a positive regular Borel measure μ having the following two properties:

(1) $\mu(E) < \infty$ if E is compact;
(2) $\mu(E + x) = \mu(E)$ for all measurable $E \subset G$ and all $x \in G$.

One proves that a Haar measure always exists and that it is unique up to multiplication by a positive constant; by abuse of language one may therefore talk about *the* Haar measure. The Haar measure of G is finite if, and only if, G is compact and it is then usually† normalized to have total mass one. If $G = \mathbf{T}$ or $G = \mathbf{T}^n$ the Haar measure is simply the normalized Lebesgue measure. If $G = \mathbf{R}$ the Haar measure is again a multiple of the Lebesgue measure. If G is discrete, the Haar measure is usually† normalized to have mass one at each point.

† Except when G is finite; it is as usual to introduce the "compact" normalization as it is the "discrete."

If G is the direct sum of G_1 and G_2, the Haar measure of G is the product measure of the Haar measures of G_1 and G_2. The Haar measure on the complete direct sum of a family of compact groups is the product of the corresponding normalized Haar measures. In particular, the Haar measure on the group $\mathbf{D}$ defined above corresponds to the well-known Lebesgue measure on the Cantor set, the homeomorphism defined above being also measure preserving.

Let G be an LCA group; we denote the Haar measure on G by dx, and the integral of f with respect to the Haar measure by $\int_G f(x)\,dx$ or simply $\int f(x)\,dx$. For $1 \leqq p \leqq \infty$ we denote by $L^p(G)$ the L^p space on G corresponding to the Haar measure. One defines convolution on G by $(f * g)(y) = \int_G f(y-x)g(x)\,dx$ and proves that if $f, g \in L^1(G)$ then $f * g \in L^1(G)$ and $\| f * g \|_{L^1(G)} \leqq \| f \|_{L^1(G)} \| g \|_{L^1(G)}$ so that $L^1(G)$ is a Banach algebra under convolution. We may define homogeneous Banach spaces on any LCA group G as we did for $\mathbf{T}$ or $\mathbf{R}$, that is, as Banach spaces B of locally integrable functions, norm invariant under translation and such that the mappings $y \to f_y$ are continuous from G to B for all $f \in B$. Remembering that for $1 \leqq p < \infty$ the continuous functions with compact support are norm dense in $L^p(G)$, it is clear that $L^p(G)$ is a homogeneous Banach space on G.

Let B be a homogeneous Banach space on an LCA group G. Using vector-valued integration we can extend the definition of convolution so that $f * g$ is defined and belongs to B for all $f \in L^1(G)$ and $g \in B$ and show that $\| f * g \|_B \leqq \| f \|_{L^1(G)} \| g \|_B$.

DEFINITION: A *summability kernel* on the LCA group G is a directed family $\{k_\alpha\}$ in $L^1(G)$ satisfying the following conditions:

(a) $\| k_\alpha \|_{L^1(G)} < \text{const}$;
(b) $\int k_\alpha(x)\,dx = 1$;
(c) if V is an neighborhood of 0 in G, $\lim_\alpha \int_{G \setminus V} | k_\alpha(x) |\,dx = 0$.

If $\{k_\alpha\}$ is a summability kernel on G and if B is a homogeneous Banach space on G, then $\lim_\alpha \| k_\alpha * g - g \|_B = 0$ for all $g \in B$.

3. CHARACTERS AND THE DUAL GROUP

A *character* on an LCA group G is a continuous homomorphism of G into the multiplicative group of complex numbers of modulus 1, that is, a continuous complex-valued function $\xi(x)$ on G satisfying:

$$|\xi(x)| = 1 \qquad \text{and} \qquad \xi(x + y) = \xi(x)\xi(y).$$

The set $\hat{G}$ of all the characters on G is clearly a commutative multiplicative group (under pointwise multiplication). We change the notation and write the group operation of $\hat{G}$ as addition and replace $\xi(x)$ by $<x, \xi>$ or sometimes by $e^{i\xi x}$.

We introduce a topology to $\hat{G}$ by stipulating that convergence in $\hat{G}$ is equivalent to uniform convergence on compact subsets of G (the elements of $\hat{G}$ being functions on G). Thus, a basis of neighborhoods of 0 in $\hat{G}$ is given by sets of the form $\{\xi; | <x, \xi> - 1 | < \varepsilon \text{ for all } x \in K\}$ where K is a compact subset of G and $\varepsilon > 0$. Neighborhoods of other points in $\hat{G}$ are translates of neighborhhods of 0. It is not hard to see that with this topology $\hat{G}$ is an LCA group; we call it *the dual group of* G.

For each $x \in G$, the mapping $\xi \to <x, \xi>$ defines a character on $\hat{G}$. The *Pontryagin duality theorem* states that every character on $\hat{G}$ has this form and that the topology of uniform convergence on compact subsets of $\hat{G}$ coincides with the original topology on G. In other words, if $\hat{G}$ is the dual group of G, then G is the dual of $\hat{G}$.

Examples: (a) For $G = \mathbf{T}$ with the usual topology every character has the form $t \to e^{int}$ for some integer n, the topology of uniform convergence on $\mathbf{T}$ is clearly the discrete topology and $\hat{\mathbf{T}} = \mathbf{Z}$. Similarly, we check $\hat{\mathbf{Z}} = \mathbf{T}$; this illustrates the Pontryagin duality theorem.

The example $G = \mathbf{T}$ hints the following general theorem: *The dual group of any compact group is discrete* (see exercise 5 at the end of this section). Also: *The dual group of every discrete group is compact.*

(b) Characters on $\mathbf{R}$ all have the form $x \to e^{i\xi x}$ for some real ξ. The dual group topology is the usual topology of the reals and $\hat{\mathbf{R}}$ is isomorphic to $\mathbf{R}$.

(c) If H is a closed subgroup of an LCA group G, the annihilator of H, denoted $H^{\perp}$, is the set of all characters of G which are equal to 1 on H. $H^{\perp}$ is clearly a closed subgroup of $\hat{G}$. If $\xi \in H^{\perp}$, ξ defines canonically a character on G/H; on the other hand, every character on G/H defines canonically (by composition with the mapping $G \to G/H$) a character on G. This establishes an algebraic isomorphism between the dual group of G/H and $H^{\perp}$. One checks that this is also a homeomorphism and the dual of G/H can be identified with $H^{\perp}$.

(d) By (c) above and the Pontryagin duality theorem: $\hat{G}/H^{\perp}$ is the dual group of H.

(e) If G_1 and G_2 are LCA groups, then $\widehat{G_1 \oplus G_2}$ can be identified with $\hat{G}_1 \oplus \hat{G}_2$ through

$$<(x_1, x_2), (\xi_1, \xi_2)> \;=\; <x_1, \xi_1> <x_2, \xi_2> .$$

(f) If G_α is a compact abelian group for every α belonging to some index set I, and if G is the complete direct sum of $\{G_\alpha\}$, then $\hat{G}$ can be identified with the direct sum of $\{\hat{G}_\alpha\}$ (with the discrete topology). The *direct sum* of a family $\{\hat{G}_\alpha\}$ of groups is the subgroup of the complete direct sum consisting of those vectors $\{\xi_\alpha\}_{\alpha \in I}$, $\xi_\alpha \in \hat{G}_\alpha$, such that $\xi_\alpha = 0$ in $\hat{G}_\alpha$ for all but a finite number of indices.

The dual group of the group of order two is again the group of order two. Consequently, the dual group of the group $\mathbf{D}$ introduced above is the direct sum of a sequence of groups of order two. If we identify the elements of $\mathbf{D}$ as sequences $\{\varepsilon_n\}$, $\varepsilon_n = 0, 1$, then $\hat{\mathbf{D}}$ is the discrete group of sequences $\{\zeta_n\}$, $\zeta_n = 0, 1$ with only a finite number of ones, and

$$<\{\varepsilon_n\}, \{\zeta_n\}> \;=\; (-1)^{\Sigma \varepsilon_n \zeta_n} .$$

4. FOURIER TRANSFORMS

Let G be an LCA group; the *Fourier transform* of $f \in L^1(G)$ is defined by

$$\hat{f}(\xi) = \int_G \overline{<x, \xi>} f(x)\, dx, \qquad \xi \in \hat{G}.$$

We denote by $A(\hat{G})$ the space of all Fourier transforms of functions in $L^1(G)$. Since we have $\widehat{(f+g)} = \hat{f} + \hat{g}$ and $\widehat{f * g} = \hat{f} * \hat{g}$, $A(\hat{G})$ is an algebra of functions on $\hat{G}$ under the pointwise operations. The functions in $A(\hat{G})$ are continuous on $\hat{G}$; in fact, an equivalent way to define the topology on $\hat{G}$ is as the weak topology determined by $A(\hat{G})$, that is, as the weakest topology for which all the functions in $A(\hat{G})$ are continuous.

With the Haar measures on G and $\hat{G}$ properly normalized one proves inversion formulas stating essentially that $f(-x)$ is the Fourier transform of $\hat{f}$ in some appropriate sense, and literally if f is continuous and $\hat{f} \in L^1(\hat{G})$. One deduces the uniqueness theorem stating that if $f \in L^1(G)$ and $\hat{f} = 0$ then $f = 0$.

From the inversion formulas one can also prove *Plancherel's theorem*. This states that the Fourier transformation is an isometry of $L^1 \cap L^2(G)$ onto a dense subspace of $L^2(\hat{G})$ and can therefore be extended to an isometry of $L^2(G)$ onto $L^2(\hat{G})$. One can now define the Fourier transform of functions in $L^p(G)$, $1 < p < 2$, by interpolation, and obtain inequalities generalizing the Hausdorff-Young theorem (as we did in VI.3 for the case $G = \mathbf{R}$).

We denote by $M(G)$ the space of (finite) regular Borel measures on G. $M(G)$ is a Banach space canonically identified with the dual of $C_0(G)$. The fact that the underlying space G is a group permits the definition of convolution in $M(G)$ (analogous to that which we introduced in I.6 for the case $G = \mathbf{T}$). With the convolution as multiplication $M(G)$ is a Banach algebra. We keep the notation $\mu * \nu$ for the convolution of the measures μ and ν. $L^1(G)$ is identified as a closed subalgebra of $M(G)$ through the correspondence $f \leftrightarrow f dx$.

The *Fourier* (*Fourier-Stieltjes*) *transform* of $\mu \in M(G)$ is defined by

$$\hat{\mu}(\xi) = \int \overline{<x, \xi>}\, d\mu(x), \qquad \xi \in \hat{G}.$$

For all $\mu \in M(G)$, $\hat{\mu}(\xi)$ is uniformly continuous on $\hat{G}$. If $\mu = f dx$ with $f \in L^1(G)$, then $\hat{\mu}(\xi) = \hat{f}(\xi)$. The mapping $\mu \to \hat{\mu}$ is clearly linear and we have $\widehat{\mu * \nu} = \hat{\mu}\hat{\nu}$ so that the family $B(\hat{G}) = \{\hat{\mu}; \mu \in M(G)\}$ of all Fourier-Stieltjes transforms is an algebra of uniformly continuous functions on $\hat{G}$ under pointwise addition and multiplication.

A function φ defined on $\hat{G}$ is called *positive definite* if, for every choice of $\xi_1, \ldots, \xi_N \in \hat{G}$ and complex numbers $z_1, \ldots, z_N$ we have $\sum_{j,k=1}^{N} \varphi(\xi_j - \xi_k) z_j \bar{z}_k \geqq 0$. Weil's generalization of Herglotz-Bochner's theorem states that a function $\varphi(\xi)$ on $\hat{G}$ is the Fourier transform of a positive measure on G if, and only if, it is continuous and positive definite.

5. ALMOST-PERIODIC FUNCTIONS AND THE BOHR COMPACTIFICATION

Let G be an LCA group. A function $f \in L^\infty(G)$ is, by definition, *almost-periodic* if the set of all translates of f, $\{f_y\}_{y \in G}$ is precompact in the norm topology of $L^\infty(G)$ (compare with VI.5.5). We denote the space of all almost-periodic functions on G by $AP(G)$. One proves that almost-periodic functions are uniformly continuous and are uniform limits of trigonometric polynomials on G (i.e., of finite linear combinations of characters). Since trigonometric polynomials are clearly almost-periodic, one obtains that $AP(G)$ is precisely the closure in $L^\infty(G)$ of the space of trigonometric polynomials.

If G is compact we have $AP(G) = C(G)$. In the general case we consider the groups $(\hat{G})_d$, the dual group of G with its topology replaced by the discrete topology, and $\bar{G}$, the dual group of $(\hat{G})_d$. $\bar{G}$ is the group of all homomorphisms of $\hat{G}$ into $\mathbf{T}$, and it therefore contains G (which is identified with the group of all continuous homomorphisms of $\hat{G}$ into $\mathbf{T}$). One proves that the natural imbedding of G into $\bar{G}$ is a continuous isomorphism and that G is dense in $\bar{G}$. Being the dual of a discrete group, $\bar{G}$ is compact; we call it *the Bohr compactification of* G. The Bohr compactification of the real line is the dual group of the discrete real line and is usually called the *Bohr group*.

Assume $f \in AP(G)$; let $\{P_j\}$ be a sequence of trigonometric polynomials which converges to f uniformly. Then, since G is dense in $\bar{G}$, $\{P_j\}$ converges uniformly on $\bar{G}$ (every character on G extends by continuity to a character on $\bar{G}$). It follows that f is the restriction to G of $\lim P_j = F \in C(\bar{G})$. Conversely, since every continuous function F on $\bar{G}$ can be approximated uniformly by trigonometric polynomials, it follows that $AP(G)$ is simply the restriction to G of $C(\bar{G})$.

EXERCISES

1. Let G be an LCA group and μ the Haar measure on G. Show that if U is a nonempty open set in G then $\mu(U) > 0$. *Hint*: Every compact set $E \subset G$ can be covered by a finite number of translates of U.

2. Let G be an LCA group and μ the Haar measure on G. Let H be a compact subgroup. Describe the Haar measure on G/H.

3. Let G_1 and G_2 be compact abelian groups and let $G = G_1 \oplus G_2$. Denote by μ, μ_1, μ_2 the normalized Haar measures on G, G_1, G_2, respectively. Considering μ_j, $j = 1, 2$, as measures on G (carried by the closed subgroups G_j), prove that

$$\mu = \mu_1 * \mu_2 .$$

4. Let G be a compact group and $\{H_n\}$ an increasing sequence of compact subgroups such that $\bigcup_1^\infty H_n$ is dense in G. Denote by μ, μ_n, respectively, the normalized Haar measure of G, H_n, respectively. Considering the μ_n's as measures on G, show that $\mu_n \to \mu$ in the weak-star topology of measures.

5. Let G be a group and let ξ_1 and ξ_2 be distinct characters on G. Show that

$$\sup_{x \in G} \left| <x, \xi_1> - <x, \xi_2> \right| \geqq \sqrt{3}.$$

Deduce that if G is a compact abelian group, then $\hat{G}$ is discrete.

6. Let G be a compact abelian group with normalized Haar measure and let $\xi \in \hat{G}$. Show that

$$\int_G <x, \xi> dx = \begin{cases} 1 & \text{if } \xi = 0 \\ 0 & \text{if } \xi \neq 0. \end{cases}$$

7. Let G be a compact abelian group. Show that the characters on G form a complete orthonormal family in $L^2(G)$.

Chapter VIII

Commutative Banach Algebras

Many of the spaces we have been dealing with are algebras. We used this fact, implicitly or semi-explicitly, but only on the most elementary level. Our purpose in this chapter is to introduce the reader to the theory of commutative Banach algebras and to show, by means of examples, how natural and useful the Banach algebra setting can be in harmonic analysis. There is no claim, of course, that every problem in harmonic analysis has to be considered in this setting; however, if a space under study happens to be either a Banach algebra, or the dual space of one, keeping this fact in mind usually pays dividends. The introduction that we offer here is by no means unbiased. The topics discussed are those that seem to be the most pertinent to harmonic analysis and some very important aspects of the theory of commutative Banach algebras (as well as the entire realm of the noncommutative case) are omitted. As further reading on the theory of Banach algebras we mention [5], [15], [19] and [21].

1. DEFINITION, EXAMPLES, AND ELEMENTARY PROPERTIES

1.1 DEFINITION: A complex *Banach algebra* is an algebra B over the field **C** of complex numbers, endowed with a norm $\|\cdot\|$ under which it is a Banach space and such that

$$\|xy\| \leqq \|x\| \, \|y\| \tag{1.1}$$

for any $x, y \in B$.

Examples: (1) The field **C** of complex numbers, with the absolute value as norm.

(2) Let X be a compact Hausdorff space and $C(X)$ the algebra of all continuous complex-valued functions on X with pointwise addition and multiplication. $C(X)$ is a Banach algebra under the supremum norm (also referred to as the sup-norm)

$$\|f\|_\infty = \sup_{x \in X} |f(x)|. \tag{1.2}$$

(3) Similarly, if X is a locally compact Hausdorff space, we denote by $C_0(X)$ the sup-normed algebra (with pointwise addition and multiplication) of all continuous functions on X which *vanish at infinity* (i.e., the functions f for which $\{x; |f(x)| \geqq \varepsilon\}$ is compact for all $\varepsilon > 0$).

(4) $C^n(\mathbf{T})$—the algebra of all n-times continuously differentiable functions on $\mathbf{T}$ with pointwise addition and multiplication and with the norm

$$\|f\|_{C^n} = \sum_0^n \frac{1}{j!} \max_t |f^{(j)}(t)|$$

(5) $HC(D)$—the algebra of all functions holomorphic in D (the unit disc $\{z; |z| < 1\}$) and continuous in $\bar{D}$, with pointwise addition and multiplication and with the sup-norm.

(6) $L^1(\mathbf{T})$—with pointwise addition and the convolution as multiplication (I.1.8) and with the norm $\| \ \|_{L^1}$. Condition (1.1) is proved in theorem I.1.7. Similarly—$L^1(\mathbf{R})$.

(7) $M(\mathbf{T})$—the space of (Borel) measures on $\mathbf{T}$ with convolution as multiplication and with the norm $\| \ \|_{M(\mathbf{T})}$ (see I.7). Similarly—$M(\mathbf{R})$.

(8) The algebra of linear operators on a Banach space with the standard multiplication and the operator norm.

(9) Let B be a Banach space; we introduce to B the trivial multiplication $xy = 0$ for all $x, y \in B$. With this multiplication B is a Banach algebra.

All the foregoing examples, except (8), have the additional property that the multiplication is commutative. In all that follows we shall deal mainly with commutative Banach algebras.

1.2 In all the examples except for (3), (6), and (9), the algebras have a unit element for the multiplication: the number 1 in (1); the function $f(x) = 1$ in (2), (4), and (5); the unit mass at the origin in (7); and the identity operator in (8). It is clear from I.1.7 that if $f \in L^1(\mathbf{T})$ were a unit element, we would have $\hat{f}(n) = 1$ for all n, which, by the Riemann-Lebesgue lemma is impossible; thus $L^1(\mathbf{T})$ does not have a unit.

Let B be a Banach algebra. We consider the direct sum $B_1 = B \oplus \mathbf{C}$, that is, the set of pairs (x, λ), $x \in B$, $\lambda \in \mathbf{C}$; and define addition, multiplication, scalar multiplication and norm in B_1 by:

$$\begin{aligned}(x_1, \lambda_1) + (x_2, \lambda_2) &= (x_1 + x_2, \lambda_1 + \lambda_2)\\ (x_1, \lambda_1)(x_2, \lambda_2) &= (x_1 x_2 + \lambda_1 x_2 + \lambda_2 x_1, \lambda_1 \lambda_2)\\ \lambda(x_1, \lambda_1) &= (\lambda x_1, \lambda\lambda_1)\\ \|(x, \lambda)\|_{B_1} &= \|x\|_B + |\lambda|.\end{aligned}$$

It is clear, by direct verification, that B_1 is a Banach algebra with a unit element (namely $(0, 1)$). We now identify B with the set of pairs of the form $(x, 0)$, which is clearly an ideal of codimension 1 in B_1. We say that B_1 is obtained from B by a formal adjoining of a unit element; this simple operation allows the reduction of many problems concerning Banach algebras without a unit to the corresponding problems for Banach algebras with unit. If B is an algebra with a unit element we often denote the unit by 1 and identify its scalar multiples with the corresponding complex numbers. Thus we write "$1 \in B$" instead of "B has a unit element," and so on. This notation will be used when convenient and may be dropped when the unit element has been identified differently.

1.3 Every normed algebra, that is, complex algebra with a norm satisfying (1.1) but under which it is not necessarily complete, can be completed into a Banach algebra. This is done in the same way a normed space is completed into a Banach space. If B_0 is a normed algebra, we denote by B the space of equivalence classes of Cauchy sequences in B_0, determined by the equivalence relation:

$$\{x_n\} \sim \{y_n\} \qquad \text{if, and only if,} \qquad \lim \|x_n - y_n\| = 0.$$

One checks immediately, and we leave it to the reader, that if $\{x_n\} \sim \{x_n'\}$ and $\{y_n\} \sim \{y_n'\}$, then $\{x_n + y_n\} \sim \{x_n' + y_n'\}$, $\{\lambda x_n\} \sim \{\lambda x_n'\}$, $\{x_n y_n\} \sim \{x_n' y_n'\}$ and $\lim_{n\to\infty} \|x_n\| = \lim \|x_n'\|$; hence we can define addition, scalar multiplication, multiplication, and norm in B as follows: for $x, y \in B$, let $\{x_n\}$ (resp. $\{y_n\}$) be a Cauchy sequence in the equivalence class x (resp. y), then $x + y$ (resp. $\lambda x, xy$) will be the equivalence class containing $\{x_n + y_n\}$ (resp. $\{\lambda x_n\}, \{x_n y_n\}$) and $\|x\|$ is, by definition, $\lim_{n\to\infty} \| x_n \|$. With these definitions, B is a Banach algebra and the mapping which associates with an element $a \in B_0$ the equivalence class of the "constant" sequence $\{x_n\}$, $x_n = a$ for all n, is an isometric embedding of B_0 in B as a dense subalgebra.

1.4 The condition (1.1) on the norm in a Banach algebra implies the continuity of the multiplication in both factors simultaneously. Conversely:

Theorem: *Let B be an algebra with unit and with a norm $\| \ \|$ under which it is a Banach space. Assume that the multiplication is continuous in each factor separately. Then there exists a norm $\| \ \|'$, equivalent to $\| \ \|$, for which* (1.1) *is valid.*

PROOF: By the continuity assumption, every $x \in B$ defines a continuous linear operator $A_x : y \to xy$ on B. If $x \neq 0$, $A_x(1) = x$, and consequently $A_x \neq 0$; also $A_{x_1 x_2}(y) = x_1 x_2 y = A_{x_1}(x_2 y) = A_{x_1} A_{x_2}(y)$, hence the mapping $x \to A_x$ is an isomorphism of the algebra B into the algebra of all continuous linear operators on B. Let $\| \ \|'$ be the induced norm, that is,

$$\| x \|' = \| A_x \| = \sup_{\|y\| < 1} \| xy \| \tag{1.3}$$

then $\| \ \|'$ is clearly a norm on B and it clearly satisfies (1.1). We also remark that

$$\| x \|' \geqq \| 1 \|^{-1} \| x \| \tag{1.4}$$

(take $y = \| 1 \|^{-1}$ in (1.3)) and, consequently, if x_n is a Cauchy sequence in $\| \ \|'$, it is also a Cauchy sequence in $\| \ \|$, and so converges to some x_0 in B. We contend now that $\lim \| x_n - x_0 \|' = 0$ which is the same as $\lim \| A_{x_n} - A_{x_0} \| = 0$. This follows from: (a) $\{A_{x_n}\}$ is a Cauchy sequence in the algebra of linear operators on B hence converges in norm to some operator A_0; (b) $A_{x_n} y = x_n y \to x_0 y = A_{x_0} y$ for all $y \in B$ (here we use the continuity of xy in x). It follows that $A_0 = A_{x_0}$ and the contention is proved. We have proved that B is complete under the norm $\| \ \|'$, and since the two norms, $\| \ \|$ and $\| \ \|'$ are comparable (1.4), they are in fact equivalent (closed graph theorem). ◄

Remark: The norm $\| \ \|'$ has the additional property that $\| 1 \|' = 1$; hence there is no loss of generality in assuming as we shall henceforth do implicitly, that whenever $1 \in B$, $\| 1 \| = 1$.

1.5 ***Theorem:*** *Let B be a commutative Banach algebra and let I be a closed ideal in B. The quotient algebra B/I endowed with the canonical quotient norm is a Banach algebra.*

PROOF: The only thing to verify is the validity of (1.1). Let $\varepsilon > 0$, let $\tilde{x}, \tilde{y} \in B/I$ and let $x, y \in B$ be representatives of the cosets $\tilde{x}$, $\tilde{y}$, respectively, such that $\|x\| < \|\tilde{x}\| + \varepsilon$, $\|y\| < \|\tilde{y}\| + \varepsilon$. We have $xy \in \tilde{x}\tilde{y}$ and consequently

$$\|\tilde{x}\tilde{y}\| \leqq \|xy\| \leqq \|x\| \, \|y\| \leqq \|\tilde{x}\| \, \|\tilde{y}\| + \varepsilon(\|\tilde{x}\| + \|\tilde{y}\|) + \varepsilon^2$$

and since $\varepsilon > 0$ is arbitrary, $\|\tilde{x}\tilde{y}\| \leqq \|\tilde{x}\| \, \|\tilde{y}\|$. ◀

EXERCISES FOR SECTION 1

1. Verify condition (1.1) in the case of $C^n(\mathbf{T})$ (example 4 above).
2. Let B be a homogeneous Banach space on $\mathbf{T}$, define multiplication in B as convolution (inherited from $L^1(\mathbf{T})$). Show that with this multiplication B is a Banach algebra.
3. Let X be a locally compact, noncompact, Hausdorff space and denote by X_∞ its one-point compactification. Show that $C(X_\infty)$ is isomorphic (though not isometric) to the algebra obtained by formally adjoining a unit to $C_0(X)$.
4. Let B be an algebra with two consistent norms (see IV.1.1), $\| \;\|_0$ and $\| \;\|_1$. Assume that both these norms are multiplicative (i.e., satisfy condition (1.1)). Show that all the interpolating norms $\| \;\|_\alpha$, $0 < \alpha < 1$ (see IV.1.2), are multiplicative. *Hint*: $\mathscr{B}$ is a normed algebra and $\mathscr{B}_\alpha$ are ideals in $\mathscr{B}$.

2. MAXIMAL IDEALS AND MULTIPLICATIVE LINEAR FUNCTIONALS

2.1 Let B be a commutative Banach algebra with a unit 1. An element $x \in B$ is *invertible* if there exists an element $x^{-1} \in B$ such that $xx^{-1} = 1$.

Lemma: *Consider a Banach algebra B with a unit* 1. *Let $x \in B$ and assume $\|x - 1\| < 1$. Then x is invertible and*

$$x^{-1} = \sum_{j=0}^{\infty} (1 - x)^j. \tag{2.1}$$

PROOF: By (1.1), $\|(1 - x)^j\| \leqq \|1 - x\|^j$; hence the series on the right of (2.1) converges in B. Writing $x = 1 - (1 - x)$ and multiplying term be term we obtain $x \sum_{j=0}^{\infty}(1 - x)^j = 1$. ◀

2.2 Lemma: *Let $x \in B$ be invertible and $y \in B$ satisfying $\| y - x \| < \| x^{-1} \|^{-1}$. Then y is invertible and*

$$(2.2) \qquad y^{-1} = x^{-1} \sum_{j=0}^{\infty} (1 - x^{-1}y)^j.$$

PROOF: $\| 1 - x^{-1}y \| \leqq \| x^{-1} \| \, \| x - y \|$. Apply lemma 2.1 to $x^{-1}y$. ◀

Corollary: *The set U of invertible elements in B is open and the function $x \to x^{-1}$ is continuous on U.*

PROOF: We only need to check the continuity, Let $x \in U$, $y \to x$; by (2.2) $y^{-1} - x^{-1} = x^{-1} \sum_{j=1}^{\infty}(1 - x^{-1}y)^j$; hence

$$\| y^{-1} - x^{-1} \| \leqq \sum_{j=1}^{\infty} \| x^{-1} \|^{j+1} \| x - y \|^j \leqq 2 \| x^{-1} \|^2 \| x - y \|$$

provided $\| x - y \| < \frac{1}{2} \| x^{-1} \|^{-1}$. ◀

2.3 DEFINITION: The *resolvent set* $R(x) = R_B(x)$ of an element x in a Banach algebra B with a unit is the set of complex numbers λ such that $x - \lambda$ is invertible.

Lemma: *For $x \in B$, $R(x)$ is open and $F(\lambda) = (x - \lambda)^{-1}$ is a holomorphic B-valued function on $R(x)$.*

PROOF: This is again an immediate consequence of lemma 2.2. If $\lambda_0 \in R(x)$ and λ is close to λ_0, it follows from (2.2) that

$$(2.3) \qquad \begin{aligned} (x - \lambda)^{-1} &= (x - \lambda_0)^{-1} \sum_{0}^{\infty} (1 - (x - \lambda_0)^{-1}(x - \lambda_0 + \lambda_0 - \lambda))^j \\ &= - \sum (x - \lambda_0)^{-j-1}(\lambda_0 - \lambda)^j. \end{aligned}$$

(2.3) is the expansion of $(x - \lambda)^{-1}$ to a convergent power series in $\lambda - \lambda_0$ with coefficients in B. ◀

2.4 Lemma: *$R(x)$ can never be the entire complex plane.*

PROOF: Assume $R(x) = \mathbf{C}$. The function $(x - \lambda)^{-1}$ is an entire B-valued function and as $|\lambda| \to \infty$

$$\|(x-\lambda)^{-1}\| = |\lambda|^{-1}\left\|\left(\frac{x}{\lambda}-1\right)^{-1}\right\| \sim |\lambda|^{-1} \to 0.$$

It follows from Liouville's theorem (see the appendix) that $(x-\lambda)^{-1} \equiv 0$, which is impossible. ◀

Theorem *(Gelfand-Mazur): A complex commutative Banach algebra which is a field is isomorphic to* $\mathbf{C}$.

PROOF: Let $x \in B$, λ a complex number $\lambda \notin R(x)$; then $x-\lambda$ is not invertible and, since the only noninvertible element in a field is zero, $x = \lambda$. Thus, having identified the unit of B with the number 1, B is canonically identified with $\mathbf{C}$. ◀

2.5 We now turn to establish some basic facts about ideals in a Banach algebra.

Lemma: *Let I be an ideal in an algebra B with a unit. Then I is contained in a maximal ideal.*

PROOF: Consider the family $\mathscr{I}$ of all the ideals in B which contain I. $\mathscr{I}$ is partially ordered by inclusion and, by Zorn's lemma, contains a maximal linearly ordered subfamily $\mathscr{I}_0$. The union of all the ideals in $\mathscr{I}_0$ is a proper ideal, since it does not contain the unit element of B, and it is clearly maximal by the maximality of $\mathscr{I}_0$. ◀

Remark: The condition $1 \in B$ in the statement of the lemma can be relaxed somewhat. For instance, if $I \subset B$ is an ideal and if $u \in B$ is such that (u, I)—the ideal generated by u and I—is the whole algebra, then u belongs to no proper ideal containing I, and the union of all the ideals in $\mathscr{I}_0$ (in the proof above) is again a proper ideal since it does not contain u.

2.6 DEFINITION: The ideal $I \subset B$ is *regular* if B has a unit mod$I$; that is, if there exists an element $u \in B$ such that $x - ux \in I$ for all $x \in B$.

If B has a unit element, every $I \subset B$ is regular. If I is regular in B and u is a unit mod$I$ then, since for every $x \in B$, $x = ux + (x - ux)$, we see that $(u, I) = B$. Using the remark 2.5 we obtain:

Lemma: *Let I be a regular ideal in an algebra B. Then I is contained in a (regular) maximal ideal.*

2.7 Lemmas 2.5 and 2.6 did not depend on the topological structure of B. If B is a Banach algebra with a unit it follows from lemma 2.1 that the distance of 1 to any proper ideal is one, and consequently the closure of any proper ideal is again a proper ideal; in particular, maximal ideals in B are closed. Our next lemma shows that the same is true even if B does not have a unit element provided we restrict our attention to regular maximal ideals.

Lemma: *Let I be a regular ideal in a Banach algebra B. Let u be a unit mod I. Then* $\operatorname{dist}(u, I) \geqq 1$.

PROOF: We show that if $v \in B$ and $\| u - v \| < 1$, then $(I, v) = B$; hence $v \notin I$. For $x \in B$ we have

$$(2.4) \quad x = \left(\sum_0^\infty (u-v)^j x - u \sum_0^\infty (u-v)^j x \right) + v \sum_0^\infty (u-v)^j x.$$

The difference $(\sum (u-v)^j x) - u(\sum (u-v)^j x)$ belongs to I since u is a unit mod I, and the third term is a multiple of v; hence $(I, v) = B$ and the lemma is proved. ◀

Corollary: *Regular maximal ideals in a Banach algebra are closed.*

2.8 DEFINITION: A *multiplicative linear functional* on a Banach algebra B is a nontrivial† linear functional $w(x)$ satisfying

$$(2.5) \qquad w(xy) = w(x)w(y), \qquad x, y \in B.$$

Equivalently, it is a homomorphism of B onto the complex numbers.

We do not require in the definition that w be continuous—we can prove the continuity:

Lemma: *Multiplicative linear functionals are continuous and have norms bounded by* 1.

PROOF: Let w be a multiplicative linear functional; denote its kernel by M. M is clearly a regular maximal ideal and is consequently closed. The mapping $x \to w(x)$ identifies canonically the quotient algebra B/M with $\mathbf{C}$, and if we denote by $\| \ \|'$ the norm induced on $\mathbf{C}$ by B/M, we clearly have $\| \lambda \|' = \| 1 \|' |\lambda|$ for all $\lambda \in \mathbf{C}$. By (1.1), $\| 1 \|' \geqq 1$ and hence $|\lambda| \leqq \| \lambda \|'$ for all $\lambda \in \mathbf{C}$; it follows that for any $x \in B$, $|w(x)| \leqq \| w(x) \|' \leqq \| x \|$.

† A linear functional which is not identically zero.

Theorem: *The mapping* $w \to \ker(w)$ *defines a one-one correspondence between the multiplicative linear functionals on* B *and its regular maximal ideals.*

PROOF: A multiplicative linear functional w is completely determined by its kernel M: if $x \in M$ then $w(x) = 0$: if $x \notin M, w(x)$ is the unique complex number for which $x^2 - w(x)x \in M$. On the other hand, if M is a regular maximal ideal in a commutative Banach algebra B, the quotient algebra B/M is a field. Since M is closed B/M is itself a complex Banach algebra (theorem 1.5), and by theorem 2.4, B/M is canonically identified with $\mathbf{C}$. It follows that the mapping $B \to B/M$ is a multiplicative linear functional on B. ◀

Corollary: *Let* B *be a commutative Banach algebra with a unit element. An element* $x \in B$ *is invertible if, and only if,* $w(x) \neq 0$ *for every multiplicative linear functional on* B.

PROOF: If x is invertible $w(x)w(x^{-1}) = 1$ for every multiplicative linear functional w, hence $w(x) \neq 0$. If x is not invertible then xB is a proper ideal which by lemma 2.5, is contained in a maximal ideal M. Since $x = x \cdot 1 \in xB \subseteq M$ it follows that $w(x) = 0$ where w is the multiplicative linear functional whose kernel is M. ◀

2.9 At this point we can already give one of the nicest applications of the theory of Banach algebras to harmonic analysis.

Theorem (*Wiener*): *Let* $f \in A(\mathbf{T})$ *and assume that* f *vanishes nowhere on* $\mathbf{T}$; *then* $f^{-1} \in A(\mathbf{T})$.

PROOF: We have seen in I.6.1 that $A(\mathbf{T})$ is an algebra under pointwise multiplication and that the norm

$$\|f\|_{A(\mathbf{T})} = \sum_{-\infty}^{\infty} |\hat{f}(n)|$$

is multiplicative. Since $A(\mathbf{T})$ is clearly a Banach space (isometric to ℓ^1), it follows that it is a Banach algebra.

Let w be a multiplicative linear functional on $A(\mathbf{T})$; denote $\lambda = w(e^{it})$ (the value of w at the function $e^{it} \in A(\mathbf{T})$). Since $\|e^{it}\|_{A(\mathbf{T})} = 1$ it follows from lemma 2.8 that $|\lambda| \leqq 1$; similarly we obtain that $\lambda^{-1} = w((e^{it})^{-1}) = w(e^{-it})$ satisfies $|\lambda^{-1}| \leqq 1$, and consequently $|\lambda| = 1$, that is $\lambda = e^{it_0}$ for some t_0. By the multiplicativity of $w, w(e^{int}) = e^{int_0}$ for all n; by the linearity, $w(P) = P(t_0)$ for every trigonometric polynomial P; and by the continuity, $w(f)=f(t_0)$ for all $f\in A(\mathbf{T})$. It follows that every multiplicative linear functional on $A(\mathbf{T})$ is an evaluation at some $t_0\in\mathbf{T}$; every $t_0\in\mathbf{T}$ clearly gives rise to such a functional and we have thus identified all multiplicative linear functionals on $A(\mathbf{T})$. Let $f\in A(\mathbf{T})$ such that $f(t)\neq 0$ for all $t\in\mathbf{T}$. By corollary 2.8, f is invertible in $A(\mathbf{T})$, that is, there exists a function $g\in A(\mathbf{T})$ such that $g(t)f(t)=1$ or equivalently $f^{-1}\in A(\mathbf{T})$. ◂

2.10 The algebra $A(\mathbf{T})$ is closely related to $A(\hat{\mathbf{R}})$ and theorem 2.9 can be used in determining the maximal ideals of $A(\hat{\mathbf{R}})$ (or, equivalently, $L^1(\mathbf{R})$); this can also be done directly and our proof below has the advantage of applying for many convolution algebras (see also exercise 4 at the end of this section).

Theorem: *Every multiplicative linear functional on $L^1(\mathbf{R})$ has the form $f\to\hat{f}(\xi_0)$ for some $\xi_0\in\hat{\mathbf{R}}$.*

PROOF: Let w be a multiplicative linear functional on $L^1(\mathbf{R})$. As any linear functional on $L^1(\mathbf{R})$, w has the form $w(f)=\int f(x)\overline{h(x)}\,dx$ for some $h\in L^\infty(\mathbf{R})$. We have

$$w(f*g) = \iint f(x-y)g(y)\overline{h(x)}\,dy\,dx = \iint f(x)g(y)\overline{h(x+y)}dxdy$$

$$w(f)w(g)=\int f(x)\overline{h(x)}\,dx\int g(y)\overline{h(y)}\,dy = \iint f(x)g(y)\overline{h(x)\,h(y)}\,dx\,dy$$

By the multiplicativity of w and the fact that the linear combinations of the form $\sum f_j(x)g_j(y)$, $f_j, g_j\in L^1(\mathbf{R})$ are dense in $L^1(\mathbf{R}\times\mathbf{R})$, it follows that $h(x+y)=h(x)h(y)$ almost everywhere in $\mathbf{R}\times\mathbf{R}$. Thus (see exercise VI.4.7) $h(x)=e^{i\xi_0 x}$ for some $\xi_0\in\hat{\mathbf{R}}$, and $w(f)=\hat{f}(\xi_0)$. ◂

2.11 We shall use the term "*function algebra*" for algebras of continuous functions on a compact or locally compact Hausdorff space with pointwise addition and multiplication. It is clear that if B is a function algebra on a space X and if $x\in X$, then $f\to f(x)$ is either a multiplicative linear functional on B, or zero, and consequently (lemma 2.8) if B is a Banach algebra under a norm $\|\ \|$, we have $|f(x)|\leq\|f\|$ for all $x\in X$ and $f\in B$.

2.12 Let B be a function algebra on a locally compact Hausdorff space X and assume that for all $x\in X$ there exists a function $f\in B$ such that $f(x)\neq 0$. Denote by w_x the multiplicative linear functional

$f \to f(x)$. Recall that B is *separating* on X if for any $x_1, x_2 \in X$, $x_1 \neq x_2$, there exists an $f \in B$ such that $f(x_1) \neq f(x_2)$; this amounts to saying that if $x_1 \neq x_2$ then $w_{x_1} \neq w_{x_2}$. Thus, if B is separating on X and not all the functions in B vanish at any $x \in X$, the mapping $x \to w_x$ identifies X as a set of multiplicative linear functionals on B. In general we obtain only part of the set of multiplicative linear functionals as w_x, $x \in X$ (see exercise 6 at the end of this section); however, in some important cases, every multiplicative linear functional on B has the form w_x for some $x \in X$. We give one typical illustration.

DEFINITION: A function algebra B on a space X is *self-adjoint on* X if whenever $f \in B$ then also $\bar{f} \in B$ (where $\bar{f}(x) = \overline{f(x)}$).

DEFINITION: A function algebra B on a space X is *inverse closed* if $1 \in B$ and whenever $f \in B$ and $f(x) \neq 0$ for all $x \in X$, then $f^{-1} \in B$.

Thus we can restate theorem 2.9 as: "$A(\mathbf{T})$ is inverse closed."

Theorem: *Let B be a separating, self-adjoint, inverse-closed function algebra on a compact Hausdorff space X. Then every multiplicative linear functional on B has the form w_x (i.e., $f \to f(x)$) for some $x \in X$.*

PROOF: If we denote $M_x = \{f; f(x) = 0\}$, or equivalently $M_x = \ker(w_x)$, then, by theorem 2.8, the assertion that we want to prove is equivalent to the assertion that every maximal ideal in B has the form M_x for some $x \in X$. We prove this by showing that every proper ideal is contained in at least one M_x. Let I be an ideal in B and assume $I \nsubseteq M_x$ for all $x \in X$. This means that for every $x \in X$ there exists a function $f \in I$ such that $f(x) \neq 0$. Since f is continuous, $f(y) \neq 0$ for all y in some neighborhood O_x of x. By the compactness of X we can find a finite number of points $x_1, \dots, x_n$ with corresponding $f_j \in I$ and neighborhoods O_j, $i = 1, \dots, n$, such that $X = \bigcup_1^n O_j$ and such that $f_j(y) \neq 0$ for $y \in O_j$. The function $\varphi = \sum_1^n \bar{f}_j f_j$ belongs to I, is positive on X, and since B is assumed to be inverse closed, φ is invertible and $1 \in I$, that is, $I = B$. ◀

Corollary: *Let X be a compact Hausdorff space. Then every multiplicative linear functional on $C(X)$ has the form w_x (i.e., $f \to f(x)$) for some $x \in X$.*

EXERCISES FOR SECTION 2

1. Use the method of the proof of 2.9 to determine all the multiplicative linear functionals on $C(\mathbf{T})$.

2. The same for $C^n(\mathbf{T})$. *Hint*: $\| e^{imt} \|_{C^n} = O(m^n)$.

3. Check the results of exercises 1 and 2 using 2.12.

4. Let G be an LCA group, and let B denote the convolution algebra $L^1(G)$. Show that every multiplicative linear functional on B has the form $f \to \hat{f}(\gamma)$ for some $\gamma \in \hat{G}$. *Hint*: Repeat the proof of 2.10.

5. Determine the multiplicative linear functionals on $HC(D)$ (see section 1, example 5).

6. Let B be the sup-norm algebra of all the continuous functions f on $\mathbf{T}$ such that $\hat{f}(n) = 0$ for all negative integers n. Show that B is a separating function algebra on $\mathbf{T}$; however, not every multiplicative linear functional on B has the form w_t for some $t \in \mathbf{T}$. *Hint*: What is the relationship between B and $HC(D)$?

7. Show that a commutative Banach algebra B may have *no* multiplicative linear functionals (*hint*: example 9 of section 1); however, if $1 \in B$, B has at least one such functional.

8. Determine the multiplicative linear functionals on $C_0(X)$, X being a locally compact Hausdorff space.

3. THE MAXIMAL-IDEAL SPACE AND THE GELFAND REPRESENTATION

3.1 Consider a commutative Banach algebra B and denote by $\mathfrak{M}$ the set of all of its regular maximal ideals. By theorem 2.8 we have canonical identification of every $M \in \mathfrak{M}$ with a multiplicative linear functional, and hence, by lemma 2.8, we can identify $\mathfrak{M}$ with a subset of the unit ball U^* of B^*—the dual space of B. This identification induces on $\mathfrak{M}$ whatever topological structure we have on U^*, and two important topologies come immediately to mind: the norm topology and the weak-star topology. We limit our discussion of the metric induced on $\mathfrak{M}$ by the norm in B^* to exercises 1–3 at the end of this section and refer to [6] for a more complete discussion. The topology induced on $\mathfrak{M}$ by the weak-star topology on B^* is more closely related to the algebraic properties of B; we shall refer to it as *the weak-star topology on* $\mathfrak{M}$.

Lemma: *$\mathfrak{M} \cup \{0\}$ is closed in U^* in the weak-star topology. If $1 \in B$ then $\mathfrak{M}$ is closed.*

PROOF: In order to prove the first statement we have to show that if $u_0 \in \overline{\mathfrak{M}}$, then $u_0(xy) = u_0(x)u_0(y)$ for all $x, y \in B$. From this it would follow that either $u_0 \in \mathfrak{M}$ or $u_0 = 0$. Let $\varepsilon > 0$, $x, y \in B$ and consider the neighborhood of u_0 in U^* defined by

(3.1) $\{u; |u(x) - u_0(x)| < \varepsilon, |u(y) - u_0(y)| < \varepsilon, |u(xy) - u_0(xy)| < \varepsilon\}$;

Since $u_0 \in \overline{\mathfrak{M}}$ there exists a $w \in \mathfrak{M}$ in (3.1), and remembering that $w(xy) = w(x)w(y)$ we obtain

$$|u_0(xy) - u_0(x)u_0(y)| \leqq \varepsilon(1 + \|x\| + \|y\|).$$

Since $\varepsilon > 0$ is arbitrary, $u_0(xy) = u_0(x)u_0(y)$.

In order to prove the second statement we have to show that if $1 \in B$, $0 \notin \overline{\mathfrak{M}}$. Since $w(1) = 1$ for all $w \in \mathfrak{M}$, it follows that $\{u; |u(1)| < \frac{1}{2}\}$ is a neighborhood of 0 disjoint from $\mathfrak{M}$ and the proof is complete. ◀

Since U^*, with the weak-star topology, is a compact Hausdorff space, it follows that the same is true for $\mathfrak{M} \cup \{0\}$ or, if $1 \in B$, for $\mathfrak{M}$. This is sufficiently important to be stated as:

Corollary: *$\mathfrak{M}$, with the weak-star topology, is a locally compact Hausdorff space. If $1 \in B$ then $\mathfrak{M}$ is compact.*

We shall see later (see theorem 3.5) that in some cases the compactness of $\mathfrak{M}$ implies $1 \in B$. However, considering example (9) of section 1, we realize that $\mathfrak{M}$ may be compact (as a matter of fact empty!) for algebras without unit. The reader who feels unconvinced by an example consisting of the empty set should refer to exercise 4 at the end of this section.

3.2 For $x \in B$ and $M \in \mathfrak{M}$ we now write $\hat{x}(M) = x \bmod M$ (i.e., the image of x under the multiplicative linear functional corresponding to M). By its definition, the weak-star topology on $\mathfrak{M}$ is the weakest topology such that all the functions $\{\hat{x}(M); x \in B\}$ are continuous.

Lemma: *If $1 \in B$, the mapping $x \to \hat{x}$ is a homomorphism of norm one of B into $C(\mathfrak{M})$.*

PROOF: The algebraic properties of the mapping are obvious. For every $M \in \mathfrak{M}$ and $x \in B$. $|\hat{x}(M)| \leqq \|x\|$ (lemma 2.8) and hence $\sup_M |\hat{x}(M)| \leqq \|x\|$. On the other hand $\hat{1}(M) = 1$ and the norm of the mapping is not smaller than one. ◀

If we do not assume $1 \in B$, the set $\{M; |\hat{x}(M)| \geqq \varepsilon\}$ is compact in $\mathfrak{M}$ for every $x \in B$ and $\varepsilon > 0$; consequently $x \to \hat{x}$ is a homomorphism of norm at most one, of B into $C_0(\mathfrak{M})$. The subalgebra $\hat{B}$ of $C(\mathfrak{M})$ (resp. $C_0(\mathfrak{M})$) obtained as the image of B under the homomorphism $x \to \hat{x}$ is called *the Gelfand representation* of B. The function $\hat{x}(M)$ is sometimes referred to as the *Fourier-Gelfand transform* of x.

3.3 In many cases we can identify the weak-star topology on $\mathfrak{M}$ concretely in virtue of the following simple fact (cf. [15], p. 6): let τ_1, τ_2 be Hausdorff topologies on a space $\mathfrak{M}$ and assume that $\mathfrak{M}$ is compact in both topologies and that the two topologies are comparable; then $\tau_1 = \tau_2$. In our case this means that if $1 \in B$ and if τ is a Hausdorff topology on $\mathfrak{M}$ in which $\mathfrak{M}$ is compact, and such that all the functions $\hat{x}(M)$ in $\hat{B}$ are τ-continuous, then, since the weak-star topology is weaker than or equal to τ, the two are equal. By a formal adjoining of a unit we obtain, similarly, that if $1 \notin B$ and τ is a Hausdorff topology on $\mathfrak{M}$ such that $\mathfrak{M}$ is locally compact and $\hat{B} \subseteq C_0(\mathfrak{M}, \tau)$, then τ is the weak-star topology on $\mathfrak{M}$.

3.4 DEFINITION: *The radical*, Rad(B), of a commutative Banach algebra B is the intersection of all the regular maximal ideals in B.

Rad(B) is clearly a closed ideal in B and is the kernel of the homomorphism $x \to \hat{x}$ of B onto $\hat{B}$. The radical of B may coincide with B (example 9 of section 1) in which case we say that B is a radical algebra; it may be a nontrivial proper ideal, or it may be reduced to zero.

3.5 DEFINITION: A commutative Banach algebra B is *semisimple* if Rad$(B) = 0$. Equivalently: B is semisimple if the mapping $x \to \hat{x}$ is an isomorphism.

3.6 DEFINITION:† *The spectral norm* of an element $x \in B$, denoted $\| x \|_{sp}$, is $\sup_{M \in \mathfrak{M}} |\hat{x}(M)|$.

The spectral norm can be computed from the B norm by:

Lemma: $$\| x \|_{sp} = \lim_{n \to \infty} \| x^n \|^{1/n}.$$

† The origin of the term is in the fact that the set of complex numbers λ for which $x - \lambda$ is not invertible (assuming $1 \in B$) is commonly called "the spectrum of x" and the spectral norm of x is defined as sup $|\lambda|$ for λ in the spectrum of x. By the corollary 2.8, the spectrum of x coincides with the range of $\hat{x}(M)$, which justifies our definition; we prefer to avoid using the much abused word "spectrum" in any sense other than that of chapter VI.

PROOF: The contention is that the limit on the right exists and is equal to $\|x\|_{sp}$. This follows from the two inequalities:

$$\text{(a)}\quad \|x\|_{sp} \leqq \liminf \|x^n\|^{1/n};$$
$$\text{(b)}\quad \|x\|_{sp} \geqq \limsup \|x^n\|^{1/n}.$$

We notice first that for every n, $\|x\|_{sp}^n = \|x^n\|_{sp} \leqq \|x^n\|$, that is, $\|x\|_{sp} \leqq \|x^n\|^{1/n}$, which proves (a).

In the proof of (b) we assume first that $1 \in B$, and consider $(x - \lambda)^{-1}$ as a function of λ. By lemma 2.3, $(x - \lambda)^{-1}$ is holomorphic for $|\lambda| > \|x\|_{sp}$. If $|\lambda| > \|x\|$ we have

$$(x - \lambda)^{-1} = -\lambda^{-1}(1 - x/\lambda)^{-1} = -\lambda^{-1}\sum x^n\lambda^{-n},$$

and if F is any linear functional on B,

$$\langle (x - \lambda)^{-1}, F\rangle = -\lambda^{-1}\sum \langle x^n, F\rangle\lambda^{-n}$$

is holomorphic, hence convergent for $|\lambda| > \|x\|_{sp}$. Pick any $\lambda > \|x\|_{sp}$; then $\langle \lambda^{-n}x^n, F\rangle$ is bounded (in fact, it tends to zero) for all $F \in B^*$, and it follows from the uniform boundedness theorem that $\lambda^{-n}\|x^n\|$ is bounded, hence $\limsup \|x^n\|^{1/n} \leqq \lambda$. Since λ is any number of modulus greater than $\|x\|_{sp}$, (b) follows.

If $1 \notin B$ we may adjoin a unit formally. Both the norm and the spectral norm of an element $x \in B$ are the same in the extended algebra and since (b) is valid in that algebra, the proof is complete. ◀

Corollary: $x \in \mathrm{Rad}(B) \Leftrightarrow \lim \|x^n\|^{1/n} = 0$.

PROOF: $x \in \mathrm{Rad}(B) \Leftrightarrow \|x\|_{sp} = 0$. ◀

3.7 Lemma 3.6 allows a simple characterization of the Banach algebras for which the spectral norm is equivalent to the original norm. Such an algebra is clearly semisimple, and the Gelfand representation identifies it with a (uniformly) closed subalgebra of the algebra of all continuous functions on its maximal ideal space.

Theorem: *A necessary and sufficient condition for the equivalence of* $\|\ \|_{sp}$ *and the original norm* $\|\ \|$ *of a Banach algebra* B *is the existence of a constant* K *such that* $\|x\|^2 \leqq K\|x^2\|$ *for all* $x \in B$.

PROOF: If $\|\ \| \leqq K_1\|\ \|_{sp}$ then $\|x\|^2 \leqq K_1^2\|x\|_{sp}^2 \leqq K_1^2\|x^2\|$; this establishes the necessity. On the other hand, if the condition above is satisfied,

$$\|x\| \leqq K^{1/2}\|x^2\|^{1/2} \leqq K^{1/2+1/4}\|x^4\|^{1/4} \leqq$$
$$\leqq \cdots \leqq K^{1/2+1/4+\cdots+2^{-n}}\|x^{2^n}\|^{2^{-n}},$$

and, by lemma 3.6, $\|x\| \leqq K\|x\|_{sp}$. ◀

3.8 DEFINITION: A commutative Banach algebra B is *self-adjoint* if $\hat{B}$ is self-adjoint on the maximal ideal space $\mathfrak{M}$.

Remark: Notice the specific reference to the maximal ideal space. The algebra of functions defined on the segment $I = [0, 1]$ which are restrictions to I of functions holomorphic on the unit disc and continuous on the boundary (i.e., of functions in $HC(D)$), is self-adjoint as a function algebra on I. As a Banach algebra it is isomorphic to $HC(D)$ which is *not* self-adjoint. (See also exercise 11 at the end of this section).

Theorem: *Let B be self-adjoint with unit and assume that there exists a constant K such that* $\|x\|^2 \leqq K\|x^2\|$ *for all* $x \in B$. *Then* $\hat{B} = C(\mathfrak{M})$.

PROOF: By the Stone-Weierstrass theorem B is dense in $C(\mathfrak{M})$, and by 3.7 it is uniformly closed. ◀

3.9 Let $F(z) = \sum a_n z^n$ be a holomorphic function in the disc $|z| < R$ and x an element in a Banach algebra B such that $\|x\|_{sp} < R$. It follows from lemma 3.6 that the series $\sum a_n x^n$ converges in B (if $1 \notin B$, we assume $a_0 = 0$) and we denote its sum by $F(x)$. If M is a maximal ideal in B, we clearly have $\widehat{F(x)}(M) = F(\hat{x}(M))$.

Instead of power series expansion, we can use the Cauchy integral formula:

Theorem: *Assume* $1 \in B$. *Let F be a complex-valued function, holomorphic in a region*† *G in the complex plane. Let* $x \in B$ *be such that the range of* $\hat{x}$ *is contained in G. Let* γ *be a closed rectifiable curve*† *in G, enclosing the range of* $\hat{x}$, *and whose index with respect to any* $\hat{x}(M)$, $M \in \mathfrak{M}$, *is one, and is zero with respect to any point outside G. Then the integral*

$$F(x) = \frac{1}{2\pi i}\int_\gamma \frac{F(z)}{z - x}\,dz \tag{3.2}$$

is a well-defined element in B and

$$\widehat{F(x)}(M) = F(\hat{x}(M)) \tag{3.3}$$

for all $M \in \mathfrak{M}$.

† Not necessarily connected!

PROOF: The integrand is a continuous B-valued function of z, hence (3.2) is well defined and (3.3) is valid. ◀

Remarks: (a) The element $F(x)$ defined by (3.2) does not depend on the particular choice of γ. Also, it can be shown, using the Cauchy integral (3.2), that for a given $x \in B$, the mapping $F \to F(x)$ is a homomorphism of the algebra of functions holomorphic in a neighborhood of the range of $\hat{x}$ into B (cf. [5], §6).

Remarks: (b) Though the integral (3.2) clearly depends upon the assumption $1 \in B$, we can "save" the theorem in the case $1 \notin B$ by formally adjoining a unit. Denoting by B' the algebra obtained by adjoining a unit to B, we notice that for $x \in B$ the range of $\hat{x}$ over $\mathfrak{M}'$ (the maximal ideal space of B') is the union of $\{0\}$ with the range of $\hat{x}$ on $\mathfrak{M}$. If we require $0 \in G$, then (3.2) can be taken as a B'-valued integral, and if $F(0) = 0$ then $F(x)$ vanishes whenever $\hat{x}$ does, and in particular $F(x) \in B$.

3.10 Theorem 3.9 states essentially that $\hat{B}$ is stable under the operation of analytic functions. For the algebra $A(\mathbf{T})$ this is Paul Levy's extension of Wiener's theorem 2.9:

Theorem: *(Wiener-Levy): Let* $f(t) = \sum \hat{f}(j) e^{ijt}$ *with* $\sum |\hat{f}(j)| < \infty$, *Let* F *be holomorphic in a neighborhood of the range of* f. *Then* $g(t) = F(f(t))$ *has an absolutely convergent Fourier series.*

3.11 As another simple application we mention that if there exists an element $x \in B$ such that $\hat{x}(M)$ is bounded away from zero on $\mathfrak{M}$, then, denoting by $F(z)$ the function which is identically 1 for $|z| > \varepsilon$ and identically zero for $|z| < \varepsilon/2$ (where $\varepsilon = \frac{1}{2}\inf|\hat{x}(M)|$), we see that $\widehat{F(x)} = 1$ on $\mathfrak{M}$. If we assume that B is semisimple it follows that $F(x)$ is a unit element in B.

The assumption that $\hat{x}$ is bounded away from zero for some $x \in B$ implies directly that $\mathfrak{M}$ is compact (see the proof of 3.1); if we assume, on the other hand, that $\mathfrak{M}$ is compact, then 0 is not a limit point of $\mathfrak{M}$ in U^* and consequently there exists a neighborhood of zero in U^*, disjoint from $\mathfrak{M}$. By the definition of the weak-star topology this is equivalent to the existence of a finite number of elements $x_1, \cdots, x_n$ in B such that $|\hat{x}_1(M)| + \cdots + |\hat{x}_n(M)|$ is bounded away from zero on $\mathfrak{M}$. Operating with functions of several complex variables one can prove again that $1 \in B$. We refer the reader to ([5], §13) for a discussion of operation by functions of several complex variables on elements in B and state the following theorem without proof:

Theorem: *If B is semisimple and $\mathfrak{M}$ is weak-star compact, then $1 \in B$.*

3.12 Another very important theorem which is proved by application of holomorphic functions of several complex variables, deals with the existence of idempotents:

Theorem: *Assume that $\mathfrak{M}$ is disconnected and let $U \subset \mathfrak{M}$ be both open and compact in the weak-star topology. Then there exists an element $u \in B$ such that*

$$\text{(a)} \quad u^2 = u$$

$$\text{(b)} \quad \hat{u}(M) = \begin{cases} 0 & M \notin U \\ 1 & M \in U. \end{cases}$$

Remarks: (i) If B is semisimple then (a) is a consequence of (b). (ii) The idempotent u allows a decomposition of B into a direct sum of ideals. Writing $I_1 = uB = \{x \in B; x = ux\}$ and $I_2 = \{x \in B;\ ux = 0\}$, it is clear that I_1 and I_2 are disjoint closed ideals and for any $x \in B \quad x = ux + (x - ux)$; thus $B = I_1 \oplus I_2$.

We refer to ([5], §14) for a proof; here we only point out that if we know that $\hat{B}$ is uniformly dense in $C(\mathfrak{M})$ (resp. $C_0(\mathfrak{M})$), and in particular if $\hat{B}$ is self-adjoint on $\mathfrak{M}$, theorem 3.12 follows from 3.9. In fact, there exists an element $x \in B$ such that $|\hat{x}(M) - 1| < \frac{1}{4}$ for $M \in U$ and $|\hat{x}(M)| < \frac{1}{4}$ for $M \in \mathfrak{M} \setminus U$. Defining $F(z)$ as 1 for $|z - 1| < \frac{1}{4}$ and 0 for $|z| < \frac{1}{4}$, we obtain u as $F(x)$.

EXERCISES FOR SECTION 3

1. Show that the distance between any two points of $\mathbf{T}$, considered as the maximal ideal space of $C(\mathbf{T})$, in the metric induced by the dual (in this case $M(\mathbf{T})$) is equal to 2; hence the norm topology is discrete.

2. We have seen that the maximal ideal space of $HC(D)$ is $\overline{D} = D \cup \mathbf{T} = \{z;\ |z| \leqq 1\}$. Show that the norm topology on $\overline{D}$ (i.e., the topology induced by the metric of the dual space on the set of multiplicative linear functionals) coincides with the topology of the complex plane on D and with the discrete topology on $\mathbf{T}$. *Hint*: Schwarz' lemma.

3. Show that the relation $\| w_1 - w_2 \|_{B^*} < 2$ is an equivalence relation in the space of maximal ideals (multiplicative linear functionals)

of a sup-normed Banach algebra B. The corresponding equivalence classes are called the "Gleason parts" of the maximum ideal space.

4. Let B be an arbitrary Banach algebra and let B_1 be a Banach algebra with trivial multiplication (example 9 of section 1). Denote by $\tilde{B}$ the orthogonal direct sum of B and B_1, that is, the set of all pairs (x, y) with $x \in B$, $y \in B_1$ with the following operations:

$$(x, y) + (x_1, y_1) = (x + x_1, y + y_1)$$

$$\lambda(x, y) = (\lambda x, \lambda y) \quad \text{for complex } \lambda$$

$$(x, y)(x_1, y_1) = (xx_1, 0)$$

and the norm $\|(x, y)\| = \|x\|_B + \|y\|_{B_1}$. Show that B is a Banach algebra without unit and with the same maximal ideal space as B.

5. Let X be a compact (locally compact) Hausdorff space. Show that the maximal ideal space of $C(X)$ (resp. $C_0(X)$), with the weak-star topology, coincides with X as a topological space.

6. We recall that a set $\{x_1, \dots, x_n\} \subset B$ is a *set of generators* in B if it is contained in no proper closed subalgebra of B, or, equivalently, if the algebra of polynomials in $x_1, \dots, x_n$ is dense in B. Show that if $\{x_1, \dots, x_n\}$ is a set of generators in B, then the mapping $M \to (\hat{x}_1(M), \dots, \hat{x}_n(M))$ identifies $\mathfrak{M}$, as a topological space, with a bounded subset of $\mathbf{C}^n$.

7. Let I be a closed ideal in B. Denote by $\boldsymbol{h}(I)$—the hull of I—the set of all regular maximal ideals containing I. Show that the maximal ideal space of B/I can be canonically identified with $\boldsymbol{h}(I)$.

8. Let I_1, I_2 be (nontrivial) closed ideals in an algebra B such that $1 \in B$. Assume that $B = I_1 \oplus I_2$. Show that $\mathfrak{M}$ is disconnected.

9. Show that *not* every multiplicative linear functional on $M(\mathbf{T})$ has the form $\mu \to \hat{\mu}(n)$ for some integer n.

10. Show that for any LCA group $G, L^1(G)$ and $M(G)$ are semisimple.

11. Let B be a Banach algebra with a unit, realized as a self-adjoint function algebra on a space X.

(a) Prove that B is self-adjoint if, and only if, $\hat{f}$ is real valued on $\mathfrak{M}$ for every $f \in B$ which is real valued on X.

(b) Prove that B is self-adjoint if, and only if, $1 + |f|^2$ is invertible in B for all $f \in B$.

12. Let $\{w_n\}_{n \in Z}$ be a sequence of positive numbers satisfying $1 \leqq w_{n+m} \leqq w_n w_m$ for all $n, m \in \mathbf{Z}$. Denote† by $A\{w_n\}$ the subspace of $A(\mathbf{T})$ consisting of the functions f for which

† Compare with exercise V.2.7.

$$\|f\|_{\{w_n\}} = \sum_{-\infty}^{\infty} |\hat{f}(n)| w_n < \infty .$$

(a) Show that with the norm so defined, $A\{w_n\}$ is a Banach algebra.
(b) Assume that for some $k > 0$, $w_n = O(|n|^k)$. Show that the maximal ideal space of $A\{w_n\}$ can be identified with $\mathbf{T}$.
(c) Under the assumption of (b), show that $A\{w_n\}$ is self-adjoint if, and only if,

$$w_n w_{-n}^{-1} = O(1) \qquad \text{as } |n| \to \infty .$$

13. Let B be a Banach algebra with norm $\| \ \|_0$ and maximal ideal space $\mathfrak{M}$. Let $B_1 \subset B$ be a dense subalgebra which is itself a Banach algebra under a norm $\| \ \|_1$. Assume that its maximal ideal space is again $\mathfrak{M}$ (this means that every multiplicative linear functional on B_1 is continuous there with respect to $\| \ \|_0$). Assume that $\| \ \|_0$ and $\| \ \|_1$ are consistent on B_1 and denote by $\| \ \|_\alpha$, $0 < \alpha < 1$, the interpolating norms (see IV.1.2 and exercise 1.4) and by B_α the completion of B_1 with respect to the norm $\| \ \|_\alpha$. Show that the maximal ideal space of B_α is again $\mathfrak{M}$. *Remark*: If B is semisimple the norm $\| \ \|_0$ is majorized by $\| \ \|_1$ (see section 4) and hence by $\| \ \|_\alpha$ for all $0 \leqq \alpha \leqq 1$.

14. Let $B_1 \subseteq B' \subseteq B$ be Banach algebras with norms $\| \ \|_1$, $\| \ \|'$ and $\| \ \|$ respectively. Assume that B_1 is dense in B and in B' in their respective norms and that B and B_1 have the same maximal ideal space $\mathfrak{M}$. Show that the maximal ideal space of B' is again $\mathfrak{M}$. *Remark*: The assumption that B_1 is dense in B' is essential; see exercise 9.11.

4. HOMOMORPHISMS OF BANACH ALGEBRAS

4.1 We have seen (lemma 2.8) that homomorphisms of any Banach algebra into the field $\mathbf{C}$ are always continuous. The Gelfand representation enables us to extend this result:

Theorem: *Let B be a semisimple Banach algebra, let B_1 be any Banach algebra and let φ be a homomorphism of B_1 into B. Then φ is continuous.*

PROOF: We use the closed graph theorem and prove the continuity of φ by showing that its graph is closed. Let $x_j \in B_1$ and assume that $x_j \to x_0$ in B_1 and $\varphi x_j \to y_0$ in B. Let M be any maximal ideal in B; the mapping $x \to \widehat{\varphi x}(M)$ is a multiplicative linear functional on

B_1 and hence, by (2.8), it is continuous. It follows that $\widehat{\varphi x_j}\,(M)$ converges to $\widehat{\varphi x_0}(M)$; on the other hand, since $\varphi x_j \to y_0$ in B we have $\widehat{\varphi x_j}(M) \to \hat{y}_0(M)$, hence $\widehat{\varphi x_0}\,(M) = \hat{y}_0(M)$. Hence $\varphi x_0 - y_0 \in M$ for all maximal ideals M in B and, by the assumption that B is semisimple, $\varphi x_0 = y_0$. Thus the graph of φ is closed and φ is continuous. ◀

Corollary: *There exists at most one norm, up to equivalence, with which a semisimple algebra can be a Banach algebra.*

4.2 Let B and B_1 be Banach algebras with maximal ideal spaces $\mathfrak{M}$ and $\mathfrak{M}_1$, respectively. Let φ be a homomorphism of B_1 into B. As we have seen in the course of the preceding proof, every $M \in \mathfrak{M}$ defines a multiplicative linear functional $w(x) = \widehat{\varphi x}\,(M)$ on B_1. We denote the corresponding maximal ideal by ϕM. It may happen, of course that w is identically zero and the "corresponding maximal ideal" is then the entire B_1; thus ϕ is a mapping from $\mathfrak{M}$ into $\mathfrak{M}_1 \cup \{B_1\}$. In terms of linear functionals, ϕ is clearly the restriction to $\mathfrak{M}$ of the adjoint of φ.

Theorem: *$\phi: \mathfrak{M} \to \mathfrak{M}_1 \cup \{B_1\}$ is continuous when both spaces are endowed with the weak-star topology. If $\varphi(B_1)$ is dense in B in the spectral norm, then ϕ is a homeomorphism of $\mathfrak{M}$ onto a closed subset of $\mathfrak{M}_1$.*

PROOF: A sub-basis for the weak-star topology on $\mathfrak{M}_1 \cup \{B_1\}$ is the collection of sets of the form $U_1 = \{M_1; \hat{x}_1(M_1) \in O\}$, O being an open set in $\mathbb{C}$ and $x_1 \in B_1$. The ϕ pre-image in $\mathfrak{M}$ of U_1 is $U = \{M; \widehat{\varphi x_1}\,(M) \in O\}$ which is clearly open. If $\widehat{\varphi(B_1)}$ is uniformly dense in $\hat{B}$, the functions $\widehat{\varphi x_1}(M)$, $x_1 \in B_1$, determine the weak-star topology on $\mathfrak{M}$ and it is obvious that ϕ is one-to-one into $\mathfrak{M}_1$ and that it is a homeomorphism. It remains to show† that if $\widehat{\varphi(B_1)}$ is uniformly dense in $\hat{\mathrm{B}}$ then $\phi(\mathfrak{M})$ is closed in $\mathfrak{M}_1$. We start with two remarks:

(a) For $M_1 \in \mathfrak{M}_1$ the mapping $\varphi x_1 \to \hat{x}_1(M_1)$ is well defined on $\varphi(B_1)$ if, and only if, for all $x_1 \in B_1$, $\varphi x_1 = 0$ implies $\hat{x}_1(M_1) = 0$.

(b) When the above-mentioned mapping is defined, it is clearly multiplicative and (assuming $\widehat{\varphi(B_1)}$ uniformly dense in $\hat{B}$) it can be extended to a multiplicative linear functional on B if, and only if, for all $x_1 \in B_1$:

† Notice that if $1 \in B$ then $\mathfrak{M}$ is compact and $\phi(\mathfrak{M})$ is therefore compact, so that in this case the rest of the proof is superfluous.

$$|\hat{x}_1(M_1)| \leqq \|\varphi x_1\|_{sp}.$$

Assume now that $M_1 \in \mathfrak{M}_1$ is in the weak-star closure of $\varphi(\mathfrak{M})$ For any $x_1 \in B_1$ and $\varepsilon > 0$ there exists an $M \in \mathfrak{M}$ such that

$$|\hat{x}_1(M_1) - \hat{x}_1(\varphi M)| = |\hat{x}_1(M_1) - \widehat{\varphi x}_1(M)| < \varepsilon.$$

Since $\varepsilon > 0$ is arbitrary and since $|\widehat{\varphi x}_1(M)| \leqq \|\varphi x_1\|_{sp}$, it follows that $|\hat{x}_1(M_1)| \leqq \|\varphi x_1\|_{sp}$, and by our remark (b) there exists an $M_0 \in \mathfrak{M}$ such that $\widehat{\varphi x}_1(M_0) = \hat{x}_1(M_1)$ for all $x_1 \in B_1$; that is, $\varphi M_0 = M_1$ and $M_1 \in \varphi(\mathfrak{M})$. ◀

4.3 From 4.2 it follows in particular that if φ is an automorphism of B, then φ is a homeomorphism of $\mathfrak{M}$, and if B is semisimple (so that we can identify it with its Gelfand representation), φ is given by $\widehat{\varphi x}(M) = \hat{x}(\varphi M)$.

In other words: *Every automorphism of a semisimple Banach algebra is given by a (homeomorphic) "change of variable" on the maximal ideal space.*

Of course not every homeomorphism of $\mathfrak{M}$ defines an automorphism of $\hat{B}$ (or B), and the question which homeomorphisms do, is equivalent to the characterization of all the automorphisms of B and can be quite difficult.

***4.4** The following lemma is sometimes helpful in determining the automorphisms of Banach algebras of functions on the line.

Lemma: *Let φ be a continuous function defined on an interval $[a,b]$ and having the following property: If $r_0, r_1, \cdots, r_N$ are real numbers such that all the 2^N points*

$$\eta_\alpha = r_0 + \sum_1^N \varepsilon_j r_j, \qquad \varepsilon_j = 0,1 \tag{4.1}$$

lie in $[a,b]$, then the numbers $\{\varphi(\eta_\alpha)\}$ are linearly dependent over the rationals. Then the set of points in a neighborhood of which φ is a polynomial of degree smaller than N is everywhere dense in $[a,b]$.

PROOF: Let I be any interval contained in $[a,b]$; we show that there exists an interval $I' \subset I$ such that φ coincides on I' with some polynomial of degree smaller than N. Without loss of generality we may assume that $I \supset [0, N+1]$ so that if $0 \leqq r_j \leqq 1$, $j = 0, 1, \ldots, N$, all the points η_α defined by (4.1) are contained in I. By the assumption of the lemma, to each choice of $(r_0, \ldots, r_N)$ such that $0 \leqq r_j \leqq 1$, corresponds at least one vector $(A_1, \ldots, A_{2N})$ with integral entries not all of which vanish, such that

$$\sum_{\alpha=1}^{2N} A_\alpha \varphi(\eta_\alpha) = 0. \tag{4.2}$$

Denote by $E(A_1, \ldots, A_{2N})$ the set of points $(r_0, \ldots, r_N)$ in the $N+1$-dimensional cube $0 \leqq r_i \leqq 1$, for which (4.2) is valid. Since φ is continuous it follows that $E(A_1, \ldots, A_{2N})$ is closed for every $(A_1, \ldots, A_{2N})$ and since $\bigcup E(A_1, \ldots, A_{2N})$ is the entire cube $0 \leqq r_j \leqq 1$, it follows from the theorem of Baire that some $E(A_1, \ldots, A_{2N})$ contains a box of the form $r_j^0 \leqq r_j \leqq r_j^1$, $j = 0, \ldots, N$.

Let $\varepsilon > 0$ be smaller than $\frac{1}{2}(r_0^1 - r_0^0)$ and let ψ_ε be an infinitely differentiable function carried by $(-\varepsilon, \varepsilon)$ such that $\int \psi_\varepsilon(\xi) d\xi = 1$. We put $\varphi_\varepsilon = \varphi * \psi_\varepsilon$ and notice that

$$\sum A_\alpha \varphi_\varepsilon(\eta_\alpha) = \psi_\varepsilon * \left(\sum A_\alpha \varphi(\eta_\alpha)\right)$$

and consequently

$$\sum_{\alpha=1}^{2N} A_\alpha \varphi_\varepsilon(\eta_\alpha) = 0 \tag{4.3}$$

for

$$r_0^0 + \varepsilon \leqq r_0 \leqq r_0^1 - \varepsilon, \quad r_j^0 \leqq r_j \leqq r_j^1, \qquad j = 1, \ldots, N. \tag{4.4}$$

Now, φ_ε is infinitely differentiable and we can differentiate (4.3) with respect to various r_j's, $j \geqq 1$. Assume that $A_{\alpha_0} \neq 0$ and that the coefficient of r_{j_0} in η_{α_0} is equal to one. Differentiating (4.3) with respect to r_{j_0} we obtain

$$\sum{}' A_\alpha \varphi_\varepsilon'(\eta_\alpha) = 0 \tag{4.5}$$

where the summation extends now only over those values of α such that the coefficient of r_{j_0} in η_α is equal to one. Also, (4.5) is nontrivial since it contains the term $A_{\alpha_0}\varphi_\varepsilon'(\eta_{\alpha_0})$. Repeating this argument with other r_j's we finally obtain a nontrivial relation $A_\alpha \varphi_\varepsilon^{(M)}(\eta_\alpha) = 0$, that is $\varphi_\varepsilon^{(M)}(\eta_\alpha) = 0$, with $M \leqq N$, and it follows that on the range of η_α corresponding to (4.4), say I', φ_ε is a polynomial of degree

$\leqq M-1<N$. As $\varepsilon \to 0$, $\varphi_\varepsilon \to \varphi$ and φ is a polynomial of degree smaller than N on I'. ◀

Corollary: *If φ, as above, is N-times continuously differentiable on $[a,b]$, then it is a polynomial of degree smaller than N on $[a,b]$.*

****4.5 Theorem*** (*Beurling-Helson*): *Let $\boldsymbol{\varphi}$ be an automorphism of $A(\hat{\mathbf{R}})$ and let φ be the corresponding change of variable on $\hat{\mathbf{R}}$ (see 4.3). Then $\varphi(\xi) = \alpha\xi + \beta$ with $\alpha, \beta \in \hat{\mathbf{R}}$ and $\alpha \neq 0$.*

PROOF: The proof is done in two steps. First we show that φ is linear on some interval on $\hat{\mathbf{R}}$, and then that the fact that φ is linear on some interval implies that it is linear on $\hat{\mathbf{R}}$.

First step: By 4.1, $\boldsymbol{\varphi}$ is a continuous linear operator on $A(\hat{\mathbf{R}})$. Let N be an even integer such that $2^N > \|\boldsymbol{\varphi}\|^4$; we claim that φ satisfies the condition of lemma 4.4 for this value of N. If we show this, it follows from 4.4 that φ is a polynomial on some interval $I' \subseteq \hat{\mathbf{R}}$. φ maps I' onto some interval I_0 and, since φ is an automorphism we can repeat the same argument for $\boldsymbol{\varphi}^{-1}$ and obtain that the inverse function of φ is a polynomial on some interval $I_0' \subseteq I_0$. Since a polynomial whose inverse function (on some interval) is again a polynomial must be linear it follows that φ is linear on I'.

The adjoint $\boldsymbol{\varphi}^*$ of $\boldsymbol{\varphi}$ maps the unit measure concentrated at ξ to the unit measure concentrated at $\varphi(\xi)$. Since $\|\boldsymbol{\varphi}^*\| = \|\boldsymbol{\varphi}\|$ we obtain that for every choice of $a_j \in \mathbf{C}$ and $\xi_j \in \hat{\mathbf{R}}$

$$\left\| \sum a_j \delta_{\varphi(\xi_j)} \right\|_{\mathscr{F}L^\infty} \leqq \|\boldsymbol{\varphi}\| \left\| \sum a_j \delta_{\xi_j} \right\|_{\mathscr{F}L^\infty}. \tag{4.6}$$

We remember also that by Kronecker's theorem (VI.9.2) if $\{\varphi(\xi_j)\}$ are linearly independent over the rationals $\left\| \sum a_j \delta_{\varphi(\xi_i)} \right\|_{\mathscr{F}L^\infty} = \sum |a_j|$.

We show now that for every choice of $r_0, r_1, \cdots, r_N \in \hat{\mathbf{R}}$, if the 2^N points η_α given by (4.1) are all distinct there exists a measure ν carried by $\{\eta_\alpha\}$ such that $\|\nu\|_{M(\hat{\mathbf{R}})} = 1$ and $\|\nu\|_{\mathscr{F}L^\infty} \leqq 2^{-N/4}$.

Put

$$\mu_j = \tfrac{1}{4}(\delta_0 + \delta_{r_{2j-1}} + \delta_{r_{2j}} - \delta_{(r_{2j-1}+r_{2j})}) \qquad j = 1, \cdots, N/2. \tag{4.7}$$

The total mass of μ_j is clearly 1 and

$$\begin{aligned} \hat{\mu}_j(x) &= \tfrac{1}{4}(1 + e^{ir_{2j-1}x} + e^{ir_{2j}x} - e^{i(r_{2j-1}+r_{2j})x}) = \\ &= \tfrac{1}{2} e^{ir_{2j-1}x/2} \cos r_{2j-1}x/2 + \frac{i}{2} e^{-i(r_{2j}+r_{2j-1}/2)x} \sin r_{2j-1}x/2 \end{aligned} \tag{4.8}$$

so that

(4.9) $$|\hat{\mu}(x)| \leqq \tfrac{1}{2}(|\cos r_{2j-1}x/2| + |\sin r_{2j-1}x/2|) \leqq 2^{-\frac{1}{2}}.$$

We now take $\nu = \delta_{r_0} * \mu_1 * \cdots * \mu_{N/2}$. ν is clearly carried by $\{\eta_\alpha\}$ and if the η_α's are all distinct the total mass of ν is 1. On the other hand

$$\| \nu \|_{\mathscr{F}L^\infty} \leqq \prod_1^{N/2} \| \mu_j \|_{\mathscr{F}L^\infty} \leqq 2^{-N/4}$$

If $\{\varphi(\eta_\alpha)\}$ are linearly independent over the rationals it follows first that the η_α's are all distinct, and by (4.6) applied to ν, and Kronecker's theorem $1 \leqq \| \varphi \| 2^{-N/4}$ which contradicts the assumption $2^N > \| \varphi \|^4$.

Second step: For all n and λ, $\| e^{in\xi} \hat{\mathbf{V}}_\lambda \|_{A(\hat{\mathbf{R}})} \leqq 3$; hence

$$\| e^{in\varphi(\xi)} \hat{\mathbf{V}}_\lambda(\varphi(\xi)) \|_{A(\hat{\mathbf{R}})} \leqq 3 \| \varphi \|.$$

As $\lambda \to \infty$, $\hat{\mathbf{V}}_\lambda(\varphi(\xi))$ becomes 1 on larger and larger intervals on $\hat{\mathbf{R}}$, which eventually cover any finite interval on $\hat{\mathbf{R}}$. By VI.2.4 (or by taking weak limits), $e^{in\varphi(\xi)}$ is a Fourier-Stieltjes transform of a measure of total mass at most $3 \| \varphi \|$. If we denote by μ_1 the measure on $\mathbf{R}$ such that $\hat{\mu}_1(\xi) = e^{i\varphi(\xi)}$, it follows that $e^{in\varphi(\xi)}$ is the Fourier-Stieltjes transform of

$$\mu_1^{*n} = \overbrace{\mu_1 * \cdots * \mu_1}^{n\,\text{times}}$$

and we have

(4.10) $$\| \mu_1^{*n} \|_{M(\mathbf{R})} \leqq 3 \| \varphi \|.$$

By the first step we know that $\varphi(\xi) = a\xi + b$ on some interval I on $\hat{\mathbf{R}}$. We now consider the measure μ obtained from μ_1 by multiplying it by e^{-ib} and translating it by a. We have $\hat{\mu}(\xi) = e^{-i(a\xi + b)}\hat{\mu}_1(\xi)$, that is $\hat{\mu}(\xi) = 1$ on I (and $|\hat{\mu}(\xi)| = 1$ everywhere). It follows from (4.10) that

(4.11) $$\| \mu^{*n} \|_{M(\mathbf{R})} \leqq 3 \| \varphi \|$$

Consider the measures $\nu_n = 2^{-n}(\delta + \mu)^{*n}$ ($\delta = \delta_0$ being the unit mass at $\xi = 0$). We have

$$\nu_n = 2^{-n} \sum_0^n \binom{n}{j} \mu^{*j}$$

and consequently $\| \nu_n \|_{M(\mathbf{R})} \leqq 3 \| \varphi \|$; also, $\hat{\nu}_n(\xi) = \left(\frac{1 + \hat{\mu}(\xi)}{2} \right)^n$ which is equal to 1 if $\hat{\mu}(\xi) = 1$ and tends to zero if $\hat{\mu}(\xi) \neq 1$. Taking a weak limit of

ν_n as $n \to \infty$ we obtain a measure ν such that $\hat{\nu}(\xi)$ is equal almost everywhere to the characteristic function of the set $\{\xi; \hat{\mu}(\xi) = 1\}$ which clearly implies $\hat{\nu}(\xi) = 1$ identically on $\hat{\mathbf{R}}$, hence $\hat{\mu}(\xi) = 1$ almost everywhere on $\hat{\mathbf{R}}$ and since $\hat{\mu}$ is continuous, $\hat{\mu}(\xi) = 1$ everywhere. It follows that $e^{i\varphi(\xi)} = e^{i(a\xi + b)}$ everywhere on $\hat{\mathbf{R}}$, and $\varphi(\xi) = a\xi + b$. ◀

Remarks: (a) The first step of the proof applies to a large class of algebras. For instance, if B is an algebra whose maximal ideal space is $\hat{\mathbf{R}}$, $B \supset A(\hat{\mathbf{R}})$ and for each constant K there exists an integer N such that every set $\{\eta_\alpha\}$ defined by (4.1) (such that the η_α's are all distinct) carries a measure ν such that $\|\nu\|_{B^*} < K^{-1} \|\nu\|_{M(\mathbf{R})}$, step 1 goes verbatim for B.

(b) If we assume that φ is continuously differentiable, step 2 is superfluous. In fact, step 1 shows that the set of points near which φ is linear is everywhere dense and if φ' exists and is continuous, the slope must be always the same and φ must be linear. This proves that for the algebras discussed in remark (a), the continuously differentiable changes of variable induced by an automorphism must be linear.

(c) We have used the fact that φ was an automorphism of $A(\hat{\mathbf{R}})$ rather than an endomorphism once, when deducing in the first step that φ was linear in some interval from the fact that it is a polynomial (on an interval) whose inverse function is also a polynomial. This part of the argument can be replaced (see exercise 12 at the end of this section), and we thereby obtain that every nontrivial endomorphism of $A(\hat{\mathbf{R}})$ is given by a linear change of variable (and consequently is an automorphism).

EXERCISES FOR SECTION 4

1. Let B be a semisimple Banach algebra with norm $\| \;\|$ and $B_1 \subset B$ a subalgebra of B which is a Banach algebra with a norm $\| \;\|_1$. Show that there exists a constant C such that $\|x\| \leq C\|x\|_1$ for all $x \in B_1$.

2. Let B be a Banach algebra of infinitely differentiable functions on $[0, 1]$, having $[0, 1]$ as its space of maximal ideals. Show that there exists a sequence $\{M_n\}_{n=0}^{\infty}$ such that $\sup|f^{(n)}(x)| \leq M_n\|f\|$ for every $f \in B$.

3. Show that the space of all infinitely differentiable functions on $[0, 1]$ cannot be normed so as to become a Banach algebra.

4. Let B be a semisimple Banach algebra with maximal ideal space $\mathfrak{M}$. Prove that a homeomorphism ψ of $\mathfrak{M}$ is induced by an endomorphism of B if, and only if, $\hat{f} \in \hat{B} \Rightarrow \hat{f} \circ \psi \in \hat{B}$, where $(\hat{f} \circ \psi)(M) = \hat{f}(\psi(M))$.

5. What condition on ψ above is equivalent to its being induced by an automorphism?

6. Construct examples of semisimple Banach algebras B and B_1 and a homomorphism $\varphi: B_1 \to B$ (such that $\widehat{\varphi B_1}$ is not dense in $\hat{B}$) and such that the corresponding mapping φ (a) is not one-to-one; (b) is one-to-one but not a homeomorphism; (c) maps $\mathfrak{M}$ onto a dense proper subset of $\mathfrak{M}_1$.

7. Show that a homeomorphism φ of $\mathbf{T}$ onto itself is induced by an automorphism of $A(\mathbf{T})$ if, and only if, $e^{in\varphi} \in A(\mathbf{T})$ for all n and $\| e^{in\varphi} \|_{A(\mathbf{T})} = O(1)$.

8. (Van der Corput's lemma): (a) Let φ be real-valued on an interval $[a, b]$, and assume that it has there a monotone derivative satisfying $\varphi'(\xi) > \rho > 0$ on $[a, b]$. Show that $\left| \int_a^b e^{i\varphi(\xi)}\, d\xi \right| \leqq 2/\rho$.

(b) Instead of assuming $\varphi' > \rho$ on $[a, b]$, assume that φ is twice differentiable and that $\varphi'' > \kappa > 0$ on $[a, b]$. Show that

$$\left| \int_a^b e^{i\varphi(\xi)}\, d\xi \right| \leqq 6\kappa^{-\frac{1}{2}}$$

Hints: For (a) write $\int e^{i\varphi(\xi)} d\xi = -i \int d\Phi(\xi)/\varphi'(\xi)$, where $\Phi(\xi) = e^{i\varphi(\xi)}$ and apply the so-called "second mean-value theorem." For (b), if $\varphi'(c) = 0$, write

$$\int_a^b = \int_a^{c-\kappa^{-\frac{1}{2}}} + \int_{c-\kappa^{-\frac{1}{2}}}^{c+\kappa^{-\frac{1}{2}}} + \int_{c+\kappa^{-\frac{1}{2}}}^b$$

the middle integral is clearly bounded by $2\kappa^{-\frac{1}{2}}$; evaluate the other two integrals by (a).

9. Let φ be twice differentiable, real-valued function on $[0, 1]$ and assume that $\varphi'' > \kappa > 0$ there. Show that

$$\left| \int_0^1 \xi e^{i\varphi(\xi)}\, d\xi \right| \leqq 12\kappa^{-\frac{1}{2}}$$

Hint: Integrate by parts and use exercise 8.

10. Let φ be twice differentiable, real-valued function on $[-1, 1]$ and assume that $\varphi'' > \eta > 0$ there. Put $\Phi_n(\xi) = (1 - |\xi|)\, e^{in\varphi(\xi)}$ for $|\xi| \leqq 1$ and $\Phi_n(\xi) = 0$ for $|\xi| \geqq 1$. Show that for all $x \in \mathbf{R}$,

$$\left| \frac{1}{2\pi} \int_{-1}^{1} \Phi_n(\xi)\, e^{i\xi x}\, d\xi \right| \leqq 4\eta^{-\frac{1}{2}} n^{-\frac{1}{2}}.$$

Hint: $\Phi_n(\xi)e^{i\xi x} = (1 - |\xi|)\, e^{i(n\varphi(\xi) - \xi x)}$. The second derivative of the exponent is $\geqq \eta n$; use exercise 9.

11. Show that for some $c > 0$, $\| \Phi_n \|_{A(\hat{\mathbf{R}})} \geqq c\sqrt{n}$; Φ_n being the function introduced in exercise 10. *Hint*: Use exercise 10, Plancherel's theorem, and the fact that $\| \Phi_n \|_{L^2(\hat{\mathbf{R}})}$ is independent of n.

*12. Prove that every nontrivial endomorphism of $A(\hat{\mathbf{R}})$ is given by a linear change of variable. *Hint*: See remark (c) of 4.5. If φ is the change of variable induced by an endomorphism $\boldsymbol{\varphi}$, φ is a polynomial on some interval and if it is not linear, $\varphi''(\xi) > \eta$ (or $\varphi''(\xi) \leqq -\eta$) for some $\eta > 0$ on some interval I. A linear change of variable allows the assumption $[-1, 1] \subset I$. As in the second step of the proof of 4.5, $\| \widehat{e^{in\varphi(\xi)} \mathbf{V}_\lambda(\varphi(\xi))} \|_{A(\hat{\mathbf{R}})} \leqq 3 \| \boldsymbol{\varphi} \|$ hence $\| \Phi_n(\xi)\hat{\mathbf{V}}_\lambda(\varphi(\xi)) \|_{A(\hat{\mathbf{R}})} \leqq 3 \| \boldsymbol{\varphi} \|$. For λ sufficiently large $\hat{\mathbf{V}}_\lambda(\varphi(\xi))\, \Phi_n(\xi) = \Phi_n(\xi)$ and by exercise 11, $\| \Phi_n \|_{A(\hat{\mathbf{R}})}$ tends to infinity with n, which gives the desired contradiction.

5. REGULAR ALGEBRAS

5.1 DEFINITION: A function algebra B on a compact Hausdorff space X is *regular* if, given a point $p \in X$ and a compact set $K \subset X$ such that $p \notin K$, there exists a function $f \in B$ such that $f(p) = 1$ and f vanishes on K. The algebra B is *normal* if, given two disjoint compact sets K_1, K_2 in X, there exists $f \in B$ such that $f \equiv 0$ on K_1 and $f \equiv 1$ on K_2.

Examples: (a) Let X be a compact Hausdorff space. Then $C(X)$ is normal. This is essentially the contents of Urysohn's lemma (see [15], p. 6).

(b) $HC(D)$, the algebra of functions holomorphic inside the unit disc and continuous on the boundary, is *not* regular.

Theorem (*partition of unity*): *Let X be a compact Hausdorff space and B a normal function algebra on X, containing the identity. Let $\{U_j\}_{j=1}^n$ be an open covering of X. Then there exist functions φ_j, $j = 1, 2, \ldots, n$, in B satisfying*

$$\text{(5.1)} \qquad \begin{cases} \text{support of } \varphi_j \subset U_j, \\ \displaystyle\sum_1^n \varphi_j = 1. \end{cases}$$

PROOF: We use induction on n. Assume $n = 2$. Let V_1, V_2 be open sets satisfying $\bar{V}_j \subset U_j$ and $V_1 \cup V_2 = X$. There exists a function

$f \in B$ such that $f \equiv 0$ on the complement of V_1 and $f \equiv 1$ on the complement of V_2. Put $\varphi_1 = f$, $\varphi_2 = 1 - f$.

Assume now that the statement of the theorem is valid for coverings by fewer than n open sets and let $U_1, \ldots, U_n$ be an open covering of X. Put $U' = U_{n-1} \cup U_n$ and apply the induction hypothesis to the covering $U_1, \ldots, U_{n-2}, U'$ thereby obtaining functions $\varphi_1, \ldots, \varphi_{n-2}, \varphi'$ in B, satisfying (5.1). Denote the support of φ' by S and let V_{n-1}, V_n be open sets such that $\bar{V}_j \subset U_j$ $(j = n-1, n)$ and $V_{n-1} \cup V_n \supset S$. Let $f \in B$ such that $f \equiv 0$ on $S \setminus V_{n-1}$ and $f \equiv 1$ on $S \setminus V_n$. Put $\varphi_{n-1} = \varphi' f$ and $\varphi_n = \varphi'(1 - f)$. The functions $\varphi_1, \ldots, \varphi_n$ satisfy (5.1). ◀

Remark: The family $\{\varphi_j\}$ satisfying (5.1) is called *a partition of unity in B*, subordinate to $\{U_j\}$. Partitions of unity are the main tool in transition from "local" properties to "global" ones. A typical and very important illustration is theorem 5.2 below.

5.2 Let $\mathscr{F}$ be a family of functions on a topological space X. A function f is said *to belong to $\mathscr{F}$ locally at a point $p \in X$*, if there exists a neighborhood U of p and a function $g \in \mathscr{F}$ such that $f = g$ in U. If f belongs to $\mathscr{F}$ locally at every $p \in X$, we say that *f is locally in $\mathscr{F}$*.

Theorem: *Let X be a compact Hausdorff space and $\mathscr{F}$ a normal algebra of functions on X. If a function f belongs to $\mathscr{F}$ locally, then $f \in \mathscr{F}$.*

PROOF: For every $p \in X$ let U be an open neighborhood of p and $g \in \mathscr{F}$ such that $g = f$ in U. Since X is compact, we can pick a finite cover of X, $\{U_j\}_{1=1}^{n}$, among the abovementioned neighborhoods. Denote the corresponding elements of $\mathscr{F}$ by g_j; that is, $g_j = f$ in U_j. Let $\{\varphi_j\}$ be a partition of unity in $\mathscr{F}$ subordinate to $\{U_j\}$. Then

$$f = \sum \varphi_j f = \sum \varphi_j g_j \in \mathscr{F}. \tag{5.2}$$

◀

Remark: It is clear from (5.2) that if $J \subset \mathscr{F}$ is an ideal, and if f belongs to J locally (i.e., $g_j \in J$), then $f \in J$.

5.3 We consider a semisimple Banach algebra B with a unit, and denote its maximal ideal space by $\mathfrak{M}$.

DEFINITION: *The hull*, $\boldsymbol{h}(I)$, of an ideal I in B, is the set of all $M \in \mathfrak{M}$ such that $I \subset M$. Equivalently: $\boldsymbol{h}(I)$ is the set of all common zeros of $\hat{x}(M)$ for $x \in I$.

Since the set of common zeros of any family of continuous functions is closed, $\boldsymbol{h}(I)$ is always closed in $\mathfrak{M}$.

DEFINITION: *The kernel*, $\boldsymbol{k}(E)$, of a set $E \subset \mathfrak{M}$, is the ideal $\bigcap_{M \in E} M$. Equivalently: $\boldsymbol{k}(E)$ is the set of all $x \in B$ such that $\hat{x}(M) = 0$ on E. $\boldsymbol{k}(E)$ is always a closed ideal in B.

5.4 If $E \subset \mathfrak{M}$, then $\boldsymbol{h}(\boldsymbol{k}(E))$ is a closed set in $\mathfrak{M}$ that clearly contains E. One can show (see [15], p. 60) that the hull-kernel operation is a proper closure operation defining a topology on $\mathfrak{M}$. Since $\boldsymbol{h}(\boldsymbol{k}(E))$ is closed in $\mathfrak{M}$, the hull-kernel topology is not finer than the weak-star topology. The two coincide if for every closed set $E \subset \mathfrak{M}$ we have $E = \boldsymbol{h}(\boldsymbol{k}(E))$ which means that if $M_0 \notin E$ there exists an element $x \in \boldsymbol{k}(E)$ such that $\hat{x}(M_0) \neq 0$. Remembering that $x \in \boldsymbol{k}(E)$ means $\hat{x}(M) = 0$ on E, we see that the hull-kernel topology coincides with the weak-star topology on $\mathfrak{M}$ if, and only if, $\hat{B}$ is a regular function algebra. In this case we say that B is regular.

DEFINITION: A semisimple Banach algebra B *is regular* (*resp. normal*) if $\hat{B}$ is regular (resp. normal) on $\mathfrak{M}$.

5.5 ***Theorem***: *Let B be a regular Banach algebra and E a closed subset of $\mathfrak{M}$. Then the maximal ideal space of $B/\boldsymbol{k}(E)$ can be identified with E.*

PROOF: The maximal ideals in $B/\boldsymbol{k}(E)$ are the canonical images of maximal ideals in B which contain $\boldsymbol{k}(E)$, that is, which belong to $\boldsymbol{h}(\boldsymbol{k}(E)) = E$. This identifies $\mathfrak{M}(B/\boldsymbol{k}(E))$ and E as sets and we claim that they can be identified as topological spaces. We notice that the Gelfand representation of $B/\boldsymbol{k}(E)$ is simply the restriction of $\hat{B}$ to E. A typical open set in a sub-base for the topology of $\mathfrak{M}(B/\boldsymbol{k}(E))$ has the form

$$U = \{M; M \in E, \widehat{x \bmod \boldsymbol{k}(E)}(M) \in O\},$$

O being an open set in the complex plane and $x \in B$. A typical open set in a sub-base for the topology of $\mathfrak{M}$ has the form $U' = \{M; \hat{x}_1(M) \in O'\}$ with O' open in $\mathbb{C}$ and $x_1 \in B$. If $O = O'$ and $x = x_1$ then $U = E \cap U'$ and the topology on $\mathfrak{M}(B/\boldsymbol{k}(E))$ is precisely the topology induced by $\mathfrak{M}$. ◀

5.6 ***Theorem***: *Let B be a regular Banach algebra, I an ideal in B, E a closed set in $\mathfrak{M}$ such that $E \cap \boldsymbol{h}(I) = \varnothing$. Then there exists an element $x \in I$ such that $\hat{x}(M) = 1$ on E.*

PROOF: The ideal generated by I and $\boldsymbol{k}(E)$ is contained in no maximal ideal since $M \supset (I, \boldsymbol{k}(E))$ implies $M \supset I$ and $M \supset \boldsymbol{k}(E)$, that is, $M \in E \cap \boldsymbol{h}(I)$. It follows that the image of I in $B/\boldsymbol{k}(E)$ is the entire algebra and consequently there exists an element $x \in I$ such that $x \equiv 1$ mod $\boldsymbol{k}(E)$, which is the same as saying $\hat{x}(M) = 1$ on E. ◀

Corollary: *A regular Banach algebra is normal.*

PROOF: If E_1 and E_2 are disjoint closed sets in $\mathfrak{M}$, apply the theorem to $I = \boldsymbol{k}(E_1)$, $E = E_2$. ◀

5.7 We turn now to some general facts about the relationship between ideals in regular Banach algebras and their hulls.

Theorem: *Let I be an ideal in a regular Banach algebra B and $x \in B$. Then $\hat{x}$ belongs to $\hat{I}$ locally at every interior point of $\boldsymbol{h}(x)$ and at every point $M \notin \boldsymbol{h}(I)$.*

PROOF: We write $\boldsymbol{h}(x)$ for $\boldsymbol{h}((x))$, that is, the set of zeros of $\hat{x}$ in $\mathfrak{M}$. If M is an interior point of $\boldsymbol{h}(x)$, $\hat{x} = 0$ in a neighborhood of M and $0 \in I$. If $M \notin h(I)$, M has a compact neighborhood E disjoint from $\boldsymbol{h}(I)$. By theorem 5.6 there exists an element $y \in I$ such that $\hat{y}(M) = 1$ on E. Now $\hat{x} = \hat{x}\hat{y}$ on E and $xy \in I$. ◀

Corollary: *Let I be an ideal in a regular semisimple Banach algebra B and $x \in B$. If the support of $\hat{x}$ is disjoint from $\boldsymbol{h}(I)$, then $x \in I$.*

PROOF: By theorem 5.7, $\hat{x}$ belongs to $\hat{I}$ locally at every point, and by corollary 5.6 and the remark following theorem 5.2 it follows that $\hat{x} \in \hat{I}$; hence $x \in I$. ◀

5.8 Let E be a closed subset of $\mathfrak{M}$. The set $I_0(E)$ of all $x \in B$ such that $\hat{x}(M)$ vanishes on a neighborhood of E is clearly an ideal and if B is regular, $\boldsymbol{h}(I_0(E)) = E$. It follows from corollary 5.7 that $I_0(E)$ is contained in every ideal I such that $\boldsymbol{h}(I) = E$. In other words: $I_0(E)$ is the smallest ideal satisfying $\boldsymbol{h}(I) = E$, and $\overline{I_0(E)}$ is the smallest closed ideal satisfying $\boldsymbol{h}(I) = E$. On the other hand, $\boldsymbol{k}(E)$ is clearly the largest ideal satisfying $\boldsymbol{h}(I) = E$.

DEFINITION: A *primary ideal* in a commutative Banach algebra is an ideal contained in only one maximal ideal.

In other words, an ideal is primary if its hull consists of a single point.

If B is a semisimple regular Banach algebra, every maximal ideal $M \subset B$ contains a smallest primary ideal, namely $I_0(\{M\})$. We simplify the notation and write $I_0(M)$ instead of $I_0(\{M\})$. Similarly, M contains a smallest closed primary ideal, namely $\overline{I_0(M)}$. In some cases $\overline{I_0(M)} = M$ and we then say that M contains no nontrivial closed primary ideals. Such is the case if $B = C(\mathbf{T})$ (trivial) and also if $B = A(\hat{\mathbf{R}})$ (theorem VI.4.11′). On the other hand, if $B = C^n(\mathbf{T})$ with $n \geqq 1$, the maximal ideal $\{f; f(t_0) = 0\}$ contains the nontrivial closed primary ideal $\{f; f(t_0) = f'(t_0) = 0\}$.

5.9 DEFINITION: A semisimple Banach algebra B *satisfies condition* (D) *at* $M \in \mathfrak{M}$ if, given any $x \in M$, there exists a sequence $\{x_n\} \subset I_0(M)$ such that $xx_n \to x$ in B. We say that B *satisfies the condition* (D) if B satisfies (D) at every $M \in \mathfrak{M}$.

If B satisfies condition (D) at $M \in \mathfrak{M}$, M contains no nontrivial closed primary ideal since $I_0(M)$ is dense in M. It is not known if the condition that M contains no nontrivial closed primary ideals is sufficient to imply (D) or not; however, if we know that there exists a constant K such that for every neighborhood U of M there exists $y \in B$ such that $\|y\| \leqq K$, $\hat{y}$ has its support in U and $\hat{y} \equiv 1$ in some (smaller) neighborhood of M, then we can deduce (D) from $\overline{I_0(M)} = M$. For $x \in M$ let $z_n \in I_0(M)$ such that $z_n \to x$. Let U_n be a neighborhood of M such that $z_n \equiv 0$ on U_n and let $y_n \in B$ such that $\|y_n\| \leqq K$, $y_n = 0$ outside U_n and $y_n \equiv 1$ near M. Put $x_n = 1 - y_n$. We have $x_n \in I_0(M)$, $x - xx_n = xy_n = (x - z_n)y_n$ (since $z_n y_n = 0$), and $\|(x - z_n)y_n\| \leqq K\|x - z_n\| \to 0$. ◄

5.10 ***Lemma***: *Let B be a regular semisimple Banach algebra satisfying condition (D) at $M_0 \in \mathfrak{M}$. Let I be a closed ideal in B and $x \in M_0$. Assume that there exists a neighborhood U of M_0 such that $x \in I$ locally at every $M \in U \setminus \{M_0\}$. Then $x \in I$ locally at M_0.*

PROOF: Let $y \in B$ such that the support of $\hat{y}$ is included in U and $\hat{y} = 1$ in some neighborhood V of M_0. yx belongs to I locally at every $M \neq M_0$ and yxx_n belongs locally to I everywhere ($\{x_n\}$ being the sequence given by (D); remember that $\hat{x}_n \equiv 0$ near M_0); hence $yxx_n \in I$ and since $xx_n \to x$ and I is closed, $yx \in I$. But $\widehat{yx} = \hat{x}$ in V and the lemma follows. ◄

Theorem (*Ditkin-Shilov*): *Let B be a semisimple regular Banach algebra satisfying (D). Let I be a closed ideal in B and $x \in \boldsymbol{k}(\boldsymbol{h}(I))$*

such that the intersection of the boundary of $\boldsymbol{h}(x)$ *with* $\boldsymbol{h}(I)$ *contains no nontrivial perfect sets. Then* $x \in I$.

PROOF: Denote by $\mathfrak{N}$ the set of $M \in \mathfrak{M}$ such that x does not belong to I locally at M. By theorem 5.7, $\mathfrak{N} \subset (\text{bdry } \boldsymbol{h}(x)) \cap \boldsymbol{h}(I)$ and by the lemma, $\mathfrak{N}$ has no isolated points; hence $\mathfrak{N}$ is perfect and since $(\text{bdry } \boldsymbol{h}(x)) \cap \boldsymbol{h}(I)$ contains no nontrivial perfects sets, $\mathfrak{N} = \varnothing$ and $x \in I$. ◀

Corollary: *Under the same assumptions on* B; *if* $E \subset \mathfrak{M}$ *is compact and its boundary contains no nontrivial perfect subsets, then* $\overline{I_0(E)} = \boldsymbol{k}(E)$.

5.11 We have been dealing so far with algebras with a unit element. The definitions and most of the results can be extended to algebras without a unit element simply by identifying the algebra B as a maximal ideal in $B \oplus \mathbf{C}$. Instead of $\mathfrak{M}$ we consider its one point compactification $\overline{\mathfrak{M}}$ and we say that B is regular on $\mathfrak{M}$ if $B \oplus \mathbf{C}$ is regular on $\overline{\mathfrak{M}}$. This is equivalent to adding to the regularity condition the following regularity at infinity: given $M \in \mathfrak{M}$, there exists $\hat{x} \in \hat{B}$ such that $\hat{x}(M) = 1$ and $\hat{x}$ has compact support. Similarly, we have to require in defining "x belongs locally to I" not only $x \in I$ locally at every $M \in \mathfrak{M}$, but also $x \in I$ at infinity, that is, the existence of some $y \in I$ such that $\hat{x} = \hat{y}$ outside of some compact set. The condition (D) at infinity is: for every $x \in B$ there exists a sequence $x_n \in B$ such that $\hat{x}_n$ are compactly supported and $xx_n \to x$.

EXERCISES FOR SECTION 5

1. Let B be a semisimple Banach algebra, let $x_1, \dots, x_n \in B$ be generators for B, and assume that

$$\int_{-\infty}^{\infty} \frac{\log \| e^{iyx_j} \|}{1 + y^2} \, dy < \infty, \qquad j = 1, \dots, n.$$

Show that B is regular.

2. Describe the closed primary ideals of $C^n(\mathbf{T})$, n being a positive integer.

6. WIENER'S GENERAL TAUBERIAN THEOREM

In this section we prove Wiener's lemma stated in the course of the proof of theorem VI.6.1, and Wiener's general Tauberian theorem. These results are obtained as more or less immediate consequences of some of the material in the preceding section; it should be kept in mind that Wiener's work preceded, and to some extent motivated, the study of general Banach algebras.

6.1 We start with the analog of Wiener's lemma for $A(\mathbf{T})$.

Lemma: *Let $f, f_1 \in A(\mathbf{T})$ and assume that f is bounded away from zero on the support of f_1. Then $f_1 f^{-1} \in A(\mathbf{T})$.*

PROOF: $A(\mathbf{T})$ is a regular Banach algebra. Denote by I the principal ideal generated by f; then $h(I) = \{t; f(t) = 0\}$ is disjoint from the support of f_1. By corollary 5.7, $f_1 \in I$, which means $f_1 = gf$ for some $g \in A(\mathbf{T})$. Thus $f_1 f^{-1} \in A(\mathbf{T})$ locally and we apply 5.2. ◀

6.2 We obtain Wiener's lemma by showing that $A(\hat{\mathbf{R}})$ is locally the same as $A(\mathbf{T})$.

Lemma: *Let $\varepsilon > 0$ and let ψ be a continuously differentiable function supported by $(-\pi + \varepsilon, \pi - \varepsilon)$. There exists a constant K depending on ψ such that for all $-1 \leqq \alpha \leqq 1$, $\| e^{i\alpha t}\psi \|_{A(\mathbf{T})} \leqq K$.*

PROOF: We clearly have $\| e^{i\alpha t}\psi \|_{C^1(\mathbf{T})}$ bounded, and the $A(\mathbf{T})$ norm is majorized by the $C^1(\mathbf{T})$ norm. ◀

Theorem: *Let f be a continuous function carried by $(-\pi + \varepsilon, \pi - \varepsilon)$. Then*

$$\int |\hat{f}(\xi)| \, d\xi < \infty \quad \Leftrightarrow \quad \sum |\hat{f}(n)| < \infty .$$

PROOF: Let ψ be continuously differentiable, carried by $(-\pi + \varepsilon/2, \pi - \varepsilon/2)$ such that $\psi(t) = 1$ on $(-\pi + \varepsilon, \pi - \varepsilon)$. Assume that $\sum |\hat{f}(n)| < \infty$; then $f \in A(\mathbf{T})$; hence $f e^{i\alpha t}\psi \in A(\mathbf{T})$ and $\| f e^{i\alpha t}\psi \|_{A(\mathbf{T})} \leqq K \| f \|_{A(\mathbf{T})}$ for $-1 \leqq \alpha \leqq 1$. Now $f e^{i\alpha t}\psi = f e^{i\alpha t}$ and its $A(\mathbf{T})$ norm is $(1/2\pi) \sum |\hat{f}(n - \alpha)|$. Integrating the inequality

$$\sum |\hat{f}(n - \alpha)| \leqq K \sum |\hat{f}(n)|$$

on $0 < \alpha \leqq 1$, we obtain

$$\int |\hat{f}(\xi)| \, d\xi \leqq K \sum |\hat{f}(n)| .$$

Conversely, assuming $\int |\hat{f}(\xi)| \, d\xi = \sum \int_0^1 |\hat{f}(n - \alpha)| \, d\alpha < \infty$, it follows from Fubini's theorem that for almost all α, $0 \leqq \alpha \leqq 1$, $\sum |\hat{f}(n - \alpha)| < \infty$, which means $e^{i\alpha t} f \in A(\mathbf{T})$. As in the first part of the proof this implies $e^{i\alpha t} f e^{-i\alpha t}\psi = f \in A(\mathbf{T})$ and $\sum |\hat{f}(n)| < \infty$. ◀

Corollary: *Identifying* $\mathbf{T}$ *with* $(-\pi, \pi]$, *a function* f *defined in a neighborhood of* $t_0 \in \mathbf{T}$ *belongs to* $A(\mathbf{T})$ *locally at* t_0 *if, and only if, it belongs to* $A(\hat{\mathbf{R}})$ *at* t_0.

6.3 ***Lemma*** (*Wiener's lemma*): *Let* f *and* $f_1 \in A(\hat{\mathbf{R}})$ *be such that the support of* f_1 *is compact and* f *is bounded away from zero on it. Then* $f_1 = gf$ *with* $g \in A(\hat{\mathbf{R}})$.

PROOF: Without loss of generality we assume that the support of f_1 is included in $(-2, 2)$. Replacing f by $f\varphi$, where $\varphi \in A(\hat{\mathbf{R}})$, $\varphi = 1$ on $(-2, 2)$ and $\varphi = 0$ outside of $(-3, 3)$, it follows from lemma 6.1 that $g = f_1 f^{-1} \in A(\mathbf{T})$; hence, by theorem 6.2, $g \in A(\hat{\mathbf{R}})$. ◀

6.4 ***Theorem***: (*Wiener's general Tauberian theorem*): *Let* $f \in A(\hat{\mathbf{R}})$ *and assume* $f(\xi) \neq 0$ *for all* $\xi \in \hat{\mathbf{R}}$. *Then* f *is contained in no proper closed ideal of* $A(\hat{\mathbf{R}})$.

PROOF: By lemma 6.3 it follows that if $f_1 \in A(\hat{\mathbf{R}})$ has compact support, then $f_1/f \in A(\hat{\mathbf{R}})$, that is, f_1 belongs to (f) (the principal ideal generated by f). By theorem VI.1.12, (f) is dense in $A(\hat{\mathbf{R}})$ and the proof is complete. ◀

Instead of considering principal ideals, one may consider any closed ideal I. If for every $\xi \in \hat{\mathbf{R}}$ there exists $f \in I$ such that $f(\xi) \neq 0$, then $I = A(\hat{\mathbf{R}})$. As a corollary we obtain again that all maximal ideals in $A(\hat{\mathbf{R}})$ have the form $\{f; f(\xi_0) = 0\}$ for some $\xi_0 \in \hat{\mathbf{R}}$. We leave the details to the reader.

6.5 The Tauberian character of theorem 6.4 is not obvious at first glance. A Tauberian theorem is a theorem that indicates conditions under which some form of summability implies convergence or, more generally, another form of summability. The first such theorem was proved by Tauber and stated that if $\lim_{x \to 1-0} \sum_{n=0}^{\infty} a_n x^n = A$ and $a_n = o(1/n)$, then $\sum a_n = A$. Hardy and Littlewood, who introduced the term "Tauberian theorem," improved Tauber's result by showing that the condition $a_n = o(1/n)$ can be replaced by $a_n = O(1/n)$, an improvement that is a great deal deeper and harder than Tauber's rather elementary result. Wiener's original statement of theorem 6.4 was much more clearly Tauberian:

Theorem (*Wiener's general Tauberian theorem*): *Let* $K_1 \in L^1(\mathbf{R})$ *and* $f \in L^\infty(\mathbf{R})$. *Assume* $\hat{K}_1(\xi) \neq 0$ *for all* $\xi \in \hat{\mathbf{R}}$ *and*

(6.1) $$\lim_{x\to\infty}\int K_1(x-y)f(y)\,dy = A\int K_1(x)\,dx.$$

Then

(6.2) $$\lim_{x\to\infty}\int K_2(x-y)f(y)\,dy = A\int K_2(x)\,dx,$$

for all $K_2 \in L^1(\mathbf{R})$.

Remark: If $f(x)$ tends to a limit when $x \to \infty$, then (6.1) is clearly satisfied with $A = \lim_{x\to\infty} f(x)$. (6.1) states that $f(x)$ tends to the limit A in the mean with respect to the kernel K_1; the theorem states that the existence of the limit with respect to the mean K_1 implies that of the limit with respect to any mean K_2, provided $\hat{K}_1$ never vanishes. We refer to [27], chapter 3, for examples of derivations of "concrete" Tauberian theorems from theorem 6.5.

PROOF OF THEOREM 6.5: Denote by I the subset of $L^1(\mathbf{R})$ of functions K_2 satisfying (6.2). I is clearly a linear subspace, invariant under translation and closed in the $L^1(\mathbf{R})$ norm, that is, a closed ideal in $L^1(\mathbf{R})$. Since $K_1 \in I$, it follows from theorem 6.4 that $I = L^1(\mathbf{R})$ and the proof is complete. ◀

7. SPECTRAL SYNTHESIS IN REGULAR ALGEBRAS

Let B be a semisimple regular Banach algebra with a unit.† Denote by $\mathfrak{M}$ its maximal ideal space and by B^* its dual.

7.1 DEFINITION: A functional $\nu \in B^*$ *vanishes on an open set* $O \subset \mathfrak{M}$ if $\langle x, \nu\rangle = 0$ for every $x \in B$ such that the support of $\hat{x}$ is contained in O.

Lemma: *If* $\nu \in B^*$ *vanishes on the open sets* O_1 *and* O_2 *then* ν *vanishes on* $O_1 \cup O_2$.

PROOF: Let $x \in B$ and assume that the support of $\hat{x}$ is contained in $O_1 \cup O_2$. Denote by O_3 the complement in $\mathfrak{M}$ of the support of $\hat{x}$ and let $\hat{\varphi}_j$, $j = 1, 2, 3$ be a partition of unity in $\hat{B}$ subordinate to O_j, $j = 1, 2, 3$. Then $x = x\varphi_1 + x\varphi_2$ and $\langle x, \nu\rangle = \langle x\varphi_1, \nu\rangle + \langle x\varphi_2, \nu\rangle = 0$, since $x\varphi_3 = 0$ and $x\varphi_j$ has its support in O_j. ◀

From the lemma it follows immediately that if $\nu \in B^*$ vanishes on every set in some finite collection of open sets it vanishes also on their union; and since $\mathfrak{M}$ is compact the same holds for arbitrary

† The standing assumption $1 \in B$ is introduced for convenience only. It is not essential and the reader is urged to extend the notions and results to the case $1 \notin B$.

unions. The union of all the open sets on which ν vanishes is the largest set having this property and we define *the support*, $\Sigma(\nu)$, of ν as the complement of this set (compare with VI.4).

7.2 For $M \in \mathfrak{M}$ we denote by δ_M the multiplicative linear functional associated with M, $\langle x, \delta_M \rangle = \hat{x}(M)$; thus δ_M is naturally identifiable with the measure of mass 1 concentrated at M.

DEFINITION: A functional $\nu \in B^*$ *admits spectral synthesis* if ν belongs to the weak-star closure of the span in B^* of $\{\delta_M\}_{M \in \Sigma(\nu)}$.

Since the subspace of B orthogonal to the span of $\{\delta_M\}_{M \in \Sigma(\nu)}$ is precisely the set of all $x \in B$ such that $\hat{x}(M) = 0$ for all $M \in \Sigma(\nu)$, that is, the ideal $\boldsymbol{k}(\Sigma(\nu))$, we see, using the Hahn-Banach theorem as we did in VI.6, that ν *admits spectral synthesis if, and only if, it is orthogonal to* $\boldsymbol{k}(\Sigma(\nu))$.

7.3 It seems natural to define a set of spectral synthesis as a set E having the property that every $\nu \in B^*$ such that $\Sigma(\nu) = E$ admits spectral synthesis. If $\mathfrak{M}$ is very large, however, there may be sets E which are the support of no $\nu \in B^*$ and we prefer to introduce the following.

DEFINITION: A closed set $E \subset \mathfrak{M}$ is a *set of spectral synthesis* if every $\nu \in B^*$ such that $\Sigma(\nu) \subseteq E$ is orthogonal to $\boldsymbol{k}(E)$.

This condition implies in particular that if $\Sigma(\nu) = E$ then ν admits spectral synthesis.

It is clear that the condition $\Sigma(\nu) \subset E$ is equivalent to the condition that ν be orthogonal to $I_0(E)$. The condition that E is of spectral synthesis is therefore equivalent to requiring that every $\nu \in B^*$ which is orthogonal to $I_0(E)$ be also orthogonal to $\boldsymbol{k}(E)$. By the Hahn-Banach theorem this means $\overline{I_0(E)} = \boldsymbol{k}(E)$. Thus: *$E$ is of spectral synthesis if and only if $I_0(E)$ is dense in* $\boldsymbol{k}(E)$. We restate corollary 5.10 as:

Theorem: *Assume that B satisfies (D) and let $E \subset \mathfrak{M}$ be closed and its boundary contain no perfect subsets. Then E is of spectral synthesis.*

7.4 In some cases every closed $E \subset \mathfrak{M}$ is of spectral synthesis and we say that *spectral synthesis is possible in B*. Such is the case if $B = C(X)$, X being a compact Hausdorff space. Another class of examples is given by theorem 7.3: B satisfying (D) with $\mathfrak{M}$ containing no perfect subsets. In particular, if G is a discrete abelian group and $B = A(G)$ (to which we formally add a unit if we want to remain within our standing assumptions), then (D) is satisfied and $\mathfrak{M}$ contains no perfect subsets. It follows that for discrete G spectral synthesis holds in $A(G)$.

We devote the rest of this section to prove:

Theorem *(Malliavin): If G is a nondiscrete LCA group then spectral synthesis fails for $A(G)$.*

The construction is somewhat simpler technically in the case $G = \mathbf{D}$ than in the general case and we do it there. For a nondiscrete LCA group G, a *Cantor set* E on G is a compact set for which there exists a sequence $\{r_j\} \subset G$ such that the finite sums $\sum_1^n \varepsilon_j r_j$, $\varepsilon_j = 0, 1$, are all distinct and form a dense subset of E. The construction we give below can be adapted to show that every Cantor set of G contains a subset which is not of spectral synthesis for $A(G)$. Notice that every nondiscrete LCA group has Cantor subsets. We mention, finally, that for $G = \mathbf{R}^n$ with $n \geqq 3$, any sphere is an example of a set which is not of spectral synthesis; this was shown by L. Schwartz (some eleven years before the general case was settled).

7.5 We state the principle on which our construction depends in the general setting of this section, that is, for a semisimple regular Banach algebra with a unit. For typographical simplicity we identify B and $\hat{B}$ and use the letters f, g and so on, for elements of B. We remind the reader that the dual B^* is canonically a B module.

Theorem: *Let $f \in B$ and $\mu \in B^*$, $\mu \neq 0$, and denote*

$$C(u) = \| e^{iuf} \mu \|_{B^*}. \tag{7.1}$$

Assume that for an integer $N \geqq 1$

$$\int_{-\infty}^{\infty} C(u) |u|^N du < \infty. \tag{7.2}$$

Then there exists a real value a_0 such that $f_0 = a_0 + f$ has the property that the closed ideals generated by f_0^n, $n = 1, \dots, N+1$ are all distinct.

PROOF: We begin with two remarks.

First: There is no loss of generality assuming that $\langle 1, \mu \rangle \neq 0$. In fact for some $h \in B$ $\langle h, \mu \rangle = \langle 1, h\mu \rangle \neq 0$ and since

$$\| e^{iuf} h\mu \|_{B*} \leqq \| h \|_B \| e^{iuf} \mu \|_{B*},$$

(7.2) remains valid if we replace μ by $h\mu$.

Second: Write $\Phi(u) = \langle 1, e^{iuf}\mu \rangle$; then $|\Phi(u)| \leqq C(u) \in L^1(R)$, $\Phi(u)$ is continuous and $\Phi(0) = \langle 1, \mu \rangle \neq 0$. It follows that $\hat{\Phi}(\xi)$ is well defined and is not identically zero so that there exists a real number a_0 for which $\hat{\Phi}(-a_0) \neq 0$. This is the a_0 we are looking for (as we shall see) and again we may simplify the typography by assuming $a_0 = 0$; we simply replace f by $a_0 + f$ and notice that $e^{iu(a_0+f)} = e^{iua_0} e^{iuf}$ so that $\| e^{iu(a_0+f)}\mu \|_{B*} = \| e^{iuf}\mu \|_{B*}$. Thus we assume

$$\int_{-\infty}^{\infty} \langle 1, e^{iuf}\mu \rangle \, du \neq 0. \tag{7.3}$$

For $p \leqq N$ the B^*-valued integral

$$\Delta_p(f, \mu) = \int_{-\infty}^{\infty} (iu)^p (e^{iuf}\mu) \, du \tag{7.4}$$

is well defined since the integrand is continuous and, by (7.2), norm integrable.

Let $q \geqq 0$ be an integer, $g_1 \in B$ and consider

$$I_{p,q} = \langle g_1 f^q, \Delta_p(f, \mu) \rangle = \int_{-\infty}^{\infty} \langle g_1 f^q, e^{iuf}\mu \rangle (iu)^p \, du. \tag{7.5}$$

Integrating (7.5) by parts we obtain

$$\begin{aligned} I_{p,q} &= -pI_{p-1,q-1} && \text{if } p > 0,\ q > 0 \\ I_{p,q} &= 0 && \text{if } p = 0,\ q > 0. \end{aligned} \tag{7.6}$$

It follows that if $q > p$ we have $I_{p,q} = 0$ no matter what is $g_1 \in B$. In other words, $\Delta_p(f, \mu)$ is orthogonal to the ideal generated by f^{p+1}.

Now, using (7.6) with $p = q$, $g_1 = 1$, we obtain

$$\langle f^p, \Delta_p(f, \mu) \rangle = (-1)^p p! \int_{-\infty}^{\infty} \langle 1, e^{iuf}\mu \rangle \, du \neq 0 \tag{7.7}$$

by (7.3). Thus f^p does not belong to the closed ideal generated by f^{p+1} and the proof is complete. ◀

Corollary: *The sets* $f^{-1}(0)$ *and* $\Sigma(\mu) \cap f^{-1}(0)$ *are not sets of spectral synthesis.*

PROOF: The hull of the ideal generated by f^p is $f^{-1}(0)$. Since we found distinct closed ideals having $f^{-1}(0)$ as hull, $f^{-1}(0)$ is not of spectral synthesis. The fact that $\Delta_p(f,\mu)$ is orthogonal to the ideal generated by f^{p+1} implies (see corollary 5.7) that $\Sigma(\Delta_p(f,\mu)) \subseteq f^{-1}(0)$.

For $g \in B$ we have

$$(7.8) \quad \langle g, \Delta_p(f,\mu)\rangle = \int_{-\infty}^{\infty} \langle g, e^{iuf}\mu\rangle (iu)^p\,du = \int_{-\infty}^{\infty} \langle g\,e^{iuf}, \mu\rangle (iu)^p\,du$$

so that if the support of g is disjoint from $\Sigma(\mu)$ then $\langle g,\Delta_p(f,\mu)\rangle = 0$. This means that $\Sigma(\Delta_p(f,\mu)) \subseteq \Sigma(\mu)$; hence

$$(7.9) \qquad \Sigma(\Delta_p(f,\mu)) \subseteq \Sigma(\mu) \cap f^{-1}(0). \qquad ◀$$

7.6 In the case $B = A(\mathbf{D})$ we show that μ and f can be chosen so that $C(u)$, defined by (7.1), goes to zero faster than any (negative) power of $|u|$. We can take as μ simply the Haar measure of $\mathbf{D}$ and we shall have f quite explicitly too, but before describing it we make a few observations.

We identify the elements of $\mathbf{D}$ as sequences $\{\varepsilon_n\}$, $\varepsilon_n = 0, 1$, the group operation being addition mod 2. Functions on $\mathbf{D}$ are functions of the infinitely many variables ε_n, $n = 1, 2, \ldots$. Denote by x_m the element in $\mathbf{D}$ all of whose coordinates except the mth are zero. Denote by E_m the subgroup of $\mathbf{D}$ generated by x_{2m-1} and x_{2m}, that is, $\{0,\ x_{2m-1},\ x_{2m},\ x_{2m-1} + x_{2m}\}$. Denote by μ_m the measure having the mass $\frac{1}{4}$ at each of the points of E_m. μ_m is the Haar measure on E_m and one checks easily that the convolutions $\mu_1 * \cdots * \mu_\nu$ converge in the weak-star topology of measures to the Haar measure μ of $\mathbf{D}$. We write this formally as $\mu = \prod_1^{\infty} * \mu_m$.

Lemma: *Let* $E_1 = \{x_1, \ldots, x_k\}$ *and* $E_2 = \{y_1, \ldots, y_l\}$ *be finite sets on a group G. Let* $E = E_1 + E_2 = \{x_p + y_q,\ p = 1, \ldots, k,\ q = 1, \ldots, l\}$ *and assume that E has kl points. Let* h_1 *and* h_2 *be functions on E such that* $h_1(x_p + y_q) = g_1(x_p)$ *and* $h_2(x_p + y_q) = g_2(y_q)$. *Then, if* μ_m *is a measure carried by* E_m, $m = 1, 2$,

$$(7.10) \qquad h_1 h_2(\mu_1 * \mu_2) = (g_1\mu_1) * (g_2\mu_2).$$

PROOF: Both sides of (7.10) are carried by E and have the mass $g_1(x_p)g_2(y_q)\mu_1(\{x_p\})\mu_2(\{y_q\})$ at $x_p + y_q$. ◀

The lemma can be generalized either by induction or by direct verification to sums of N sets E_m. The flaw in notation of denoting by E_m first specific sets and then, in the lemma, variable sets (the same for μ_m) is forgivable in view of the fact that we use the lemma precisely for the sets E_m and the measures μ_m introduced above. Thus, if h_m, $m = 1, 2, \ldots$, are functions on $\mathbf{D}$ and if h_m depends only on the variables ε_{2m-1} and ε_{2m}, we have

$$\left(\prod_1^N h_m\right) \mu_1 * \cdots * \mu_N = (h_1\mu_1) * \cdots * (h_N\mu_N). \tag{7.11}$$

We shall have $f = \sum a_m\varphi_m$ with $\varphi_m \in A(\mathbf{D})$, φ_m depending only on the variables ε_{2m-1} and ε_{2m}, and the series converging in the $A(\mathbf{D})$ norm. Using (7.11) and taking weak-star limits, we obtain the convenient formula:

$$e^{iuf}\mu = \prod_1^\infty * (e^{iua_m\varphi_m}\mu_m). \tag{7.12}$$

We recall that the norm of a measure in $A(\mathbf{D})^*$ is the supremum of its Fourier transform, and that the Fourier transform of a convolution is the product of the transforms of the factors; thus we obtain

$$\| e^{iuf}\mu \|_{A(\mathbf{D})^*} \leqq \prod_1^\infty \| e^{iua_m\varphi_m}\mu_m \|_{A(\mathbf{D})^*}. \tag{7.13}$$

The functions φ_m are defined by:

$$\varphi_m(x) = \varepsilon_{2m-1}\varepsilon_{2m} \qquad \text{for } x = \{\varepsilon_j\}. \tag{7.14}$$

If we denote by ξ_m the character on $\mathbf{D}$ defined by

$$<x, \xi_m> = (-1)^{\varepsilon_m} \qquad \text{for } x = \{\varepsilon_j\}, \tag{7.15}$$

then

$$\varphi_m(x) = \tfrac{1}{4}(1 + \xi_{2m-1}\xi_{2m} - \xi_{2m-1} - \xi_{2m}) \tag{7.16}$$

so that $\varphi_m \in A(\mathbf{D})$ and $\| \varphi_m \|_{A(\mathbf{D})} = 1$.

The Fourier transform of the measure $e^{i\alpha\varphi_m}\mu_m$ can be computed explicitly: if $\xi = \{\zeta_j\} \in \hat{\mathbf{D}}$ then

$$\int <x, \xi> e^{i\alpha\varphi_m(x)} d\mu_m(x) = \tag{7.17}$$

$$= \tfrac{1}{4}(1 + (-1)^{\zeta_{2m-1}} + (-1)^{\zeta_{2m}} + e^{i\alpha}(-1)^{\zeta_{2m-1}+\zeta_{2m}}),$$

which assumes only the three values: $\dfrac{3 + e^{i\alpha}}{4}$, $\dfrac{1 - e^{i\alpha}}{4}$, $\dfrac{e^{i\alpha} - 1}{4}$.

It follows that if $|\alpha - \pi| \leqq \pi/3 \pmod{2\pi}$, then

(7.18) $$\| e^{i\alpha\varphi_m}\mu_m \|_{A(\mathbf{D})^*} \leqq \tfrac{3}{4}.$$

Theorem: *Denoting by μ the Haar measure on $\mathbf{D}$, there exists a real-valued function $f \in A(\mathbf{D})$ such that $C(u) = \| e^{iuf}\mu \|_{A(\mathbf{D})^*}$ vanishes, as $|u| \to \infty$, faster than any power of $|u|$.*

PROOF: Let N_k be a sequence of integers such that $\sum N_k 2^{-k} < \infty$. Write $a_m = \pi/3.2^{k-1}$ for $\sum_1^{k-1} N_j < m \leqq \sum_1^k N_j$. We clearly have $\sum a_m = (2\pi/3)\sum N_k 2^{-k} < \infty$, so that writing $f = \sum a_m \varphi_m$ as in (7.14), we have $f \in A(\mathbf{D})$. For $2^k \leqq u \leqq 2^{k+1}$ we have $2\pi/3 \leqq u a_m \leqq 4\pi/3$ for the N_k values of m such that $a_m = \pi/3.2^{k-1}$. For these values it follows from (7.18) that

$$\| e^{iua_m\varphi_m}\mu_m \|_{A(\mathbf{D})^*} \leqq \tfrac{3}{4}$$

and consequently, using (7.13),

(7.19) $$C(u) \leqq \prod_1^\infty \| e^{iua_m\varphi_m}\mu_m \|_{A(\mathbf{D})^*} \leqq (\tfrac{3}{4})^{N_k}$$

since all the factors in (7.19) are bounded by 1 and at least N_k of them by $\frac{3}{4}$. If we take $N_k = 2^k k^{-2}$ we obtain $C(u) \leqq (\frac{3}{4})^{u/\log^2 u}$ for $u \to \infty$, and since for real-valued f, $C(-u) = C(u)$, the proof is complete. ◀

Corollary: *There exists a real-valued $f \in A(\mathbf{D})$ such that the closed ideals generated by f^n, $n = 1, 2, \cdots$, are all distinct.*

EXERCISES FOR SECTION 7

1. Prove that for every function $\alpha(u)$ such that $u^{-1}\alpha(u)$ is monotonic and $\sum 2^{-k}\alpha(2^k) < \infty$, there exists a real-valued function $f \in A(\mathbf{D})$ for which $C(u) = O(e^{-\alpha(|u|)})$.

2. Denote by B_α, $0 < \alpha < 1$, the algebras obtained from $A(\mathbf{D})$ and $C(\mathbf{D})$ by the interpolation procedure described in IV.1. Show that spectral synthesis fails in B_α.

8. FUNCTIONS THAT OPERATE IN REGULAR BANACH ALGEBRAS

8.1 We again consider regular semisimple Banach algebras with unit.

DEFINITION: A function F, defined in a set Ω in the complex plane, *operates in B* if $F(\hat{x}) \in \hat{B}$ for every $\hat{x} \in \hat{B}$ whose range is included in Ω.

Theorem 3.9 can be stated as: a function F defined and analytic in an open set Ω operates in (any) B. Saying that B is self-adjoint is equivalent to saying that $F(z)=\bar{z}$ operates in B. If B is self-adjoint and regular, we can prove theorem 3.9 and a great deal more without the use of Cauchy's integral formula. We first prove:

Lemma: *Let B be a regular, self-adjoint Banach algebra with maximal ideal space $\mathfrak{M}$. Let $M_0 \in \mathfrak{M}$ and U be a neighborhood of M_0. Then there exists an element $e \in B$ such that $\hat{e}$ has its support within U, $\hat{e}=1$ on some neighborhood V of M_0 and $0 \leqq \hat{e} \leqq 1$ on $\mathfrak{M}$.*

PROOF: By the regularity of B there exists an $x \in B$ such that $\hat{x}$ has its support in U and $\hat{x} \equiv 1$ in some neighborhood V of M_0. Take $\hat{e}=\sin^2 \pi\hat{x}\bar{\hat{x}}/2$. (Notice that $\sin^2 \pi\hat{x}\bar{\hat{x}}/2$ is well defined by means of power series.) ◀

Theorem: *Let $x \in B$ and let f be a continuous function on $\mathfrak{M}$ such that in a neighborhood of each $M_0 \in \mathfrak{M}$, f can be written as $F(\hat{x})$, where $F(\zeta)=F(\xi+i\eta)$ is real-analytic in ξ and η in a neighborhood of $\hat{x}(M_0)$. Then $f \in \hat{B}$.*

Remark: The two points in which this result is more general than theorem 3.9 are:

(a) We allow real-analytic functions.

(b) We allow many-valued functions (provided $F(\hat{x}(M))$) can be defined as a continuous function.)

PROOF: We show that $f(M) \in \hat{B}$ locally at every point. Let $M_0 \in \mathfrak{M}$ and, replacing x by $x-\hat{x}(M_0)$, we may assume that $\hat{x}(M_0)=0$. In a neighborhood of M_0 we have $f=F(\hat{x})$ where near zero, say for $|\xi|<1$ and $|\eta|<1$, $F(\xi+i\eta)=\sum a_{n,m}\xi^n\eta^m$. Let U be a smaller neighborhood of M_0 such that $|\hat{x}(M)|<\frac{1}{2}$ in U, let $\hat{e} \in \hat{B}$ have the properties listed in the lemma and write $\hat{x}_1=\mathrm{Re}(\hat{e}\hat{x})=\frac{1}{2}(\hat{e}\hat{x}+\overline{\hat{e}\hat{x}})$ and $\hat{x}_2=\mathrm{Im}(\hat{e}\hat{x})$. By lemma 3.6 the series $\sum a_{nm}x_1^n x_2^m$ converges in B and we denote $y=\sum a_{nm}x_1^n x_2^m$; then $y(M)=F(\hat{x}(M))$ in V. ◀

8.2 It is not hard to see that operation by analytic (or real-analytic) functions, even in the setup of theorem 8.1 which allows many valued functions, is continuous. This follows from the (local) power series expansion. There is no reason to assume, however, that whenever a function F operates in a regular semisimple algebra B, the operation is continuous (see exercise 2 at end of this section). Still, the regularity of B

makes it easy to "condense singularities" which allows us to show that the "bad" behavior of the operation is localized on $\mathfrak{M}$ to the neighborhood of a finite set. The notions, arguments, and results that follow are typical of regular algebras.

8.3 We assume that B is self-adjoint, regular, and with a unit, and we assume for simplicity that F is a continuous function, defined on the real line, and operates in B.

DEFINITION: *F operates boundedly* if there exist constants $\varepsilon > 0$ and $K > 0$ such that if $\hat{x}(M)$ is real valued and $\|x\| < \varepsilon$, then $\|F(x)\| \leqq K$.† *F operates boundedly at $M \in \mathfrak{M}$*, if there exists a neighborhood U_M of M and constants $\varepsilon > 0$ and $K > 0$ such that if the support of $\hat{x}$ is contained in U_M and $\|x\| < \varepsilon$, then $\|F(x)\| \leqq K$.

Lemma: *F operates boundedly if, and only if, it operates boundedly at every $M \in \mathfrak{M}$.*

PROOF: Replacing F by $F - F(0)$ we may assume $F(0) = 0$. It is clear that if F operates boundedly, it does so locally at each $M \in \mathfrak{M}$. Assume that the operation is bounded locally, and pick $M_1, M_2, \ldots, M_n$ such that the corresponding neighborhoods $U_{M_1}, \ldots, U_{M_n}$ cover $\mathfrak{M}$. Let $V_1, \ldots, V_n$ be open sets such that $\bar{V}_j \subset U_{M_j}$ and such that $\{V_j\}$ cover $\mathfrak{M}$. Let $\psi_j \in \hat{B}$ be real valued with support inside U_{M_j} and $\psi_j \equiv 1$ on V_j, and let $\{\varphi_j\}$ be a partition of the unity in $\hat{B}$ relative to $\{V_j\}$. Let ε_j and K_j be the constants corresponding to U_{M_j} and now take $\varepsilon > 0$ so that $\varepsilon \|\psi_j\| < \varepsilon_j$ for all j, and $K = \sum K_j \|\varphi_j\|$. Assume that $\hat{x} \in \hat{B}$ is real valued and $\|x\| < \varepsilon$; then $\|\hat{x}\psi_j\| \leqq \|x\| \, \|\psi_j\| < \varepsilon_j$ and $\hat{x}\psi_j$ is supported by U_{M_j}, hence $\|F(\hat{x}\psi_j)\| \leqq K_j$. But $F(\hat{x}) = \sum \varphi_j F(\hat{x}\psi_j)$ so that $\|F(x)\| \leqq \sum \|\varphi_j\| K_j = K$. ◂

8.4 ***Lemma:*** *Let B be a regular, self-adjoint Banach algebra and F a function defined on the real line and operating in B. Then there exists at most a finite number of points of $\mathfrak{M}$ at which F does not operate boundedly.*

PROOF: Again we assume, with no loss of generality, that $F(0) = 0$. Assume that F operates unboundedly at infinitely many points in $\mathfrak{M}$ and pick a sequence of such points $\{M_j\}$ having pairwise disjoint neighborhoods V_j. We now pick a neighborhood W_j of M_j such that

† We denote by $F(x)$ the element in B whose Gelfand transform is $F(\hat{x})$.

$\overline{W_j} \subset V_j$. Saying that F does not operate boundedly at M_j means that, given any neighborhood W_j of M_j and any constants $\varepsilon_j > 0$ and $K_j > 0$, there exists a real-valued $f_j \in \hat{B}$ carried by W_j and such that $\|f_j\| < \varepsilon_j$, $\|F(f_j)\| > K_j$. We take $\varepsilon_j = 2^{-j}$ and $K_j = 2^j \|\varphi_j\|$, where $\varphi_j \in \hat{B}$ is carried by V_j and $\varphi_j \equiv 1$ on W_j. We now consider $f = \sum f_j$ and $F(f)$. By the choice of ε_j the series defining f converges and consequently $f \in \hat{B}$ and $F(f) \in \hat{B}$. Now

$$\|F(f)\| \geqq \frac{1}{\|\varphi_j\|} \|\varphi_j F(f)\| = \frac{1}{\|\varphi_j\|} \|F(f_j)\| \geqq 2^j \quad \text{for all } j,$$

which gives the desired contradiction. ◀

8.5 For some Banach algebras, lemma 8.4 takes us as far as we can go; for others it can be improved. Consider, for instance, an automorphism $\boldsymbol{\sigma}$ of B inducing the change of variables σ on $\mathfrak{M}$. If $f \in \hat{B}$, then $F(\boldsymbol{\sigma} f) = F(f(\sigma M)) = \boldsymbol{\sigma}(F(f))$, which means that the operations by F (on the function) and by σ (on the variables) commute. Since $\boldsymbol{\sigma}$ is a bounded, invertible operator, it follows that F operates boundedly at a point $M \in \mathfrak{M}$ if and only if it operates boundedly at σM. From this remark and lemma 8.4 it follows that if F does not operate boundedly at $M \in \mathfrak{M}$, the set of images of M under all the automorphisms of B is finite. In particular, if for every $M \in \mathfrak{M}$ the set $\{\sigma M\}$, $\boldsymbol{\sigma}$ ranging over all the automorphisms of B, is infinite, then every function that operates in B does so boundedly at every $M \in \mathfrak{M}$, and consequently, operates boundedly. In particular:

Theorem: *Let G be a compact abelian group and F a continuous function defined on the real line. If F operates in $A(G)$, it does so boundedly.*

PROOF: The maximal ideal space of $A(G)$ is G. For every $y \in G$ the mapping $f \to f_y$† is an automorphism of $A(G)$ which carries the maximal ideal corresponding to y to that corresponding to $0 \in G$. If G is infinite the statement of the theorem follows from the discussion above. If G is finite the operation by F is clearly continuous. ◀

Remark: Since the operation of a function on a Banach algebra is not linear, we cannot usually deduce continuity from boundedness, nor boundedness in one ball in B from boundedness in another (see exercise 3 at the end of this section).

† $f_y(x) = f(x - y)$.

8.6 For some algebras theorem 8.1 is far from being sharp. For instance, if $B = C(\mathfrak{M})$ every continuous function operates in B; if $B = C^n(\mathbf{T})$ every n-times continuously differentiable function operates. For group algebras of infinite LCA groups theorem 8.1 is sharp. We shall prove now that for the algebra $B = A(\mathbf{T})$, only analytic functions operate. This is a special case of the following:

Theorem: *Let G be a nondiscrete LCA group and let F be a function defined on an interval I of the real line. Assume that F operates in A(G). Then F is analytic on I.*

Remark: If G is not compact, one of our standing assumptions, namely $1 \in B$, is not satisfied. Since in this case all the functions in $A(G)$ tend to zero at infinity, we have to add to the statement of the theorem the assumption $0 \in I$ since otherwise every function defined on I operates trivially (the condition of operation being void). The theorem can be extended to infinite discrete groups: we have to assume $0 \in I$ (since discrete and compact implies finite) and the conclusion is that F is analytic at zero (see exercise 1 at the end of this section). As mentioned above we prove the theorem for $G = \mathbf{T}$; the proof of the general case runs along the same lines (see [24], chapter 6).

PROOF OF THE THEOREM ($G = \mathbf{T}$): Let b be an interior point of I and consider the function $F_1(x) = F(x + b)$. F_1 is defined on $I - b$ and clearly operates in $A(\mathbf{T})$. If we prove that $F_1(x)$ is analytic at $x = 0$ it would follow that $F(x)$ is analytic at b, so that, in order to prove that F is analytic at every interior point of I we may assume $0 \in \text{int}(I)$ and prove the analyticity of F at 0. Once we know that functions that operate are necessarily analytic at the interior points of I we obtain the analyticity at the endpoints as follows (we assume, for simplicity, that $I = [0, 1]$ and we prove that $F(x)$ is analytic at $x = 0$): consider $F_1(x) = F(x^2)$. F_1 is defined on $[-1, 1]$ and clearly operates in $A(\mathbf{T})$ so that near $x = 0$, $F_1(x) = \sum b_j x^j$. Now, since $F_1(x) = F(x^2)$ is even, $b_{2j-1} = 0$ for all j, so that $F_1(x) = \sum b_{2j} x^{2j}$ and $F(x) = \sum b_{2j} x^j$. The proof will therefore be complete if, assuming $0 \in \text{int}(I)$, we prove that F is analytic at 0.

By theorem 8.5, F operates boundedly which means that there exist constants $\varepsilon > 0$ and $K > 0$ such that if $f \in A(\mathbf{T})$ is real valued and $\|f\| < \varepsilon$, then $\|F(f)\| < K$, Pick a positive α so small that (i) $[-\alpha, \alpha] \subset I$, and (ii) $\alpha e^5 < \varepsilon$, and consider $F_1(x) = F(\alpha \sin x)$. By (i), F_1 is well defined and it clearly is 2π-periodic and operates in $A(\mathbf{T})$. Now if $f \in A(\mathbf{T})$ is real valued and $\|f\| < 5$, then $\alpha \sin f$ is real valued, and, by (ii), $\|\alpha \sin f\| < \varepsilon$ so that $\|F_1(f)\| < K$. In particular, if $\varphi \in A(\mathbf{T})$, $\|\varphi\| \leqq 1$, $\tau \in \mathbf{R}$, $|\tau| \leqq \pi$, then

(8.1) $$\|F_1(\varphi + \tau)\| < K.$$

Now $\alpha \sin x \in A(\mathbf{T})$; hence $F_1 \in A(\mathbf{T})$ and we can write

(8.2) $$F_1(x) = \sum A_n e^{inx};$$

in particular, F_1 is continuous. For real-valued f in $A(\mathbf{T})$, $F_1(f) \in A(\mathbf{T})$ and therefore can be written as

(8.3) $$F_1(f(t)) = \sum a_n(f) e^{int}, \qquad \sum |a_n(f)| < \infty.$$

Since F_1 is uniformly continuous on $\mathbf{R}$ it follows that

$$a_n(f) = 1/2\pi \int F_1(f(t)) e^{-int} dt$$

depends continuously on f and therefore, for each N, the mapping

(8.4) $$f \rightarrow \sum_{-N}^{N} a_n(f) e^{int}$$

is continuous from the real functions in $A(\mathbf{T})$ into $A(\mathbf{T})$. We conclude from (8.3) that $F_1(f)$ is a pointwise limit of continuous functions on $A(\mathbf{T})$, that is, is a Baire function on $A(\mathbf{T})$, and in particular: $F_1(\varphi + \tau)$ considered as a function of τ on $[-\pi, \pi]$ is a measurable vector-valued function which is bounded by K if $\|\varphi\| \leqq 1$. It follows that

(8.5) $$\left\| \frac{1}{2\pi} \int F_1(\varphi + \tau) e^{-in\tau} d\tau \right\| \leqq K;$$

however,

(8.6) $$\frac{1}{2\pi} \int F_1(\varphi + \tau) e^{-in\tau} d\tau = A_n e^{in\varphi},$$

as can be checked by evaluating both sides of (8.6) for every $t \in \mathbf{T}$, and we rewrite (8.5) as

(8.5′) $$\|A_n e^{in\varphi}\| \leqq K.$$

Let us write

(8.7) $$\mathcal{N}(u) = \sup_{\substack{f \text{ real} \\ \|f\| \leqq u}} \|e^{if}\|_{A(\mathbf{T})};$$

then it follows from (8.5′) that

(8.8) $$|A_n| \leqq K(\mathcal{N}(n))^{-1}$$

and if we show that $\mathcal{N}(u)$ grows exponentially with u, it would follow that (8.2) converges not only on the real axis, but in a strip around it, so that F_1 is analytic on $\mathbf{R}$ and, finally, F is analytic at 0. All that we need in order to complete the proof is:

Lemma: *Let $\mathcal{N}(u)$ be defined by* (8.7). *Then*

(8.9) $$\mathcal{N}(u) = e^u .$$

PROOF: It is clear from the power series expansion of e^{if} that for any Banach algebra

$$\mathcal{N}(u) \leqq e^u.$$

The proof that, for $A(\mathbf{T})$, $\mathcal{N}(u) \geqq e^u$ us based on the following two remarks:

(a) Let $f, g \in A(\mathbf{T})$, then

(8.10) $$\| f(t)g(\lambda t) \| \to \| f \| \, \| g \| \qquad \text{as } \lambda \to \infty$$

(λ being integer). We prove (8.10) by noticing that if f is a trigonometric polynomial and λ is greater than twice the degree of f, then $\| f(t)g(\lambda t) \| = \| f \| \, \| g \|$. For arbitrary $f \in A(\mathbf{T})$ and $\varepsilon > 0$ we write $f = f_1 + f_2$ where f_1 is a trigonometric polynomial and $\| f_2 \| < \varepsilon \| f \|$. If $\lambda/2$ is greater than the degree of f_1 we have

$$\| f(t)g(\lambda t) \| \geqq \| f_1(t)g(\lambda t) \| - \| f_2(t)g(\lambda t) \| \geqq (1 - 2\varepsilon) \| f \| \, \| g \|.$$

(b) If α is positive, then $e^{i\alpha \cdot \cos t} = 1 + i\alpha \cos t + \cdots$ so that

(8.11) $$\| e^{i\alpha \cos t} \|_{A(\mathbf{T})} = 1 + \alpha + O(\alpha^2).$$

Let $u > 0$; we pick a large N and write $f = \sum_1^N (u/N) \cos \lambda_j t$ where the λ_j's increase fast enough to ensure

(8.12) $$\left\| \prod_{j=1}^{N} e^{i(u/N)\cos \lambda_j t} \right\| > \left(1 - \frac{1}{N}\right) \prod_{j=1}^{N} \left\| e^{i(u/N)\cos t} \right\|$$

f is clearly real valued, $\| f \| = u$, and, by (8.11) and (8.12),

$$\| e^{if} \| > \left(1 - \frac{1}{N}\right) \left(1 + \frac{u}{N} + O\left(\frac{u^2}{N^2}\right)\right)^N > (1 - \varepsilon)e^u$$

if N is large enough. ◀

This completes the proof of the theorem (for $G = \mathbf{T}$). ◀

The lemma is not accidental: the exponential growth of $\mathcal{N}(u)$ is the real reason for the validity of the theorem. The function $\mathcal{N}(u)$ as defined by (8.7) can be considered for any Banach algebra B and if, for some B, $\mathcal{N}(u)$ does not have exponential growth at infinity, then there exist nonanalytic functions which operate in B. As an example we can take any $F(x) = \sum A_n e^{inx}$ such that A_n does *not* vanish exponentially as $|n| \to \infty$ but such that for all $k > 0, \sum |A_n| \mathcal{N}(k|n|) < \infty$; F operates in B since for any real-valued $f \in B$, $F(f) = \sum A_n e^{inf}$ and the series converges in norm.

8.7 We finish this section with some remarks concerning the so-called "individual symbolic calculus" in regular semisimple Banach algebras. Inasmuch as "symbolic calculus" is the study of functions that operate in an algebra and of their mode of operation, individual symbolic calculus is the study of the functions that operate on a fixed element in the algebra. Let us be more precise. We consider a regular, semisimple Banach algebra (identify it with its Gelfand transform) and say that a function F *operates on an element* $f \in B$ if the domain of F contains the range of f and $F(f) \in B$. It is clear that a function F operates in B if it operates on every $f \in B$ with range contained in the domain of F. It is also clear that for each fixed $f \in B$, the set of functions that operate on f is a function algebra on the range of f; we denote this algebra by $[f]$. For $F \in [f]$ we write $\|F\|_{[f]} = \|F(f)\|_B$ and with this norm $[f]$ is a normed algebra. If we denote by $[[f]]$ the subalgebra of B consisting of the elements $F(f)$, $F \in [f]$, it is clear that the correspondence $F \leftrightarrow F(f)$ is an isometry of $[f]$ onto $[[f]]$. Since $[[f]]$ consists of all $g \in B$ which respect the level lines of f (i.e., such that $f(M_1) = f(M_2) \Rightarrow g(M_1) = g(M_2)$), $[[f]]$ and $[f]$ are Banach algebras. We say that $[[f]]$ is the subalgebra *generated formally* by f; it clearly contains the subalgebra generated by f (which corresponds to the closure of the polynomials in $[f]$).

It should be noted that the "concrete" algebra $[f]$ depends on f more than $[[f]]$. The latter depends only on the level lines of f and is the same, for example, if we replace a real-valued f by f^3. Even if the ranges of f and f^3 are the same we usually have $[f] \neq [f^3]$.

If f is real valued, $[f]$ always contains nonanalytic functions. In fact, since $\|e^{if}\|_{sp} = 1$, it follows from lemma 3.6 that

$$\lim_{n\to\infty} \| e^{inf} \|^{1/n} = 1$$

and there exists therefore a sequence $\{A_n\}$ that does not vanish exponentially such that $\sum A_n e^{inf}$ converges in norm; hence $F(x) = \sum A_n e^{inx}$ belongs to $[f]$. The fact that $\{A_n\}$ does not vanish exponentially implies that $F(x)$ is not analytic on the entire real line but it can still be analytic on portions thereof which may contain the range of f. So we impose the additional condition that $A_n = 0$ unless $n = m!$, $m = 1, 2, \ldots$, which implies that $\sum A_n e^{inx}$ is analytic nowhere on $\mathbf{R}$.

8.8 Individual symbolic calculus is related to the problem of spectral synthesis in B. Assume for instance that the range of f is $[-1,1]$ and that $[f] \subseteq C^m([-1,1])$, $m \geqq 1$. Since in C^m, $F(x) = x$ does not belong to the ideal generated by x^2, and since (theorem 4.1) the imbedding of $[f]$ in C^m is continuous, f does not belong to the ideal generated by f^2 in $[[f]]$. This does not mean a-priori that the same is true in B. We do have a linear functional v on $[[f]]$ which is orthogonal to (f^2) and such that $\langle f, v\rangle \neq 0$ and we can extend it by the Hahn-Banach theorem to a functional on B; there is no reason, however, to expect that the support of the extended functional should always be contained in $f^{-1}(0)$. If v can be extended to B with $\Sigma(v) \subseteq f^{-1}(0)$, spectral synthesis fails in B.

Going back to C^m one identifies immediately a functional orthogonal to the ideal generated by x^2 but not to x; for instance, δ', the derivative (in the sense of the theory of distributions) of the point mass at zero, which assigns to every $F \in C^m$ the value of its derivative at the origin. In $[[f]]$ the corresponding functional can be denoted by $\delta'(f)$ and remembering that the Fourier transform of δ' is $\widehat{\delta'}(u) = -iu$ one may try to extend $\delta'(f)$ to B using the Fourier inversion formula $\delta'(f) = 1/2\pi \int (iu)\, e^{iuf}\, du$. Strictly speaking this is meaningless, but it provided the motivation for theorem 7.5.

EXERCISES FOR SECTION 8

1. Let B be a semisimple Banach algebra without unit and with discrete maximal ideal space. Show that every function F analytic near zero and satisfying $F(0) = 0$, operates in B.

2. As in chapter I, $\mathrm{Lip}_1(\mathbf{T})$ denotes the subalgebra of $C(\mathbf{T})$ consisting of the functions f satisfying $\sup_{t_1 \neq t_2} \left| \dfrac{f(t_1) - f(t_2)}{t_1 - t_2} \right| < \infty$.

(a) Find the functions that operate in $\mathrm{Lip}_1(\mathbf{T})$.
(b) Show that every function which operates in $\mathrm{Lip}_1(\mathbf{T})$ is bounded in every ball.
(c) Show that $F(x) = |x|$ does not operate continuously at $f = \sin t$.

3. Assume that F is defined on $\mathbf{R}$ and operates in $A(\mathbf{T})$. Assume that for every $r > 0$ there exists $K = K(r)$ such that if f is real valued and $\| f \|_{A(\mathbf{T})} \leqq r$, then $\| F(f) \|_{A(\mathbf{T})} < K(r)$. Show that F is the restriction to $\mathbf{R}$ of an entire function.

4. Let B be a regular, semisimple, self-adjoint Banach algebra with a unit. Assume that $F(x) = \sqrt{|x|}$ operates boundedly in B. Prove that $B = C(\mathfrak{M})$. *Hint*: Use theorem 3.8.

5. Use the construction of section 7 to show that for the algebra $B = A(\mathbf{D})$, $\mathcal{N}(u)$ has exponential growth at infinity; hence prove theorem 8.6 for the case $G = \mathbf{D}$.

6. Let $\alpha(u)$ be a positive function, $0 < u < \infty$, such that $\alpha(u) \leqq \frac{1}{2}$ and $\alpha(u) \to 0$ as $u \to \infty$. Show that there exists a real-valued $f \in A(\mathbf{T})$ such that

$$\| e^{iuf} \| \geqq e^{u\alpha(u)} .$$

7. (a) Show that if $\varphi(t, \tau) \in A(\mathbf{T}^2)$, then for every $\tau \in \mathbf{T}$, $\psi_\tau(t) = \varphi(t, \tau)$, considered as a function of t alone, belongs to $A(\mathbf{T})$ and $\| \psi_\tau(t) \|_{A(\mathbf{T})} \leqq \| \varphi \|_{A(\mathbf{T}^2)}$. Furthermore: $\psi_\tau(t)$ is a continuous $A(\mathbf{T})$-valued function of τ.
(b) Prove: for every function $\alpha(u)$ as in exercise 6 above, there exists a real-valued $g \in A(\mathbf{T}^2)$ whose range contains $[-\pi, \pi]$ and such that if $F(x) = \sum A_n e^{inx} \in [g]$, then $A_n = O(e^{-|n|\alpha(|n|)})$. *Hint*: Take $g(t, \tau) = f(t) + 5 \sin \tau$, where f is a function constructed in exercise 6 above. Apply part (a) and the argument of 8.6.
(c) Deduce theorem 8.6 for the case $G = \mathbf{T}^2$ from part (b).

9. THE ALGEBRA $M(\mathbf{T})$ AND FUNCTIONS THAT OPERATE ON FOURIER-STIELTJES COEFFICIENTS

In this section we study the Banach algebra of measures on a non-discrete LCA group. We shall actually be more specific and consider $M(\mathbf{T})$; this in order to avoid some (minor) technical difficulties while presenting all the basic phenomena of the general case.

9.1 We have little information so far about the Banach algebra $M(\mathbf{T})$. We know that for every $n \in \mathbf{Z}$, the mapping $\mu \to \hat{\mu}(n)$ is a multiplicative linear functional on $M(\mathbf{T})$; this identifies $\mathbf{Z}$ as part of the maximal ideal space $\mathfrak{M}$ of $M(\mathbf{T})$. How big a part of $\mathfrak{M}$ is $\mathbf{Z}$? We have one negative indication: since $\mathfrak{M}$ is compact the range of every $\hat{\mu}$ on $\mathfrak{M}$ is compact and therefore contains the closure of the sequence $\{\hat{\mu}(n)\}_{n\in\mathbf{Z}}$, which may well be uncountable (e.g., if $\hat{\mu}(n) = \cos n$, $n \in \mathbf{Z}$). Thus $\mathfrak{M}$ is uncountable and is therefore much bigger than $\mathbf{Z}$. But we also have a positive indication: a measure μ is determined by its Fourier-Stieltjes coefficients, that is, if $\hat{\mu} = 0$ on $\mathbf{Z}$ then $\mu = 0$ and therefore $\hat{\mu} = 0$ on $\mathfrak{M}$. This proves that $M(\mathbf{T})$ is semisimple and may suggest the following question:

(a) Is $\mathbf{Z}$ dense in $\mathfrak{M}$?

Other natural questions are:

(b) Is $M(\mathbf{T})$ regular?
(c) Is $M(\mathbf{T})$ self-adjoint?

***Theorem*:** *There exists a measure $\mu \in M(\mathbf{T})$ such that $\hat{\mu}$ is real-valued on $\mathbf{Z}$ but is not real valued on $\mathfrak{M}$.*

***Corollary*:** *The answer to all three questions above is "no."*

PROOF OF THE COROLLARY: It is clear that the theorem implies that $\mathbf{Z}$ is not dense in $\mathfrak{M}$. If $M \in \mathfrak{M}$ is not in the closure of $\mathbf{Z}$, there is no $\mu \in M(\mathbf{T})$ such that $\hat{\mu} = 0$ on $\mathbf{Z}$ while $\hat{\mu}(M) \neq 0$ (since $\hat{\mu} = 0$ on $\mathbf{Z}$ implies $\mu = 0$); so $M(\mathbf{T})$ is not regular. Finally: if μ is a measure with real-valued Fourier-Stieltjes coefficients and if for some $\nu \in M(\mathbf{T})$, $\hat{\nu} = \bar{\hat{\mu}}$ on $\mathfrak{M}$, we have $\hat{\nu} = \hat{\mu}$ on $\mathbf{Z}$, hence $\nu = \mu$ and $\hat{\mu} = \bar{\hat{\mu}}$ on $\mathfrak{M}$ which means that $\hat{\mu}$ is real valued on $\mathfrak{M}$. Thus, if μ has the properties described in the theorem, then $\bar{\hat{\mu}} \notin \widehat{M(\mathbf{T})}$. ◀

9.2 In the proof of theorem 9.1 we shall need

***Lemma*:** *Let G_1 and G_2 be disjoint*† *subgroups of $\mathbf{T}$ and let $E_j \subset G_j$, $j = 1, 2$, be compact. Let μ_j be carried by E_j, $j = 1, 2$. Then*

$$\| \mu_1 * \mu_2 \|_{M(\mathbf{T})} = \| \mu_1 \|_{M(\mathbf{T})} \| \mu_2 \|_{M(\mathbf{T})}. \tag{9.1}$$

PROOF: Let $\varepsilon > 0$ and let φ_j be continuous on E_j, satisfying $|\varphi_j| \leqq 1$ and $\int \varphi_j \, d\mu_j \geqq \| \mu_j \| - \varepsilon$. The function $\psi(t + \tau) = \varphi_1(t)\varphi_2(\tau)$ is well defined and continuous on $E_1 + E_2$ (this is where we use the fact that E_j are contained in disjoint subgroups) and

† That is: $G_1 \cap G_2 = \{0\}$.

$$\int \psi \, d(\mu_1 * \mu_2) = \iint \psi(t + \tau) \, d\mu_1 \, d\mu_2 = \int \varphi_1 \, d\mu_1 \int \varphi_2 \, d\mu_2$$

which implies $\| \mu_1 * \mu_2 \| \geqq (\| \mu_1 \| - \varepsilon)(\| \mu_2 \| - \varepsilon)$. ◀

9.3 PROOF OF 9.1: We construct a measure $\mu \in M(\mathbf{T})$ with real Fourier-Stieltjes coefficients and such that

(9.2) $$\| e^{in\mu} \|_{M(\mathbf{T})} = e^n \qquad \text{for } n \geqq 0 .$$

By lemma 3.6 it follows that the spectral norm of $e^{i\mu}$ is equal to e which means that $\operatorname{Im}(\hat{\mu}) = -1$ somewhere on $\mathfrak{M}$.

Let E be a perfect independent set on $\mathbf{T}$ (see VI.9.4). Let ν be a continuous measure carried by E and $\nu^{\#}$ the symmetric image of ν, defined by $\nu^{\#}(F) = \overline{\nu(-F)}$ for all measurable sets F. $\nu^{\#}$ is clearly carried by $-E$ and if we write $\mu = \nu + \nu^{\#}$ we have $\hat{\mu}(n) = 2 \operatorname{Re}(\hat{\nu}(n))$ for all $n \in \mathbf{Z}$. We claim that for such μ

(9.2′) $$\| e^{i\mu} \| = e^{\|\mu\|} .$$

Let N be a large integer and write E as a union of N disjoint closed subsets E_j such that the norm of the portion of μ carried by $E_j \cup -E_j$, call it μ_j, is precisely $\| \mu \| N^{-1}$. (Here we use the fact that μ is continuous.) Now $e^{i\mu_j} = \delta + i\mu_j + [$a measure whose norm is $O(N^{-2})]$ where δ is the identity in $M(\mathbf{T})$, that is, the unit mass concentrated at the origin. We have $\| \delta + i\mu_j \| = 1 + \| \mu \| N^{-1}$ and

$$e^{i\mu} = \prod_1^N * e^{i\mu_j} = \prod_1^N * (\delta + i\mu_j) + \rho$$

where ρ is a measure whose norm is $O(N^{-1})$. Since E is independent the E_j generate disjoint subgroups of $\mathbf{T}$ and, by lemma 9.2, $\| e^{i\mu} \| = (1 + \| \mu \| N^{-1})^N + O(N^{-1})$; as $N \to \infty$, (9.2′) follows. It is now clear that if we normalize μ to have norm 1 and apply (9.2′) to $n\mu$ we have (9.2). ◀

Remark: Since the measure μ, described above, has norm 1, its spectrum lies in the disc $|z| \leqq 1$. The only point in the unit disc whose imaginary part is -1 is $z = -i$. It follows that -1 is in the spectrum of μ^2 which means that $\delta + \mu^2$ is not invertible. The Fourier-Stieltjes coefficients of $\delta + \mu^2$ are $1 + (\hat{\mu}(n))^2 \geqq 1$ (since $\hat{\mu}(n)$ is real-valued) and yet $(1 + (\hat{\mu}(n))^2)^{-1}$ are not the Fourier-Stieltjes coefficients of any measure on $\mathbf{T}$. This phenomenon was discovered by Wiener and Pitt.

9.4 DEFINITION: A function F, defined in some subset of $\mathbf{C}$, *operates on Fourier-Stieltjes coefficients* if $\{F(\hat{\mu}(n))\}_{n\in\mathbf{Z}}$ is a sequence of Fourier-Stieltjes coefficients for every $\mu \in M(\mathbf{T})$ such that $\{\hat{\mu}(n)\}_{n\in\mathbf{Z}}$ is contained in the domain of definition of F.

Since $\mathbf{Z}$ is not the entire maximal ideal space of $M(\mathbf{T})$ there is no reason to expect that if F is holomorphic on its domain, it operates on Fourier-Stieltjes coefficients; and by the remark above, the function defined on $\mathbf{R}$ by $F(x) = (1 + x^2)^{-1}$ does not operate on Fourier-Stieltjes coefficients. This is a special case of

Theorem: *Let F be defined in an interval $I \subseteq \mathbf{R}$ and assume that it operates on Fourier-Stieltjes coefficients. Then F is the restriction to I of an entire function.*

The theorem can be proved along the same lines as 8.6. The main difference is that one shows that if F operates on Fourier-Stieltjes coefficients, the operation is bounded in *every* ball of $M(\mathbf{T})$ (rather than *some* ball, as in the case of $A(G)$), that is, for all $r > 0$ there exists a $K = K(r)$ such that if $\|\mu\| \leqq r$ and $\hat{\mu}(n) \in I$ for all $n \in \mathbf{Z}$, then $F(\hat{\mu}(n))$ are the Fourier-Stieltjes coefficients of a measure of norm $\leqq K$. We refer to [24], chapter 6 and to exercises 6 through 9 at the end of this section for further details.

9.5 The individual symbolic calculus on $M(\mathbf{T})$ is also more restrictive than an individual symbolic calculus can be in a Banach algebra considered as function algebra on the entire maximal ideal space. There exist measures μ in $M(\mathbf{T})$ with real-valued Fourier-Stieltjes coefficients such that every continuous function which operates on μ must be the restriction to $\mathbf{R}$ of some function analytic in a disc (see exercise 10 at the end of this section). This suggests that portions of the maximal ideal space of $M(\mathbf{T})$ may carry analytic structure.

EXERCISES FOR SECTION 9

1. Let $\mu \in M(\mathbf{T})$ be such that $\|e^{\alpha\mu}\| = e^{|\alpha|\|\mu\|}$ for all $\alpha \in \mathbf{C}$. Show that $\{\mu^n\}$, $n = 0, 1, 2, \ldots$, are mutually singular.

2. Let E be a linearly independent compact set on $\mathbf{T}$ and let μ be a continuous measure carried by $E \cup -E$. Show that $\{\mu^n\}$, $n = 0, 1, 2, \ldots$, are mutually singular.

3. Show that if $\mu \in M(\mathbf{T})$, $\hat{\mu}(n)$ is real for all $n \in \mathbf{Z}$ and μ^n are mutually singular for $n = 1, 2, \ldots$, then μ is continuous.

4. Deduce theorem 9.1 from theorem 9.4.

5. Let r_j, $j = 1, 2, \ldots$, be positive numbers such that $r_j/r_{j-1} < \frac{1}{3}$ and $r_j/r_{j-1} \to 0$ as $j \to \infty$. Show that $\varphi(n) = \prod_1^\infty \cos r_j n$, $n \in \mathbf{Z}$ are the Fourier-Stieltjes coefficients of a measure μ, and that μ^n are mutually singular, $n = 1, 2, \ldots$. *Hint*: Show that if $r_{j-1}/r_j > M$ for all j, then $\{\mu^n\}_{n=1}^M$ are mutually singular.

6. Let F be continuous on $[-1, 1]$ and $F(0) = 0$. Show that the following two conditions are equivalent:
(i) If $\mu \in M(\mathbf{T})$, $\|\mu\|_{M(\mathbf{T})} < r$ and $-1 \leqq \hat{\mu}(n) < 1$ for all n, then $\{F(\hat{\mu}(n))\}$ are the Fourier-Stieltjes coefficients of some measure $F(\mu)$ $F(\mu) \in M(\mathbf{T})$ such that $\|F(\mu)\|_{M(\mathbf{T})} < K$.
(ii) If $-1 \leqq a_n \leqq 1$ and $P = \sum a_n e^{int}$ is a polynomial satisfying $\|P\|_{L^1(\mathbf{T})} < r$, then $\|\sum F(a_n) e^{int}\|_{L^1(\mathbf{T})} < K$.
Also show that in (ii) we may add the assumption that the a_n are rational numbers without affecting the equivalence of (i) and (ii).

7. (For the purpose of this exercise) we say that a measure μ contains a polynomial P if for appropriate m and M, $\hat{P}(n) = \hat{\mu}(m + nM)$ for all integers n which are bounded in absolute value by twice the degree d of P. Notice that if μ contains P then

$$P(Mt) = \mathbf{V}_d(Mt) * (e^{-imt}\mu)$$

and consequently $\|P\|_{L^1(\mathbf{T})} \leqq 2\|\mu\|_{M(\mathbf{T})}$. Show that there exists a measure μ with real Fourier-Stieltjes coefficients, $\|\mu\| \leqq 2$, and μ contains every polynomial P, with rational coefficients, such that $\|P\|_{L^1(\mathbf{T})} \leqq 1$. *Hint*: Show that for every sequence of integers $\{N_j\}$ there exists a sequence of integers $\{\lambda_j\}$ such that, writing $\Lambda_j = \{k\lambda_j\}_{k=1}^{N_j}$ and $\Lambda = \bigcup_j \Lambda_j$, every function $f \in C_\Lambda$ (see chapter V for the notation C_Λ) can be written $f = \sum f_j$, with $f_j \in C_{\Lambda_j}$, and $\sum \|f_j\|_\infty \leqq 2\|f\|$. Deduce, using the Hahn-Banach theorem, that if the numbers $a_{j,k}$ are such that for each j, $\|\sum_{k=1}^{N_j} a_{j,k} e^{ikt}\|_{L^1(\mathbf{T})} \leqq 1$, then there exists a measure $\mu \in M(\mathbf{T})$ such that $\|\mu\|_{M(\mathbf{T})} < 2$ and $\hat{\mu}(k\lambda_j) = a_{j,k}$ for appropriate λ_j and $1 \leqq k \leqq N_j$. If the numbers $a_{j,k}$ above are real, one can replace μ by $\frac{1}{2}(\mu + \mu^\#)$.

8. Let F be defined and continuous on $\mathbf{R}$ and assume that it operates on Fourier-Stieltjes coefficients. Prove that the operation is bounded on every ball of $M(\mathbf{T})$. *Hint*: Use exercises 6 and 7 above; show that if $\|\nu\|_{M(\mathbf{T})} < r$ then $\|F(\nu)\|_{M(\mathbf{T})} \leqq 2\|F(r\mu)\|_{M(\mathbf{T})}$.

9. Prove theorem 9.4.

10. Show that if F is defined and continuous on $\mathbf{R}$ and if $F(\hat{\mu}(n))$ are Fourier-Stieltjes coefficients, μ being the measure introduced in

exercise 7, then F is analytic at the origin. If $F(k\hat{\mu}(n))$ are Fourier-Stieltjes coefficients for all k, then F is entire.

11. Let $\mu \in M(\mathbf{T})$ be carried by a compact independent set and assume that $\hat{\mu}(n) \to 0$ as $|n| \to \infty$. (Such measures exist: see [25].) Let B be the closed subalgebra of $M(\mathbf{T})$ generated by $L^1(\mathbf{T})$ and μ.

(a) Check that theorems 9.1 and 9.4 are valid if we replace in their statement $M(\mathbf{T})$ by B.

(b) Notice that the restriction of $\hat{B}$ to $\mathbf{Z}$ is a function algebra on $\mathbf{Z}$, intermediate between $A(\mathbf{Z})$ and c_0 both of which have $\mathbf{Z}$ as maximal ideal space, and yet its maximal ideal space is larger than $\mathbf{Z}$.

10. THE USE OF TENSOR PRODUCTS

In this final section we prove a theorem concerning the symbolic calculus and the failing of spectral synthesis in some quotient algebras of $A(\mathbf{R})$. The theorem and its proof are due to Varopoulos and serve here as an illustration of a general method which he introduced. We refer to [26] for a systematic account of the use of tensor algebras in harmonic analysis.

10.1 Let $E \subset \mathbf{R}$ be compact. We denote by $A(E)$ the algebra of functions on E which are restrictions to E of elements of $A(\mathbf{R})$. $A(E)$ is canonically identified with the quotient algebra $A(\mathbf{R})/k(E)$ (where $k(E) = \{f; f \in A(\mathbf{R}) \text{ and } f = 0 \text{ on } E\}$) and is therefore a Banach algebra with E as the space of maximal ideals (see 5.5). The main theorem of this section is:

Theorem: *Let E_1, E_2 be nonempty perfect subsets of $\mathbf{R}$ such that $E_1 \cup E_2$ is a Kronecker set. Put $E = E_1 + E_2$*†. *Then*: (a) *Every function F, defined on $\mathbf{R}$, which operates in $A(E)$ is analytic.* (b) *Spectral synthesis fails in $A(E)$.*

Remarks: (i) We place E on $\mathbf{R}$ for the sake of technical simplicity and in accordance with the general trend of this book. Only minor modifications are needed in order to place E in an arbitrary nondiscrete LCA group, obtaining thereby a proof of Malliavin's theorem 7.4 in its full generality.

(ii) We shall actually prove more, namely: $A(E)$ is isomorphic to a

† $E_1 + E_2 = \{x; x = x_1 + x_2 \text{ with } x_j \in E_j\}$.

fixed Banach algebra (subsections 10.2, 10.3, and 10.4) for which (a) and (b) are valid (subsection 10.5).

10.2 Let X and Y be compact Hausdorff spaces and $X \times Y$ their cartesian product. We denote by $V = V(X, Y)$ the projective tensor product of $C(X)$ and $C(Y)$; that is, the space of all continuous functions φ on $X \times Y$ that admit a representation of the form

$$\varphi(x, y) = \sum f_j(x) g_j(y) \tag{10.1}$$

with $f_j \in C(X)$, $g_j \in C(Y)$, and

$$\sum \|f_j\|_\infty \|g_j\|_\infty < \infty. \tag{10.2}$$

We introduce the norm

$$\|\varphi\|_V = \inf \sum \|f_j\|_\infty \|g_j\|_\infty \tag{10.3}$$

where the infimum is taken with respect to all possible representations of φ in the form (10.1). It is immediate to check that the norm $\| \ \|_V$ is multiplicative and that V is complete; thus V is a Banach algebra.

***Lemma*:** *The maximal ideal space of V can be identified canonically with $X \times Y$.*

PROOF: Denote by V_1 (resp. V_2) the subalgebra of V consisting of the functions $\varphi(x, y)$ which depend only on x (resp. only on y). It is clear that V_1 and V_2 are canonically isomorphic to $C(X)$ and $C(Y)$ respectively. A multiplicative linear functional w on V induces, by restriction multiplicative linear functionals w_1 on V_1 and w_2 on V_2. By corollary, 2.12, w_1 has the form $f \to f(x_0)$ for some $x_0 \in X$; w_2 has the form $g \to g(y_0)$ for some $y_0 \in Y$, and it follows that if

$$\varphi(x, y) = \sum f_j(x) g_j(y),$$

then

$$w(\varphi) = \sum f_j(x_0) g_j(y_0) = \varphi(x_0, y_0).$$ ◀

***Corollary*:** *V is semisimple, self-adjoint, and regular.*

10.3 We assume now that X is homeomorphic to a compact abelian group G (more precisely, to the underlying topological space of G) and that Y is homeomorphic to a compact abelian group H. We denote both homeomorphisms $X \to G$ and $Y \to H$ by σ. σ induces canonically a homeomorphism of $X \times Y$ onto $G \oplus H$, and hence an isomorphism of $C(G \oplus H)$ onto $C(X \times Y)$.

Lemma: *The canonical isomorphism of $C(G \oplus H)$ onto $C(X \times Y)$ maps $A(G \oplus H)$ into V.*

PROOF: Let χ be a character on $G \oplus H$ and let $\boldsymbol{\chi}$ be its image under the canonical isomorphism, namely

$$\boldsymbol{\chi}(x, y) = \chi(\sigma x, \sigma y). \tag{10.4}$$

Since $\chi(\sigma x, \sigma y) = \chi(\sigma x, 0)\chi(0, \sigma y)$ we have

$$\boldsymbol{\chi}(x, y) = \boldsymbol{\chi}(x, 0)\boldsymbol{\chi}(0, y) \tag{10.5}$$

so that $\boldsymbol{\chi} \in V$ and $\|\boldsymbol{\chi}\|_V = 1$. If $\varphi \in A(G \oplus H)$ then $\varphi = \sum a_\chi \chi$, the summation extending over $\widehat{G \oplus H}$ and $\|\varphi\|_A = \sum |a_\chi|$. The image of φ under the canonical isomorphism is $\boldsymbol{\varphi} = \sum a_\chi \boldsymbol{\chi}$ and therefore $\|\boldsymbol{\varphi}\|_V \leqq \sum |a_\chi| = \|\varphi\|_A$. ◀

If $\varphi \in A(G \oplus H)$ depends only on the first variable, that is, if $\varphi(x, y) = \psi(x)$, then $\psi \in A(G)$. Assuming G to be infinite we have $A(G) \neq C(G)$ and it follows that the image of $A(G \oplus H)$ in V does not contain V_1 and is therefore a proper part of V.

The connection between $A(G \oplus H)$ and V is only that of (canonical) inclusion, which is too loose for obtaining information for one algebra from the other. A closer look reveals, however, that the structure needed for the lemma is not the group structure on G or on H but only the cartesian structure of $G \oplus H$, while in order to show that the image of $A(G \oplus H)$ is not the entire V we use the group structure of G. The idea now is to keep the useful structure and obliterate the hampering one; this is the reason for the appeal to Kronecker sets.

10.4 ***Theorem:*** *Let E_1, E_2, and E be as in the statement of theorem* 10.1. *Let X and Y be homeomorphic to the (classical) Cantor set. Then $A(E)$ is isomorphic to $V(X, Y)$.*

PROOF: We begin by noticing that E_1 and E_2, being portions of a Kronecker set, are clearly totally disconnected and, being perfect and nonempty, are homeomorphic to the Cantor set. Thus, X, Y, E_1, and E_2 are all homeomorphic and we simplify the typography by identifying X with E_1 and Y with E_2. Since $E_1 \cup E_2$ is independent (being a Kronecker set) the mapping $(x, y) \to x + y$ is a homeomorphism of $X \times Y$ on E. We now show that the induced mapping of $C(E)$ onto $C(X \times Y)$ maps $A(E)$ onto V. The fact that $A(E)$ is mapped into V (and that the mapping is of norm 1) is a verbatim repetition of 10.3. We therefore have only to prove that the mapping is surjective (i.e. that it maps $A(E)$ *onto* V). We shall need:

Lemma: *Let $\tilde{E}$ be a Kronecker set. Then every $f \in C(\tilde{E})$ can be written as $f(x) = \sum a_n e^{i\lambda_n x}$ with $\sum |a_n| < 3 \| f \|_{C(\tilde{E})}$. In particular: $A(\tilde{E}) = C(\tilde{E})$ and $\| f \|_{A(\tilde{E})} \leqq 3 \| f \|_\infty$.*

PROOF OF THE LEMMA: It is enough to show that if $f \in C(\tilde{E})$ is real valued, then $f = \sum a_n e^{i\lambda_n x}$ with $\sum |a_n| < \frac{3}{2} \| f \|_\infty$. Let f be real valued and assume, for simplicity, that $\| f \|_\infty = 1$; define $g(x) = \sqrt{1-(f(x))^2}$, then g is continuous and $\Phi = f + ig$ has modulus 1 on $\tilde{E}$. Let λ_1 be such that $|G - e^{i\lambda_1 x}| < \frac{1}{10}$ on $\tilde{E}$; this implies $|f - \cos \lambda_1 x| < \frac{1}{10}$ on $\tilde{E}$.

If $\| f \|_\infty$ is not 1 we consider $\| f \|_\infty^{-1} f$ and obtain λ_1 such that

$$|f - \| f \|_\infty \cos \lambda_1 x| < \tfrac{1}{10} \| f \|_\infty \text{ on } \tilde{E}.$$

We now proceed by induction. Define $a_1 = \| f \|_\infty$, λ_1 as above, and $f_1 = f - a_1 \cos \lambda_1 x$; once we have $a_1, \ldots, a_n, \lambda_1, \ldots, \lambda_n$, and f_n, define $a_{n+1} = \| f_n \|_\infty$, λ_{n+1} by the condition $|f_n - a_{n+1} \cos \lambda_{n+1}| < a_{n+1}/10$, and $f_{n+1} = f_n - a_{n+1} \cos \lambda_{n+1} = f - \sum_1^{n+1} a_j \cos \lambda_j x$.

We clearly have $a_{n+1} < a_n/10 < \| f \|_\infty 10^{-n}$ and it follows that $f = \sum_1^\infty a_n \cos \lambda_n x$ with $\sum |a_n| = \sum a_n < \| f \|_\infty (\sum_0^\infty 10^{-n}) < \frac{3}{2} \| f \|_\infty$. Writing $\cos \lambda_n x = \frac{1}{2}(e^{i\lambda_n x} + e^{-i\lambda_n x})$ we obtain f as a series of exponentials. ◀

Remark. $A(\tilde{E})$ is actually isometric to $C(\tilde{E})$; see exercise 2 at the end of this section.

PROOF OF THE THEOREM, COMPLETED: We identify $X \times Y$ with E, and V with the subalgebra of $C(E)$ consisting of the functions φ which admit a representation

$$\varphi(x+y) = \sum f_j(x) g_j(y), \qquad x \in E_1,\ y \in E_2 \tag{10.6}$$

where $f_j \in C(E_1)$, $g_j \in C(E_2)$ such that (10.2) is valid. All that we need to show is that if $\varphi \in V$ then $\varphi \in A(E)$.

Let $\varphi \in V$ and consider a representation of the form (10.6) such that

$$\sum \| f_j \|_\infty \| g_j \|_\infty \leqq 2 \| \varphi \|_V. \tag{10.7}$$

Using the lemma we write each f_j as an exponential series (E_1, being a portion of a Kronecker set, is itself one) and similarly for the g_j.

Denoting the frequencies appearing in the f's by λ and those appearing in the g's by ν, and taking account of (10.6) and (10.7), we obtain

$$\varphi(x+y) = \sum_{\lambda,\nu} a_{\lambda,\nu} e^{i\lambda x} e^{i\nu y}, \qquad x \in E_1,\ y \in E_2 \tag{10.8}$$

where

$$\sum |a_{\gamma,\nu}| \leqq 18 \| \varphi \|_V. \tag{10.9}$$

We now use the fact that $E_1 \cup E_2$ is a Kronecker set: let (λ, ν) be a pair which appears in (10.8) and define

$$h(x) = \begin{cases} e^{i\lambda x} & x \in E_1, \\ e^{i\nu x} & x \in E_2; \end{cases}$$

h is clearly continuous and of modulus 1 on $E_1 \cup E_2$ and it follows that there exists a real number ξ such that

$$|e^{i\xi x} - h(x)| < 1/200 \qquad \text{on } E_1 \cup E_2. \tag{10.10}$$

We have (for $x \in E_1$, $y \in E_2$),

$$e^{i\xi(x+y)} - e^{i\lambda x} e^{i\nu y} = (e^{i\xi x} - e^{i\lambda x}) e^{i\xi y} + e^{i\lambda x}(e^{i\xi y} - e^{i\nu y})$$

which means that (with the canonical identifications)

$$\| e^{i\xi(x+y)} - e^{i\lambda x} e^{i\nu y} \|_V < 1/100.$$

We can now write $\varphi = \vartheta_1 + \varphi_1$ where

$$\vartheta_1(x+y) = \sum a_{\lambda,\nu} e^{i\xi(x+y)}$$

and

$$\varphi_1(x+y) = \sum a_{\lambda,\nu}(e^{i\lambda x} e^{i\nu y} - e^{i\xi(x+y)})$$

and notice that $\vartheta_1 \in A(E)$ and, by (10.9),

$$\| \vartheta_1 \|_{A(E)} \leqq \sum |a_{\lambda,\nu}| \leqq 18 \| \varphi \|_V;$$

also

$$\| \varphi_1 \|_V \leqq \frac{18}{100} \| \varphi \|_V < \tfrac{1}{5} \| \varphi \|_V.$$

Repeating, we obtain inductively

$$\varphi_n = \vartheta_{n+1} + \varphi_{n+1}$$

where

$$\vartheta_{n+1} \in A(E)$$

$\| \vartheta_{n+1} \|_{A(E)} \leqq 18 \| \varphi_n \|_V$, and $\| \varphi_{n+1} \|_V < \frac{1}{5} \| \varphi_n \|_V$. It follows that $\varphi = \sum \vartheta_n \in A(E)$ and $\| \varphi \|_{A(E)} \leqq 18(\sum_0^\infty 5^{-n}) \| \varphi \|_V \leqq 25 \| \varphi \|_V$. ◀

10.5 We now show that statements (a) and (b) of theorem 10.1 are valid for $V = V(X, Y)$ (X and Y being both homeomorphic images of the Cantor set). This is obtained as a consequence of the fact that (a) and (b) are valid for the algebra $A(\mathbf{D})$ (see exercise 8.5 for (a) and theorem 7.4 for (b)).

Since X and Y are homeomorphic to $\mathbf{D}$ we may consider V as a function algebra on $\mathbf{D} \times \mathbf{D}$. Using the group structure and the Haar measure on $\mathbf{D}$ we now define two linear operators $\boldsymbol{M}$ and $\boldsymbol{P}$ as follows:

(10.11) $$\text{for } f \in C(\mathbf{D}), \text{ write } \boldsymbol{M}f(x, y) = f(x + y);$$

(10.12) $$\text{for } \varphi \in C(\mathbf{D} \times \mathbf{D}), \text{ write } \boldsymbol{P}\varphi(x) = \int_{\mathbf{D}} \varphi(x - y, y)\, dy.$$

$\boldsymbol{M}$ maps $C(\mathbf{D})$ into $C(\mathbf{D} \times \mathbf{D})$, $\boldsymbol{P}$ maps $C(\mathbf{D} \times \mathbf{D})$ into $C(\mathbf{D})$, and, since for $f \in C(\mathbf{D})$;

$$\boldsymbol{PM}f(x) = \int_{\mathbf{D}} \boldsymbol{M}f(x - y, y)\, dy = \int_{\mathbf{D}} f(x)\, dy = f(x),$$

it follows that $\boldsymbol{PM}$ is the identity maps of $C(\mathbf{D})$.

Lemma: *$\boldsymbol{M}$ maps $A(\mathbf{D})$ into V and its norm as such is* 1. *$\boldsymbol{P}$ maps V into $A(\mathbf{D})$ and its norm as such is* 1.

PROOF: If $f = \sum a_\chi \chi$ with $\sum |a_\chi| = \|f\|_{A(\mathbf{D})} < \infty$ then

$$\boldsymbol{M}f = \sum a_\chi \chi(x)\chi(y) \in V \quad \text{and} \quad \|\boldsymbol{M}f\|_V \leqq \sum |a_\chi| = \|f\|_{A(\mathbf{D})}.$$

If $\varphi(x, y) = f(x)g(y)$ then

$$\boldsymbol{P}\varphi = \int f(x - y) g(y)\, dy = f * g;$$

hence

(10.13) $$\|\boldsymbol{P}\varphi\|_{A(\mathbf{D})} = \sum |\hat{f}(\chi)\hat{g}(\chi)| \leqq (\sum |\hat{f}(\chi)|^2)^{1/2} (\sum |\hat{g}(\chi)|^2)^{1/2} = \|f\|_{L^2} \|g\|_{L^2} \leqq \|f\|_\infty \|g\|_\infty .$$

By (10.13) and the definition (10.3) of the norm in V it follows that for arbitrary $\varphi \in V$, $\|\boldsymbol{P}\varphi\|_{A(\mathbf{D})} \leqq \|\varphi\|_V$. ◀

***Corollary*:** *Let* $\varphi \in C(\mathbf{D} \times \mathbf{D})$ *and assume that for some* $\psi \in C(\mathbf{D})$ *we have* $\varphi(x, y) = \psi(x + y)$. *Then* $\varphi \in V$ *if, and only if,* $\psi \in A(\mathbf{D})$, *and then* $\| \varphi \|_V = \| \psi \|_{A(\mathbf{D})}$.

In other words, $\boldsymbol{M}$ is an isometry of $A(\mathbf{D})$ onto the closed subalgebra V_3 of V of all the functions $\varphi(x, y)$ which depend only on $x + y$.

Remark: The subalgebra V_3 is determined by the "level lines": $x + y = \text{const}$, which clearly depend on the group structure of $\mathbf{D}$. Instead of $\mathbf{D}$ we can take any group G whose underlying topological space is homeomorphic to the Cantor set; for every such G, V has a closed subalgebra isometric to $A(G)$.

We are now ready to prove:

***Theorem*:** (a) *Every function F, defined on* $\mathbf{R}$, *which operates in V is analytic.* (b) *Spectral synthesis is not always possible in V.*

PROOF: (a) If F operates in V, so it does in V_3 (since the operation by F conserves the level lines), hence in $A(\mathbf{D})$ and by theorem 8.6 (rather, exercise 8.5), F is analytic.

(b) Let $H \subset \mathbf{D}$ be a closed set which is not a set of spectral synthesis for $A(\mathbf{D})$ (see theorem 7.4). Define:

$$H^* = \{(x, y); x + y \in H\} \subset \mathbf{D} \times \mathbf{D}.$$

We contend that H^* is not a set of spectral synthesis in V. By 7.3 we have a function $f \in A(\mathbf{D})$ which vanishes on H and which cannot be approximated by functions in $I_0(H)$ (i.e., functions that vanish in a neighborhood of H). Thus for some $\delta > 0$, $\| f - g \|_{A(\mathbf{D})} > \delta$ for every $g \in I_0(H)$. We show that H^* is not of spectral synthesis for V by showing that $\boldsymbol{M}f$, which clearly vanishes on H^*, satisfies $\| \boldsymbol{M} f - \varphi \|_V > \delta$ for every $\varphi \in V$ which vanishes in a neighborhood of H^*. For this we notice that if φ vanishes on a neighborhood of H^*, $\boldsymbol{P}\varphi$ vanishes on a neighborhood of H; now

$$\| \boldsymbol{M}f - \varphi \|_V \geqq \| \boldsymbol{PM}f - \boldsymbol{P}\varphi \|_{A(\mathbf{D})} = \| f - \boldsymbol{P}\varphi \|_{A(\mathbf{D})} > \delta. \qquad \blacktriangleleft$$

Theorems 10.4 and 10.5 clearly imply theorem 10.1.

EXERCISES FOR SECTION 10

1. Let f be a continuous complex-valued function on a topological space X. Assume $0 < |f(x)| \leqq 1$ on X. Show that f admits a unique representation in the form $f = \frac{1}{2}(g_1 + g_2)$ such that $|g_j(x)| = 1$ on

X and g_j are continuous, $j = 1, 2$. Deduce that if X is homeomorphic to the Cantor set, every continuous complex-valued function f on X admits representations of the form $f = \frac{1}{2}\| f \|_\infty(g_1 + g_2) + f_1$ where g_j are continuous, $|g_j(x)| = 1$ on X, and $\| f_1 \|_\infty$ is arbitrarily small.

2. Let B be a Banach space with the norm $\| \ \|_0$ and let $B_1 \subset B$ be a subspace which is a Banach space under a norm $\| \ \|_1$ such that the imbedding of B_1 in B is continuous. Assume that there exist constants $K > 0$ and $0 \leqq \eta < 1$ such that for every $f \in B$ there exist $g \in B_1$ and $f_1 \in B$ satisfying $f = g - f_1$, $\| g \|_1 \leqq K \| f \|_0$, and $\| f_1 \|_0 = \eta \| f \|_0$. Show that $B_1 = B$ and $\| \ \|_1 \leqq K(1 - \eta)^{-1} \| \ \|_0$. Use this and exercise 1 to prove remark 10.4. *Hint*: See either proof in 10.4.

Appendix

Vector-Valued Functions

1. RIEMANN INTEGRATION

Consider a Banach space B and let F be a B-valued function, defined and continuous on a compact interval $[a,b] \subset \mathbf{R}$. We define the (Riemann) integral of F on $[a,b]$ in a manner completely analogous to that used in the case of numerical functions, namely:

DEFINITION: $\int_a^b F(x)\,dx = \lim \sum_{i=0}^{N} (x_{j+1} - x_j)\, F(x_j)$ where

$$a = x_0 < x_1 < \cdots < x_{N+1} = b,$$

and the limit is taken as the subdivision $\{x_j\}_{j=1}^{N}$ becomes finer and finer, that is: as $N \to \infty$ and $\max_{0 \leqq j \leqq N}(x_{j+1} - x_i) \to 0$.

The existence of the limit is proved, as in the case of numerical functions, by showing that if $\{x_j\}$ and $\{y_k\}$ are subdivisions of $[a,b]$ which are fine enough to ensure that $\| F(\alpha) - F(\beta) \| < \varepsilon$ whenever α and β belong to the same interval $[x_j, x_{j+1}]$ (or $[y_k, y_{k+1}]$), then

$$\left\| \sum_{j=0}^{N} (x_{j+1} - x_j)\, F(x_j) - \sum_{k=0}^{M} (y_{k+1} - y_k)\, F(y_k) \right\| < 2(b-a)\varepsilon .$$

This is done most easily by comparing either sum to the sum corresponding to a common refinement of $\{x_j\}$ and $\{y_k\}$.

The following properties of the integral so defined are obvious:
(1) If F and G are both continuous B-valued functions on $[a,b]$, and $c_1, c_2 \in \mathbf{C}$, then:

$$\int_a^b (c_1 F(x) + c_2 G(x))\,dx = c_1 \int_a^b F(x)\,dx + c_2 \int_a^b G(x)\,dx .$$

(2) If $a < c < b$ then

$$\int_a^b F(x)\,dx = \int_a^c F(x)\,dx + \int_c^b F(x)\,dx.$$

(3)
$$\left\| \int_a^b F(x)\,dx \right\| \leqq \int_a^b \| F(x) \| \,dx.$$

(4) If μ is a continuous linear functional on B, then:

$$\left\langle \int_a^b F(x)\,dx, \mu \right\rangle = \int_a^b \langle F(x), \mu \rangle \,dx.$$

2. IMPROPER INTEGRALS

Let F be a B-valued function, defined and continuous in a non-closed interval (open or half-open; finite or infinite) say (a, b).

The (improper) integral $\int_a^b F(x)\,dx$ is, by definition, the limit of $\int_{a'}^{b'} F(x)\,dx$ where $a < a' < b' < b$ and the limit is taken as $a' \to a$ and $b' \to b$. As in the case of numerical functions the improper integral need not always exist. A sufficient condition for its existence is

$$\int_a^b \| F(x) \| \,dx < \infty.$$

3. MORE GENERAL INTEGRALS

Once in this book (in VIII.8) we integrate a vector-valued function which we do not know a-priori to be continuous. It is, however, the pointwise limit of a sequence of continuous functions and is therefore Bochner-integrable. We refer the reader to [10], chapter 3, §1, for details on the Bochner integral; we point out also that for the purpose of VIII.8, as well as in other situations where the integral is used mainly to evaluate the norm of a given vector, one can obviate the vector-valued integration by applying linear functionals to the integrand before the integration.

4. HOLOMORPHIC VECTOR-VALUED FUNCTIONS

A B-valued function $F(z)$, defined in a domain $\Omega \subseteq \mathbf{C}$ is *holomorphic in* Ω if for every continuous linear functional μ on B, the numerical function $h(z) = \langle F(z), \mu \rangle$ is holomorphic in Ω.

This condition is equivalent to the apparently stronger one stating that for each $z_0 \in \Omega$, F has the representation $F(z) = \sum_{n=0}^{\infty} a_n(z - z_0)^n$ in some neighborhood of z_0; the coefficients a_n being vectors in B and the series converging in norm. One proves that, as in the case of complex-valued functions, the power series expansion converges in the largest disc, centered at z_0, which is contained in Ω. These results are consequences of the uniform boundedness theorem (see [10], chapter 3, §2).

Many theorems about numerical holomorphic functions have their generalizations to vector-valued functions. The generalizations of theorems dealing with "size and growth" such as the maximum principle, the theorem of Phragmèn-Lindelöf, and Liouville's theorem are almost trivial to generalize. For instance: the form of Liouville's theorem which we use in VIII.2.4 is

Theorem: *Let F be a bounded entire B-valued function. Then F is a constant.*

PROOF: If $F(z_1) \neq F(z_2)$ there should exist $\mu \in B^*$ such that $\langle F(z_1), \mu \rangle \neq \langle F(z_2), \mu \rangle$. However, for all $\mu \in B^*$, $\langle F(z), \mu \rangle$ is a bounded (numerical) entire function and hence, by Liouville's theorem, is a constant. ◀

Another theorem which we use in IV.1.3 is an immediate consequence of the power series expansion. We refer to

Theorem: *Let F be a B-valued function holomorphic in a domain Ω. Let Ψ be a B^*-valued function in Ω, holomorphic in $\bar{z}$. Then $h(z) = \langle F(z), \Psi(z) \rangle$ is a holomorphic (numerical) function in Ω.*

PROOF: Let $z_0 \in \Omega$; in some disc around it $F(z) = \sum a_n(z - z_0)^n$, $\Psi(z) = \sum b_n(\overline{z - z_0})^n$, hence $h(z) = \sum \left(\sum_{k=0}^{n} \langle a_k, b_{n-k} \rangle\right)(z - z_0)^n$, and the series converges in the same disc.

Remark: Φ of IV.1.3 corresponds to $\overline{\Psi}$ here.

Bibliography

1. Bary, N. K., *A treatise on trigonometric series*. Macmillan, 1964.
2. Bohr, H., *Almost periodic functions*. Chelsea, 1947.
3. Carleman, T., *L'integrale de Fourier et questions qui s'y rattachent*. Almqvist and Wiksell, Uppsala, 1944.
4. Carleson, L., On convergence and growth of partial sums of Fourier series. *Acta Math.* **116** (1966), 135–157.
5. Gelfand, I. M., Raikov, D. A., and Shilov, G. E., *Commutative normed rings*. Chelsea, 1964.
6. Gleason, A., Function algebras. *Seminars on analytic functions*, Vol. II. Institute for Advanced Study, Princeton, N.J., 1967, pp. 213–226.
7. Goldberg, R., *Fourier transforms*. Cambridge Tracts, No. 52, 1962.
8. Hardy, G. H., and Rogozinski, W. W., *Fourier series*. Cambridge Tracts, No. 38, 1950.
9. Hewitt, E., and Ross, K. A., *Abstract harmonic analysis*. Springer-Verlag, 1963.
10. Hille, E., and Phillips, R. S., *Functional analysis and semigroups*. A.M.S. Colloquium Publications, Vol. XXXI, 1957.
11. Hoffman, K., *Banach spaces of analytic functions*. Prentice-Hall, 1962.
12. Hunt, R. A., On the convergence of Fourier series. *Proceedings of the Conference on Orthogonal Expansions and Their Continuous Analogues*. Southern Illinois University Press, 1968, pp. 234–255.
13. Kahane, J.-P., and Salem, R., *Ensembles parfaits et series trigonometriques*. Hermann, 1963.
14. Kaufman, R., A functional method for linear sets. *Israel J. Math.* **5** (1967), 185–187.
15. Loomis, L., *Abstract harmonic analysis*. Van Nostrand, 1953.
16. Lorch, Lee, The principal term in the asymptotic expansion of the Lebesgue constants. *Am. Math. Monthly* **61** (1954), 245–249.
17. Mandelbrojt, S., *Series Adherentes, regularisation* . . . . Gauthier-Villars, 1952.
18. Mikusinski, J. G., Une simple demonstration du theoreme de Titchmarsh . . . . *Bull. Acad. Polonaise des Sci.* **7** (1959), 715–717.
19. Naimark, M. A., *Normed rings*. P. Noordhoff N. V., 1964.
20. Paley, R.E.A.C., and Wiener, N., *Fourier transforms in the complex domain*. A.M.S. Colloquium Publications, Vol. XIX, 1934.
21. Rickart, C. E., *General theory of Banach algebras*. Van Nostrand, 1960.
22. Royden, H. L., *Real analysis*. Macmillan, 1963.
23. Rudin, W., *Real and complex analysis*. McGraw-Hill, 1966.

24. Rudin, W., *Fourier analysis on groups*. Interscience, 1962.

25. Rudin, W., Fourier-Stieltjes transforms of measures on independent sets. *Bull. A.M.S.* **66** (1960), 199–202.

26. Varopoulos, N. T., Tensor algebra and harmonic analysis. *Acta Math.* **119** (1967) 51–111.

27. Wiener, N., *The Fourier integral and certain of its applications*. Cambridge (1967), 1933.

28. Zygmund, A., *Trigonometric series*, 2nd ed. Cambridge University Press, 1959.

Index

A CATALOG OF SELECTED

DOVER BOOKS

IN SCIENCE AND MATHEMATICS